Teratological Testing

Advances in the Study of Birth Defects

VOLUME 2

Teratological Testing

EDITED BY

T. V. N. Persaud

MTP PRESS LIMITED
International Medical Publishers

Published by
MTP Press Limited
Falcon House
Lancaster, England

British Library Cataloguing in Publication Data

Advances in the study of birth defects
Vol. 2: Teratological testing
1. Abnormalities, Human
2. Abnormalities (Animals)
I. Persaud, T V N
616'.043 QL691
ISBN-13: 978-94-011-6653-9 e-ISBN-13: 978-94-011-6651-5
DOI: 10.1007/978-94-011-6651-5

Preface

The study of birth defects has assumed an importance even greater now than in the past because mortality rates attributed to congenital anomalies have declined far less than those for other causes of death, such as infectious and nutritional diseases. It is estimated that as many as 50% of all pregnancies terminate as miscarriages. In the majority of cases this is the result of faulty development. Major congenital malformations are found in at least 2% of all liveborn infants, and 22% of all stillbirths and infant deaths are associated with severe congenital anomalies.

Teratological studies of an experimental nature are neither ethical nor justifiable in humans. Numerous investigations have been carried out in laboratory animals and other experimental models in order to improve our understanding of abnormal intra-uterine development. In less than two decades the field of experimental teratology has advanced phenomenally. As a result of the wide range of information that is now accumulating, it has become possible to obtain an insight into the causes, mechanisms and prevention of birth defects. However, considerable work will be needed before these problems can be resolved.

The contributions in this volume deal primarily with the areas of teratological evaluation and the use of selected animal models for the study of congenital anomalies. It is not only a documentation of the latest experimental work, but it also indicates new and important areas for future research.

I am most grateful to the distinguished panel of contributors. Their enthusiasm and cooperation have made this volume possible. My sincere thanks are due to the publishers, especially Mr D. G. T. Bloomer, Managing Director, MTP Press Limited, for their encouragement and for extending to me every kindness. Finally, I am much indebted to my secretary, Mrs Barbara Clune, who has lightened the burden of editing this book.

Winnipeg, Canada
September, 1978

T. V. N. Persaud

Contents

List of contributors

A. R. BEAUDOIN
Department of Anatomy
4614 Medical Science II
The University of Michigan
Ann Arbor, Michigan 48109, USA

P. E. BINKERD
California Primate Research Center
University of California
Davis, California 95616, USA

M. CHECIU
Laboratory of Embryology
Center of Hygiene and Public Health
1900 Timisoara
Bv. Mihai Viteazul 24, Romania

H.-G. EIBS
Institut für Toxikologie und Embryonal-Pharmakologie
Freie Universität Berlin
Garystrasse 9
D-1000 Berlin 33, West Germany

J. FANGHÄNEL
Anatomisches Institut
Ernst-Moritz-Arndt-Universität
Friedrich-Loeffler-Strasse 23c
DDR-22 Greifswald
German Democratic Republic

M. HÄGELE
Institut für Toxikologie und Embryonal-Pharmakologie
Freie Universität Berlin
Garystrasse 9
D-1000 Berlin 33, West Germany

A. G. HENDRICKX
California Primate Research Center
University of California
Davis, California 95616, USA

B. P. HILDEBRAND
Engineering Physics Department
Battelle
Pacific Northwest Laboratory
Richland, Washington 99352, USA

R. D. HOOD
Department of Biology
The University of Alabama
P.O. Box 1927
University, Alabama 35486, USA

K. HOSHINO
Department of Anatomy
Kyoto University
Faculty of Medicine
Sakyo-ku, Kyoto 606, Japan

U. JACOB-MÜLLER
Institut für Toxikologie und Embryonal-Pharmakologie
Freie Universität Berlin
Garystrasse 9
D-1000 Berlin 33, West Germany

R. JELÍNEK
Institute of Experimental Medicine
Czechoslovak Academy of Sciences
Legerova 61, 120 00 Praha 2,
Czechoslovakia

L. A. KENNEDY
Department of Anatomy
University of Manitoba
730 William Avenue
Winnipeg, Manitoba, Canada R3E 0W3

R. W. KLASSEN
Department of Anatomy
Faculties of Medicine and Dentistry
University of Manitoba
730 William Avenue
Winnipeg, Manitoba, Canada R3E 0W3

TERATOLOGICAL TESTING

J. KLOSE
Institut für Toxikologie und Embryonal-Pharmakologie
Freie Universität Berlin
Garystrasse 9
D-1000 Berlin 33, West Germany

A. B. G. LANSDOWN
Department of Pathology
Wyeth Laboratories Limited
Taplow, Berkshire, England

T. V. N. PERSAUD
Department of Anatomy
Faculties of Medicine and Dentistry
University of Manitoba
730 William Avenue
Winnipeg, Manitoba, Canada R3E 0W3

W. D. B. POPE
Department of Anaesthesia
Faculty of Medicine
University of Manitoba
700 William Avenue
Winnipeg, Manitoba, Canada R3E 0Z3

Z. RYCHTER
Department of Histology
Faculty of General Medicine
Charles University
Albertov 4, 128 00 Praha 2
Czechoslovakia

S. SANDOR
Laboratory of Embryology
Center of Hygiene and Public Health
1900 Timisoara
Bv. Mihai Viteazul 24, Romania

G.-H. SCHUMACHER
Anatomisches Institut, Bereich Medizin
der Wilhelm-Pieck-Universität Rostock
Gertrudenstrasse 9, DDR 25 Rostock
German Democratic Republic

R. M. SHAH
Faculty of Dentistry
The University of British Columbia
Vancouver, B.C., Canada V6T 1W5

M. R. SIKOV
Biology Department, Batelle, P.O. Box 999
Richland, Washington 99352, USA

H. SPIELMANN
Institut für Toxikologie und Embryonal-Pharmakologie
Freie Universität Berlin, Garystrasse 9
D-1000 Berlin 33, West Germany

E. F. STULA
Haskell Laboratory for
Toxicology and Industrial Medicine
E.I. du Pont de Nemours and Company Inc.
Wilmington, Delaware 19898, USA

1
Primate teratology: selection of species and future use*

A. G. HENDRICKX AND P. E. BINKERD

INTRODUCTION

The devastating effects of the thalidomide episode of several years ago has led to a wide search for suitable animal models in which to study the aetiology of human birth defects. With the discovery that several species of nonhuman primates showed unusual similarities to the thalidomide malformation syndrome in man at comparable doses and similar exposure periods during early development, this order of mammals was scrutinized more closely for its potential as an animal model for teratological studies.

The nonhuman primate appears to be an especially appropriate model for testing environmental agents that may be teratogenic to man because of its ranking on the evolutionary scale. Its major role is that of serving as an additional species in cases of questionable results in commonly used laboratory animals (e.g. rats, mice and rabbits) and in the screening of selected drugs and other agents[1,2]. Until recently, species availability has been the primary criterion for selection of a particular nonhuman primate as an animal model for the study of pathogenesis and mechanisms of commonly occurring malformations. However, conservation measures taken by many countries in which nonhuman primates occur naturally have begun to limit both the numbers and the varieties of these valuable research animals. This has led to more judicious use of primates in biomedical research, and more importantly, to the development of breeding programmes which will assure the continued availability of these species as long as they are appropriately used. The use of nonhuman primates in teratological research will remain of high priority; therefore, the use of the most appropriate species for each particular problem becomes of paramount importance.

The purpose of this chapter is to discuss some of the most salient features that may be of value in species selection for teratological studies, namely, reproductive physiology, embryology, teratogenicity, metabolism, and the

* Research conducted by authors cited herein, supported by NIH grants and contracts RR00169, HD08658, DE03927, and N01-HD-1-2088

Table 1.1 Classification of nonhuman primates used in research

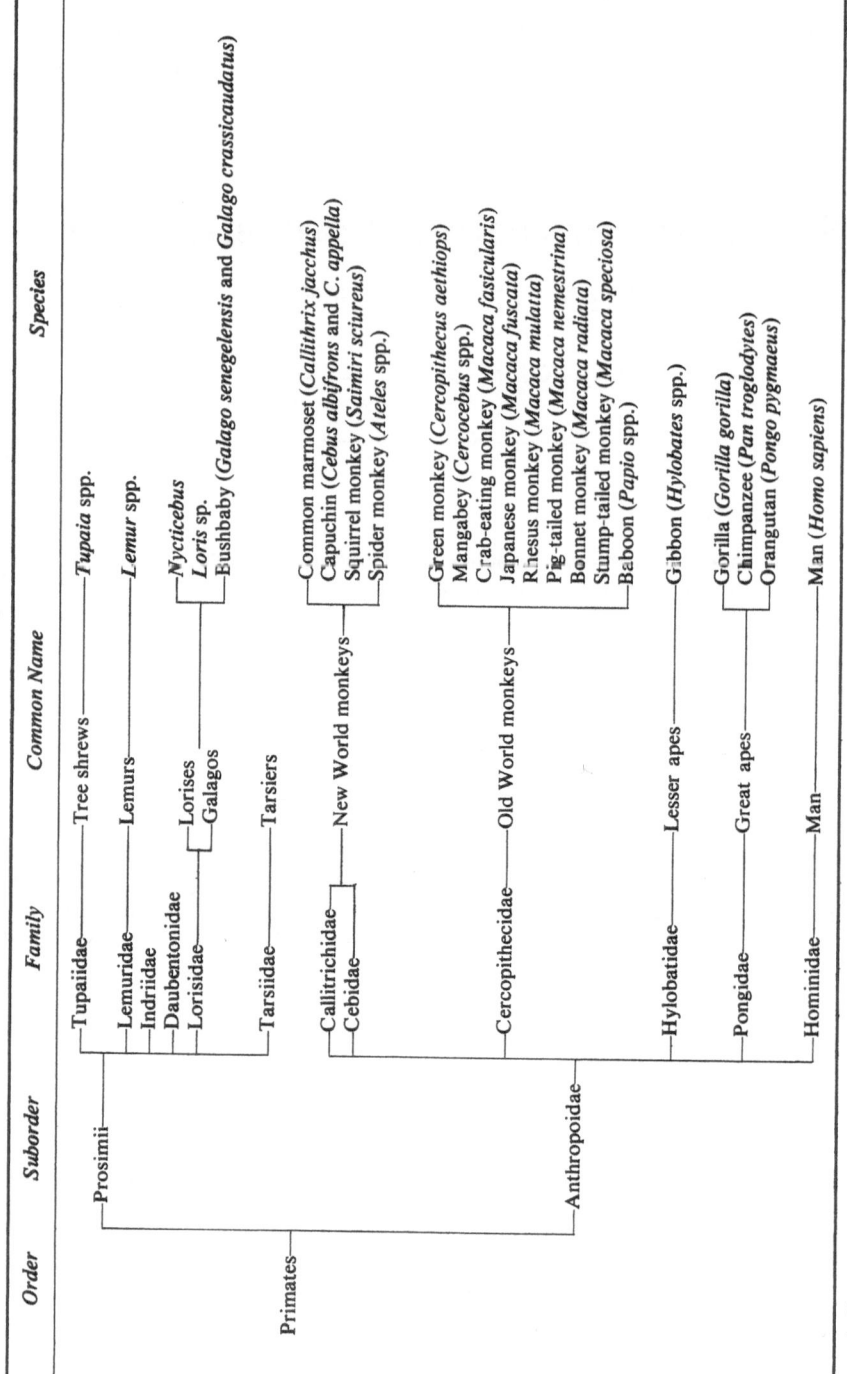

Order	*Suborder*	*Family*	*Common Name*	*Species*
Primates	Prosimii	Tupaiidae	Tree shrews	***Tupaia*** **spp.**
		Lemuridae	Lemurs	***Lemur*** **spp.**
		Indriidae		
		Daubentonidae		
		Lorisidae	Lorises	***Nycticebus***
			Galagos	***Loris*** **sp.**
				Bushbaby (*Galago senegelensis* and *Galago crassicaudatus*)
		Tarsiidae	Tarsiers	
	Anthropoidae	Callitrichidae	New World monkeys	Common marmoset (*Callithrix jacchus*)
		Cebidae		Capuchin (*Cebus albifrons* and *C. appella*)
				Squirrel monkey (*Saimiri sciureus*)
				Spider monkey (*Ateles* spp.)
		Cercopithecidae	Old World monkeys	Green monkey (*Cercopithecus aethiops*)
				Mangabey (*Cercocebus* spp.)
				Crab-eating monkey (*Macaca fasicularis*)
				Japanese monkey (*Macaca fuscata*)
				Rhesus monkey (*Macaca mulatta*)
				Pig-tailed monkey (*Macaca nemestrina*)
				Bonnet monkey (*Macaca radiata*)
				Stump-tailed monkey (*Macaca speciosa*)
				Baboon (*Papio* spp.)
		Hylobatidae	Lesser apes	Gibbon (*Hylobates* spp.)
		Pongidae	Great apes	Gorilla (*Gorilla gorilla*)
				Chimpanzee (*Pan troglodytes*)
				Orangutan (*Pongo pygmaeus*)
		Hominidae	Man	Man (*Homo sapiens*)

From the literature

effect of the more restricted availability of nonhuman primate species.

Table 1.1 illustrates the classifications of primates that have been used in reproductive and teratological research. There are about 200 species of primates which can be classified into two suborders, the Prosimii and the Anthropoidea. Each suborder can be subdivided into six families. The primate species can be organized into a scale according to increasing complexity as follows: tree shrew, lemur, loris, tarsier (members of the prosimian families), followed by the New World monkey, Old World monkey, ape, and man (members of the anthropoid families).

REPRODUCTIVE PHYSIOLOGY

Analysis of the comparative features of the reproductive functions of representative species from the four major nonhuman primate groups (prosimians, New World monkeys, Old World monkeys, and the apes) must include a discussion of the primary characteristics of the reproductive cycle as well as the events of pregnancy: implantation and placentation in addition to hormone production and metabolism.

Reproductive cycles

Table 1.2 presents information on reproductive cycle characteristics of the more commonly studied primate species and man. All of the prosimians and some New World species exhibit oestrous cycles which are characterized by overt sexual receptivity and distinct changes in vaginal cytology coincident with the periovulatory period[3]. Menstrual cycles that are comparable to the human reproductive cycle are common to some New World monkeys, all Old World species, and the greater and lesser apes[3,4]. Observations in the marmoset have failed to identify an overt oestrous or menstrual cycle in this species[5]. The literature contains various conflicting reports regarding the length of the cycle as well as the presence or absence of definitive menstruation in the squirrel monkey[3,6]. The more clearly defined manifestation of the ovarian cycle in primates that menstruate may make them better animal models for teratology experiments that require accurate timing of conception and the subsequent teratogenic susceptible period.

Studies in nonprimate mammals indicate that the hormonal controls for ovulation are comparable in animals that exhibit either an oestrous or a menstrual cycle. Specific information on the endocrine parameters of the oestrous cycles of lower primate species is limited. However, studies on the bushbaby[7] and three species of lemur[8] indicate low plasma oestrogen titres throughout the cycle except for rising concentrations of variable duration which coincide with oestrus and rising titres of plasma progesterone which peak midway through the luteal phase. Studies on selected species (mangabey, crab-eating monkey, rhesus monkey, bonnet monkey, baboon, and chimpanzee[9–14]) that exhibit menstrual cycles indicate certain quantitative and/or qualitative similarities to the well-documented human profiles for steroid (progesterone and oestrogen) and/or gonadotrophin (FSH and LH) patterns during the cycle.

Table 1.2 Comparative reproductive characteristics in primates

	Endocrinology of pregnancy				*Reproductive cycle*		
Species	*Plasma hormone patterns*	*Major steroid metabolites*	*CG* secretion* (days)	*Gestation length* (days)	*Type*	*Plasma hormone patterns*	*Length* (days)
PROSIMIANS							
Tree shrew				46–50	Oestrous		9–12
Lemur				120–135	,,		33–40
Loris				160–175	,,		42
Bushbaby				133	,,		44,32
Tarsier				180	,,		23–24
NEW-WORLD MONKEYS							
Marmoset	+	Oestradiol, 6-β-hydroxy-pregnanolone	14–126	144–152	Not definitive	+	16
Squirrel			79–112	165	Not definitive		7–8 12 25
Capuchin				153	Menstrual		21
Spider				139	Menstrual		24–27
OLD-WORLD MONKEYS							
Green				155	Menstrual		31
Mangabey	+			161–174	,,	+	26–76
Crab-eating	–			165	,,	+	25–39
Japanese				150–170	,,		24
Rhesus	–	Oestrone, C_{19} steroids	12–38	165	,,	+	27–28
Pig-tailed				162–186	,,		32–33
Bonnet	–			165	,,	+	30
Stump-tailed			14–45	182	,,		28
Baboon	–	Oestrone, C_{19} steroids, pregnanediol	12–42	175	,,	+	33
APES							
Gorilla		Oestriol, pregnanediol	21–245	255	Menstrual		25–35
Chimpanzee		Oestriol, pregnanediol	9-term	228	,,	+	37
Orangutan		Oestriol, pregnanediol	30–240	275	,,		29–32
Gibbon				195–215	,,		30
MAN		Oestriol, pregnanediol	12–term	270	Menstrual		28–29

+ similar to man; – dissimilar to man
* CG – chorionic gonadotrophin

From the literature

Evolutionary trends are evident for cycle characteristics and other reproductive parameters that have been defined in several primate species. Progression up the phylogenetic scale is characterized by a later onset of menarche and sexual activity, year-round rather than seasonal breeding patterns, longer gestation periods, and reduced frequency of twinning. These parameters are pertinent when selecting an experimental animal primarily because of practical considerations. Clearly, the lower primate species have the advantage of being more prolific due to an early puberty, shorter gestation period, and the potential for twinning or multiple births. However, these advantages are partially offset by the seasonality of their breeding behaviour. Whether the constant environmental conditions of captivity would affect this condition is speculative. On the other hand, the close phylogenetic relatedness of the Old World monkeys and great apes to man renders them desirable models due to comparable menstrual cycle features which allow accurate timing of pregnancy.

Pregnancy

Endocrinology

The production of gonadotrophins and sex steroids which are essential in preparing the implantation site and in sustaining the developing conceptus within the maternal organism represents a complex relationship of synthetic and metabolic mechanisms between mother, placenta and fetus.

Comparative data on the endocrine patterns of pregnancy for several primate species and man is presented in Table 1.2. Analysis of plasma hormone levels and urinary metabolites throughout pregnancy indicates that despite qualitative differences in the nature of urinary steroid metabolites, the marmoset exhibits gonadotrophin, oestrogen and progesterone profiles that are qualitatively and quantitatively similar to humans[5,15]. The sooty mangabey mimics the pattern observed in humans of steadily rising serum progesterone levels until term, although the absolute hormonal levels are less than those in humans[3,9].

The macaque species in general does not exhibit the same pattern of gonadotrophin and steriod production or steroid metabolism as the human. Blood progesterone levels in crab-eating monkeys are low and do not increase near term as in humans[10]. Although the bonnet monkey has a surge of progesterone during the last two weeks of gestation, the hormonal level is very low in comparison with measured progesterone levels in humans[12]. The rhesus monkey, which has been studied extensively, differs from humans in many respects. The chorionic gonadotrophin profile indicates a very limited period of hormone production in very low titres during early gestation compared with an extended period of larger titres of placental hormone production in humans[16]. In addition, the urinary metabolites and the blood levels of both oestrogen and progesterone during pregnancy in the rhesus monkey are very different from the human endocrine profile[11].

The limited number of studies that have been done in the baboon indicate that, as in the rhesus, detectable chorionic gonadotrophin secretion by the placenta in the baboon is limited to the period of early embryogenesis. The plasma progesterone levels as well as the titres and the types of urinary steroid metabolites are not comparable with those of humans[15,16].

The three species of great apes that have been investigated exhibit endocrine profiles during pregnancy that are more similar to those in humans than any other primate species studied. Although data are limited, experimental results indicate that the gorilla, chimpanzee and orangutan all secrete placental chorionic gonadotrophin for an extended period of time during pregnancy, as do humans[16]. In addition, the analysis of urinary metabolites indicates that synthesis and metabolism of oestrogen and progesterone during pregnancy closely resemble those in humans[15].

Based on endocrine data alone, it is evident that several species of great apes and the marmoset, which most closely mimic human hormonal patterns during pregnancy, would be the nonhuman primate species of choice in a teratogenic or reproductive study.

Placentation

The characteristics of implantation and placentation are important in determining the relative transport mechanisms of embryotoxic substances and their metabolites from the maternal to the fetal circulation. Events surrounding the stage of initial trophoblast attachment to the uterine endometrium are closely related to the subsequent developmental patterns of the conceptus.

Information on specific implantation mechanisms of early nonhuman primate embryos is limited; however, based on ontogenetic and comparative information derived from definitive placentation characteristics of later stages, certain evolutionary trends in early implantation patterns have been noted[17]. It has been speculated that the non-invasive, superficial, and central or eccentric attachment of the blastocyst, as occurs in the lemur and loris, represents the primitive eutherian condition[17]. The bushbaby and tarsier exhibit a minor degree of trophoblastic invasion that causes some disruption of the uterine epithelium, which is subsequently replaced. Although some species of bushbabies undergo an initial interstitial implantation, it is not homologous to the condition in the apes or man, since it becomes secondarily superficial by midgestation[17,18].

There is an increase in the proliferative activity and the invasiveness of the trophoblast proceeding up the phylogenetic scale. The New and Old World monkeys undergo an invasive, superficial implantation, with initial attachment to the uterine endometrium taking place at the embryonic pole. This latter situation represents a more precocious condition and may be adaptively advantageous for the early establishment of the chorioallantoic placenta. The interstitial implantation exhibited by the apes most closely resembles the process in humans, which is characterized by a maximum level of trophoblastic invasiveness into the uterine mucosa[17].

The establishment of a choriovitelline placenta prior to definitive chorioallantoic placentation is a primitive condition which is common to the tree shrew, loris, and bushbaby[17]. Morphological differences in chorioallantoic placenta formation which distinguish various primate species are based upon two characteristics: the pattern of villous distribution which determines the structure of the fetal circulation, and the extent of apposition between the fetal chorion and the maternal vasculature. The definitive chorioallantoic placenta in several prosimian species (lemur, loris, bushbaby) is epitheliochorial in which the fetal chorion comes into contact with the uterine epithelium; the villous distribution is diffuse, with chorionic villi distributed over the entire chorionic surface in apposition to the uterine tissue[17,19]. In contrast, the tarsier, all New and Old World monkeys, the apes and humans possess a discoidal haemochorial placenta characterized by a limited area of villous attachment with the uterine wall and the direct contact of the fetal chorion (chorionic villi) with the maternal blood[17]. The tree shrews represent an intermediate stage in placentation between these two groups with a discoidal and endotheliochorial placenta in which the chorion becomes contiguous with the endothelium of the maternal capillaries[20].

Based on these structural differences in placental structure which reflect differences in fetomaternal exchange mechanisms, the lemur, loris, bushbaby, and tree shrew would not be ideal animal models for studies involving the placental transfer of teratogens since the results could not be directly extrapolated to humans.

Little is known about the function of the different types of placentae in human and nonhuman primates during the period of organogenesis when the embryos are teratogenically most vulnerable. Data on the dynamics of transport of the early placentae in these species are limited. Such comparative studies would provide valuable information regarding the validity of extrapolating experimental data from monkey to man. (See below for further discussion on this subject.)

EMBRYOLOGY

One of the most important criteria for selection of a nonhuman primate species is comparable developmental processes between the test species and man during early gestation. Examination of early phases of development from the bilaminar disc stage through the period of organogenesis has been carried out in representative species of the four major nonhuman primate groups (prosimians, New World monkeys, Old World monkeys and the apes). A survey of the morphological and temporal aspects of early development indicates a phylogenetic pattern of embryogenesis with a graded series of major developmental events that become more similar to humans with each step up the evolutionary ladder[21,22]. The available data on prosimians indicate that the lemur, loris, and in some cases the tree shrew, exhibit numerous primitive embryonic features which show great similarities to the lower mammals. These include differentiation of the blastocyst, exposure of the embryonic ectoderm following the disappearance of the covering trophoblast, differentiation of the mesoderm, formation of the amnion by folding, and development of the allantois as a free vesicle. However, the more advanced feature of a reduced yolk sac as observed in higher primates and man is also evident in these species. Although very limited information is available for the early embryonic events in the bushbaby, two primitive features which have been noted are allantois development as a blunt projection from the caudal end of the embryo into the extraembryonic coelom[23] and formation of the amnion by the folding method[19]. The tarsier, also a member of the prosimian family, represents a transitional stage in developmental evolution manifested by a combination of primitive (exposure of the embryonic ectoderm and formation of the amnion by folding) and intermediate (precocious differentiation of the extraembryonic mesoderm, coelom, and chorion, and the replacement of the vesicular allantois by the body stalk) embryonic features in addition to advanced placentation features discussed in a previous section.

The New and Old World monkeys exhibit similar developmental patterns which are further advances on early embryogenesis. In these species the amnion forms by delamination and there is precocious development of the extraembryonic mesoderm, coelom, and mesoderm of the body stalk. These characteristics are very similar to those observed in early human embryos.

There is little information regarding the early development in the ape families; however, the evidence derived from mechanisms of implantation and placentation suggest that they represent a further specialization of early developmental events observed in the Old World monkeys which brings them in close proximity to the patterns of development in the human. Advanced developmental features that characterize early human embryos include precocious development of the primitive streak, amnion formation by delamination, a rudimentary allantois, and a vestigial yolk sac[22].

Definition of the period of organogenesis in a species used in teratological protocols is crucial since it has been determined that the period of greatest teratogenic susceptibility coincides with the embryonic period. Comparative data on the major events of organogenesis are limited to a few primate species, with the most extensive descriptive work carried out in the baboon and rhesus monkey[24–26]. Based on a description of internal and external embryonic characteristics, these studies indicate very similar developmental trends in these species in relation to that in humans. Table 1.3 shows the timespan of the embryonic period, which is defined as the period from the establishment of the primitive streak to the initial ossification of the humerus, in relation to the length of the gestation period for several species and man. The baboon and three macaque species have embryonic periods of similar length which begin and end on the same temporal scale, although the gestation period for the baboon is longer. The green monkey and the bushbaby exhibit different

Table 1.3 Temporal aspects of embryogenesis

	Embryonic period (days)		
Species	*Establishment of primitive streak*	*Ossification of humerus*	*Gestation period* (days)
Bushbaby	23	50	133
Green monkey	21	48	155
Rhesus monkey	17	46	165
Bonnet monkey	17	46	165
Crab-eating monkey	17	46	165
Baboon	17	46	180
Human	19	54	270

From reference 27

degrees of a delayed embryonic period in a shortened gestation period. Although they begin and end at varying times, the embryonic periods for all these species fall within the length of the corresponding period in humans[27].

Analysis of the relationships between embryonic age, developmental stage and embryonic size has been used as a further measure of embryological comparisons[27,28]. Based on information obtained from human specimens, Streeter[29–32] and Nishimura and colleagues[33] have provided valuable information on the temporal factors of human embryonic development. However, there are some discrepancies regarding the rates of development which are based on methods of timing pregnancies; consequently, the data provided by Nishimura *et al.*[33] are used for comparative purposes.

The relationship between embryonic age and developmental stage, the most important criteria for teratological and other developmental studies, indicates that bushbaby and green monkey embryos are more similar to human embryos than are the macaque, mangabey, and baboon embryos, which are all four to five days younger than human embryos at each stage. When size (crown–rump length) and age are correlated, it is evident that the bushbaby and the green monkey increase in length at a slower rate, and the macaques (rhesus, bonnet and crab-eating) and baboons increase at a faster rate in comparison with human embryos. When embryonic size is correlated with developmental stage, embryos of the green monkey, the three macaque species, and the baboon are similar to human embryos; however, the correlation in bushbabies indicates that their embryos are slightly smaller at each stage than humans. While green monkey embryos are similar to human embryos in two developmental relationships, the bushbaby, baboon, rhesus, bonnet, and crab-eating monkey embryos are similar in one of three correlations.

Definition of the period of organogenesis as well as recognition of comparable developmental phenomena between various primate species and man not only aids in the selection of a teratological test species but is essential in the accurate interpretation of teratological test results.

TERATOGENICITY

The primary species of nonhuman primates that have been investigated in teratological test protocols include the bushbaby, marmoset, green monkey, baboon, and several macaque species (rhesus, bonnet, crab-eating, stump-tailed, Japanese, and pig-tailed monkeys). The purpose of this section is not to present a comprehensive review of all teratogenic studies carried out in primates, but to evaluate the results of selected studies in relation to the current status of embryotoxicity in humans. Such a comparative approach will provide further insight into the subject of species selection.

As indicated in Table 1.4, four categories have been used in classifying drugs and other environmental agents according to their embryotoxic potential in man. The criteria used in this system of classification are based on clinical case reports and human epidemiological surveys in addition to definitive teratogenic studies carried out in several animal species[1,2]. These categories are: (1) established embryotoxicity (+ + +), (2) suspected embryo-

Table 1.4 Primate embryotoxicity studies

	Embryotoxicity in primates				
Substance	*Human**	*Nonhuman*	*Outcome†*	*Reference*	*Comments*
DRUGS					
Thalidomide	+++	Bushbaby	–	35, 37	
		Marmoset	+	2, 35, 36	Malformation syndrome similar to humans
		Green monkey	+		
		Japanese monkey	+		
		Stump-tailed monkey	+		
		Crab-eating monkey	+		
		Rhesus monkey	+		
		Bonnet monkey	+		
		Baboon	+		
Androgenic hormones	+++	Rhesus monkey	+	34, 35	Pseudohermaphroditic females; cryptorchidism in males
Antineoplastic agents and folic acid antagonists	+++	Rhesus monkey	+	2, 34, 35	Low level of malformations; high level of embryolethality
Anticonvulsants	++	Rhesus monkey	+	39	Low level of embryotoxic effects
			–	34, 38	
		Crab-eating monkey	+	2	Embryolethality in large doses only
Anorexic drugs	++	Rhesus monkey	–	34	
Oral hypoglycaemics	++	Rhesus monkey	+	39	Low level of embryolethality and minor anomalies (inconclusive)
Synthetic oestrogens	+	Rhesus monkey	+	44	Genital defects in both males and females
Aspirin	+	Rhesus monkey	+	34	Embryolethal and teratogenic in excessive doses only
Tranquilizers	+	Rhesus monkey	–	45	
Antibiotics	+	Rhesus monkey	–	46	
Antihistamines	–	Rhesus monkey	–	2, 34	

Imipramine	–	Rhesus monkey Bonnet monkey	–	47	
LSD	–	Rhesus monkey	–	2	
Adrenocortical steroids	–	Rhesus monkey Bonnet monkey Baboon	+	35, 48	Craniofacial, brain and lymphoid malformations
Retinoic acid (vitamin A)	–	Marmoset Pig-tailed monkey Rhesus monkey	+	35, 49	Craniofacial and appendicular skeletal malformations
ENVIRONMENTAL CHEMICALS					
Herbicides (2,4,5-T)	–	Rhesus monkey	–	55	
Fungicides (captan, folpet, difolatan)	–	Rhesus monkey Stump-tailed monkey	–	54	
Methyl mercury	+++	Stump-tailed monkey	+	50	Ataxia
Sodium cyclamate	–	Rhesus monkey	–	34	
INFECTIOUS DISEASES					
Rubella virus	+++	Crab-eating monkey Baboon	+	34	High level of embryolethality; low level of malformation
Mumps, encephalitis, influenza virus	++	Rhesus monkey	+	35, 57, 58	Intra-uterine death; cerebal and ocular defects
ENVIRONMENTAL CONDITIONS					
X-ray	+++	Rhesus monkey Bonnet monkey Marmoset	+	35, 59	Ovarian damage; head, limb, CNS defects
Hyperthermia	++	Marmoset Bonnet monkey	+	61, 62	Diverse abnormalities and intra-uterine death

* +++ known human teratogen; ++ suspected human teratogen; + suspected human teratogen under some conditions; – not teratogenic in humans under normal conditions (From reference 2)

† + embryotoxic; – not embryotoxic

toxicity (+ +), (3) possible embryotoxicity under some conditions (+), and (4) no embryotoxicity under any known conditions of human usage (–).

Drugs

Three types of drugs that have been positively implicated as teratogenic in man are: thalidomide, steroid hormones with androgenic activity, and the folic acid antagonists (Table 1.4). The most widely tested of these, thalidomide, has produced a recognizable malformation syndrome (limb defects) in several primate species[2, 34–36]. The bushbaby is the only nonhuman primate that exhibits no embryotoxic response following prenatal exposure to the drug[27, 37]. Virilization of human female offspring has been observed following intra-uterine exposure to substances with androgenic properties. Studies in rhesus monkeys indicate a similar effect in females following administration of norethindrone and testosterone propionate during pregnancy. Norethindrone additionally causes some degree of cryptorchidism in male offspring[34, 35]. Antineoplastic drugs and folic acid antagonists cause intra-uterine death as well as structural and functional defects following prenatal exposure in humans[2, 34]. Attempts to duplicate the high level of developmental toxicity of these agents (e.g. aminopterin, methotrexate) in rhesus monkeys have shown this species to be more susceptible to the lethal than the teratogenic effects[34].

Several human therapeutic drugs that are suspected of possessing teratogenic properties include anticonvulsants, anorexic drugs, and oral hypoglycaemics[2]. Experiments carried out in rhesus and crab-eating monkeys indicate a variable level of toxicity associated with these agents. Anticonvulsants produce no effect[34, 38], a low level of embryotoxicity[39], or embryolethality when administered in excessive doses[2]. In addition, the teratogenic potential of the anorexic drug methamphetamine[34] and the oral hypoglycaemic tolbutamide[39] has not been confirmed in rhesus monkeys.

Drugs which have been designated as possible human teratogens only under some conditions (e.g. synthetic oestrogens, aspirin, tranquillizers, and antibiotics), have also been tested in nonhuman primates. The nonsteroidal synthetic oestrogen diethylstilbestrol (DES) administered to human females with threatened spontaneous abortions has been implicated as a causative factor in the appearance of minor genital abnormalities in both female offspring (vaginal adenosis, cervical hooding and vaginal ridging)[40, 41] and male offspring (external genital and urinary tract abnormalities)[42, 43]. A recent report indicates that DES causes similar developmental anomalies when administered prenatally in rhesus monkeys[44].

Studies in rhesus monkeys to test several other drugs in this category indicate no embryotoxic effects (e.g. several tranquilizers and antibiotics)[34, 45, 46] or embryolethal and teratogenic effects when administered only at levels in excess of human therapeutic doses (e.g. aspirin)[34].

Agents which are believed not to be embryotoxic in humans under likely conditions of usage, including the antihistamine, meclizine, the antidepressant, imipramine, and LSD, are similarly nontoxic in pregnant rhesus and bonnet monkeys[2, 34, 47].

Substances which have not been proven as embryotoxic agents in humans but have been verified as potent embryolethal and teratogenic substances in primates are the corticosteroid triamcinolone acetonide (TAC) and vitamin A or its acidic form, retinoic acid (Table 1.4). Studies in rhesus and bonnet monkeys and baboons indicate a well-defined syndrome exhibiting dose–response characteristics of craniofacial and lymphoid defects associated with prenatal exposure to human therapeutic doses of TAC[35,48]. Sensitivity to vitamin A and retinoic acid exposure during critical periods of organogenesis in marmosets and pig-tailed and rhesus monkeys has also been well-documented. Embryolethality, growth retardation, and malformations of the skeleton, craniofacial region, and certain viscera are manifestations of the syndrome in these species[34,35,49].

Environmental chemicals

Several natural and man-made chemicals which have been examined for embryotoxic effects are listed in Table 1.4. Only one of these substances, methyl mercury, has been firmly established as deleterious to prenatal development in man[2]. Neonatal ataxia is a minor teratogenic manifestation of *in utero* exposure to methyl mercury in stump tailed monkeys[50]. The embryotoxicity of 'blighted potatoes' has not been verified in humans; similarly, conflicting experimental results associated with the prenatal administration of infected potatoes to marmosets and rhesus monkeys have failed to substantiate frank toxicity in these species[51–53]. Epidemiological surveys and laboratory studies indicate that as a group, both fungicides and herbicides are relatively non-embryotoxic in humans at ordinary use levels[2]. The embryotoxic potential of the herbicide 2,4,5-trichlorophenoxyacetic acid and the fungicides captan, folpet and difolatan has not been substantiated following administration to pregnant rhesus and stump-tailed monkeys[54]. Attention has been focused on the possible toxicity of several food additives due to implicative retrospective studies in humans. Teratogenic studies using the artificial sweetener sodium cyclamate and its major metabolite, cyclohexylamine, did not increase malformation or abortion rates in rhesus monkeys[34].

Infectious diseases

Several infectious agents have an established (e.g. rubella and herpes virus) or suspected (e.g. mumps and influenza virus) teratogenic effect in humans[56]. Prenatal exposure to rubella virus in crab-eating monkeys and baboons results in high levels of intra-uterine death and minor malformations[34]. In addition, definitive teratogenic effects (i.e. cataracts and hydrocephalus) have been experimentally produced in rhesus monkeys with encephalitis, influenza, and mumps viruses (Table 1.4)[35,57,58].

Environmental conditions

X-irradiation was one of the first environmental agents to be identified as teratogenic in man and several other mammalian species[2]. Prenatal exposure to ionizing radiation in rhesus and bonnet monkeys and marmosets produces growth retardation and malformations ranging from head, limb, and central nervous system defects to severe ovarian damage[39, 59].

Recent epidemiological evidence implicates hyperthermia as a causative factor in developmental anomalies in the human fetus[60]. The teratogenic potential of this environmental condition has been confirmed in rhesus and bonnet monkeys and marmosets by the experimental induction of elevated core temperatures during organogenesis. The toxic insult results in diverse developmental abnormalities as well as intra-uterine death (Table 1.4)[61, 62].

Several comments can be made regarding the results of teratological experiments as they relate to selection of the most suitable primate model. The use of the bushbaby in teratological screening protocols is questionable due to its lack of response to thalidomide. With few exceptions, the marmoset, bonnet, crab-eating and stump-tailed monkeys mimic the human response to the potentially toxic agents tested in these species. Although the number of studies carried out in the green, Japanese and pig-tailed monkeys is too limited to draw any conclusions, the available test results warrant further use of these species in teratology protocols. The rhesus monkey, and to a slightly lesser extent the baboon, have been the most widely used species and their general effectiveness as a primate model for human teratogens is well-documented.

METABOLISM

Interspecies differences in maternal response may arise from differences in rates or routes of metabolism of drugs and toxic substances, and these differences should be considered in species selection. Although information on comparative metabolism of nonhuman and human primates is limited, there are sufficient data to show trends in metabolic patterns to identify potentially useful groups or species. Reviews by Smith[63], Smith and Williams[64], Smith and Caldwell[65], and Clifford[66] have focused on the comparative metabolism of primate and nonprimate species and have provided the basis for making further comparisons. The purpose of this section is to discuss selected metabolic reactions which have been studied to demonstrate the value of comparative metabolism in selection of species for teratological studies.

Nonhuman primates as metabolic models for man

There are two aspects of a metabolic comparison, namely rate and routes of drug metabolism. Metabolic rates of twelve drugs in man, the rhesus monkey, the dog, and the rat are presented in Table 1.5. The plasma half-lives of the drugs, with the exception of indomethacin, isoniazid and caffeine, are generally shorter in the rhesus monkey, dog, and rat compared with man. However, these differences in half-lives could be attributable to other factors besides species differences in metabolism, including variations in tissue distribution, plasma protein binding, and excretion. It is significant to note that when using drugs with a half-life of 30 h or more in man, none of the three species studied provided suitable kinetic models. It is possible that animal models may be found only for drugs having a relatively short half-life (i.e. 4 h). From the

Table 1.5 Metabolic models for man: rates of drug metabolism in man, rhesus monkey, rat and dog expressed as plasma half-life (hours)

Drug	*Man*	*Rhesus monkey*	*Dog*	*Rat*
Indomethacin	2.0	1.5	1.5	4.0
Isoniazid	5.0	5.0	5.0	10.0
Caffeine	3.5	2.4	5.0	2.0
Pethidine	5.5	1.2	0.9	–
Antipyrine	12.0	1.8	1.7	–
Halofenate	24.0	16.0	24.0	26.0
Diazoxide	29.0	19.0	–	–
Myalex	31.0	3.0	30.0	29.0
Oxisuran	55.0	21.0	12.0	10.0
Phenylbutazone	72.0	8.0	6.0	–
Oxyphenylbutazone	72.0	8.0	0.5	–
Chlorphentermine*	92.0	14.0	–	95.0

* Excretion half-life

From reference 65

standpoint of an animal model system for man, differences in rates of drug metabolism are less important than differences in metabolic pathways; the former condition can be compensated for by different dosage schedules, whereas it is not possible to compensate for actual species differences in the latter circumstance[65].

A comparison of metabolic pathways for some representative compounds in the rhesus monkey and several nonprimate species is shown in Table 1.6. A 'good' rating indicates close similarity to man while an 'invalid' rating implies different pathways. When the pathways are similar but there are significant interspecies variations in the amounts of metabolites produced by the different pathways, a 'fair' rating is given. It is apparent that the rhesus monkey provides a good model for man in five of the seven cases. In contrast, the rat provides a good model in only one case and is a poor or invalid model for the five remaining compounds[65].

Table 1.6 Metabolic models for man: routes of drug metabolism in rat, other nonprimates, and rhesus monkey

	Species		
Compound	*Rat*	*Other* nonprimate*	*Rhesus monkey*
Chlorphentermine	Invalid	Good (G)	Good
Sulphadimethoxine	Poor	Poor (D)	Good
Phenylacetic acid	Invalid	Invalid (D)	Good
Isoniazid	–	Poor (D)	Good
Indomethacin	Poor	Poor (D)	Fair
Halofenate	Poor	Poor (D)	Good
Oxisuran	Good	Fair (D)	Fair

* D = dog; G = guinea pig

From reference 65

Metabolic reactions in primates

The biphasic pattern of drug metabolism in primates is similar to that for other species. The first phase consists of an oxidative, reductive, or hydrolytic reaction which is followed by a second phase consisting of synthetic reactions.

Smith and Caldwell[65] have provided an excellent review of Phase I metabolic reactions in primates; consequently, only a very brief account will be given here.

Phase I reactions: oxidative, reductive and hydrolytic reactions

Most studies of oxidative reactions occurring at C- centres have been in the rhesus monkey, with a limited number of studies in other New World and Old World species. Studies involving oxidation at N- and S- functions in drugs are limited to the rhesus monkey and to a lesser extent the marmoset, i.e. N-oxidation of both N-acetyl-aminofluorene and chlorphentermine occurs in man and the rhesus monkey, but this reaction is apparently defective in the marmoset. O- and N- dealkylation reactions occur in man and the rhesus monkey, the only nonhuman primate thus far studied. The deamination reaction is one of the few Phase I metabolic reactions that has been studied in several primate species, e.g. amphetamine undergoes extensive deamination in the rhesus and squirrel monkeys, the marmoset and man. Among the hydrolytic reactions, ester, amide, and carbamate hydrolyses have been shown to occur in man and the rhesus monkey.

Phase II reactions: the conjugation mechanisms

Several conjugation mechanisms which are classified as Phase II synthetic reactions will be briefly discussed in terms of species differences: glucuronic acid conjugation, sulphate conjugation, mercapturic acid formation, methylation, acetylation and amino acid conjugation.

Glucuronic acid conjugation

This is among the most common and important of the metabolic conjugation mechanisms. These conjugates can be formed with compounds containing hydroxyl (alcoholic, phenolic, hydroxylaminic), carboxyl, amino, sulphonamido, and sulphydryl groups. It is significant to note that the greatest amount of information on these reactions is available for the rhesus monkey, and this species, like man, forms the five main types of glucuronic acid conjugates. Information is scanty for many other primate species; however, the available data do indicate that Old World, New World, and prosimian species synthesize esters and N-glucuronides.

Sulphate conjugation

This consists of the combination of a hydroxy compound with sulphate to produce an acid ester of sulphuric acid; it has not been well-investigated in primates, but taxonomic group differences have been demonstrated with the simple phenols, common hydroxy compounds. In man, the rhesus monkey and the crab-eating monkey the main conjugate of phenol is phenylsulphate; in the capuchin and the marmoset it is phenylglucuronide[64].

Mercapturic acid formation and methylation reactions

These have not been widely studied; however, the investigations that have been done which involve these metabolic reactions demonstrate a similarity in the New and Old World monkey species.

Acetylation

This is a common biochemical reaction of compounds containing a primary amino group. Acetylation of the five types of amino groups, the aromatic amino group, the aliphatic amino group, the sulphonamide group (-SO_2NH_2), the hydrazine group, and the L-amino group of S-arylcysteines in the formation of mercapturic acids, occurs in both man and the rhesus monkey. Acetylation of the aromatic amino group of sulphadimethoxine, a sulphonamide, takes place in man and representative species of the Old and New World monkeys and prosimians. In addition, sulphamethazine and dapsone undergo acetylation in man, the rhesus monkey, and the squirrel monkey, but the acetylation of sulphamethazine is much lower in the squirrel monkey than in the rhesus monkey and man. Furthermore, the squirrel monkey is capable of deacetylating N^4-acetylsulphamethazine, unlike the rhesus monkey and man. However, the rhesus monkey shows much greater acetylation and deacetylation of dapsone than the squirrel monkey or man. Thus, the acetylation of certain sulphonamide drugs suggests interesting differences among the three species studied; however, there is no distinct pattern indicating greater similarity to man in either the rhesus or the squirrel monkey.

Amino acid conjugation

In primates three amino acids, glycine, glutamine and taurine, are utilized in the metabolic reaction in which certain carboxylic acids are combined with the amino acid to form an amide bond. Glycine conjugation occurs in all the primate species studied thus far and in many nonprimate species. On the other hand, glutamine, and to some extent taurine, appear to be more restricted to primates. More specifically, glutamine conjugation of phenylacetic acid occurs in man and various species of Old and New World monkeys but not in prosimians and nonprimates. The changeover from glycine to glutamine conjugation apparently has occurred at the level of the New World monkeys. It is significant to note that the chimpanzee also converts phenylacetic acid to phenylacetylglutamine, indicating that glutamine conjugation also occurs in the apes.

Drug metabolism and primate taxonomy

Analysis of phylogenetic trends in patterns of drug metabolism offers an opportunity to make general comparisons between major primate groups as well as to determine the metabolic relationship of various species to man. That some drug metabolic reactions are restricted to primate species in general or to particular groups of primates can be correlated with their evolutionary status.

The incidence of metabolic reactions of five drugs in many Old and New World monkeys and prosimians is indicated in Table 1.7. It becomes apparent that the capacity to form glucuronides with sulphadimethoxine appears to be a well-developed feature of primate species; this reaction is either absent or at a low level in nonprimates (see Table 1.6). The capacity to utilize glutamine for the conjugation of simple arylacetic acids such as phenylacetic acid

Table 1.7 Primate taxonomy and drug metabolism reactions

Metabolic reaction	*Man*	*Old World monkeys*	*New World monkeys*	*Prosimians*
N[1]-Glucuronide conjugation of sulphadimethoxine	+	+	+	+
Glutamine conjugation of simple arylacetic acids	+	+	+	−
0-Methylation of 4-hydroxy-3,5-diiodobenzoic acid	+	+	+	?
Aromatization of quinic acid to benzoic acid	+	+	−	−
Sulphate conjugation of phenol and 1-naphthol	high	high	low	?

From reference 65

appears to be a biochemical feature of anthropoid apes and monkeys but does not occur in prosimian species and nonprimates. Information of this type provides biochemical evidence for relationships between different primate groups in addition to structural and behavioural data and aids in the selection of appropriate species for pharmacological, toxicological and teratological investigations.

Drug kinetics during pregnancy

Information on the metabolism of drugs and other toxic substances in the nonpregnant animal is valuable in terms of selection of an appropriate species; however, the true test of any substance must come in the pregnant animal in order to assess potential teratogenicity. The physiological responses in pregnancy, which occur mainly in the cardiovascular system, the lungs, the kidneys, the gastrointestinal tract and the water spaces, cause functional changes that influence the pharmacokinetic properties of drugs[67] and most probably other toxic substances, especially with respect to absorption, distribution, and elimination.

Information is gradually being accumulated on drug kinetics during pregnancy, but for the most part these investigations have been done during the fetal period rather than the teratogen-sensitive embryonic period.

DISTRIBUTION OF TERATOGENIC DRUGS

Only a very limited number of investigations has been done on the distribution of drugs with teratogenic activity. Nishimura[68] compared the placental transport of thiopental and sulphamethopyrazine in early pregnancy of rodents and man, and Wilson *et al.*[69–71] have compared the distribution of hydroxyurea and aspirin in maternal plasma and in the embryos of rats and rhesus monkeys. The results with hydroxyurea indicate that the rhesus monkey embryo is somewhat less sensitive to teratogenesis than the rat embryo because a much longer exposure at equivalent levels was required to produce comparable embryotoxic effects. In a comparative study with embryotoxic doses of aspirin, pregnant rats consistently showed higher total plasma levels of salicylic acid than did the rhesus monkeys during the eight-hour period following the last treatment[70,71]. Further species differences were noted: in the rat, half of the total level remained free in the plasma and the remainder was bound to plasma proteins, while in the rhesus monkey only 25% or less of the salicylic acid remained free. This disparity in maternal plasma levels was also observed in the embryo. The rat embryo was exposed to much higher concentrations, i.e. two to four times higher than the monkey embryo, which most probably accounts for the greater degree of embryotoxicity (teratogenicity) in the rat. Although the differences in concentration in maternal plasma and in the embryo were significant, the time to reach peak concentration was similar for both species. In the rat embryo the concentration of salicylic acid followed very closely the free salicylic acid in maternal plasma, suggesting a close equilibrium between the two levels. However, there was less agreement between concentration in the embryo and the level of free salicylic acid in the maternal plasma of the rhesus monkey. This difference may be accounted for by the difference in placental transfer between the two species. During the time of treatment, encompassing day 12, the rat placenta consists of the inverted yolk-sac placenta and newly formed chorioallantoic placenta; in contrast, the rhesus monkey by 32 days has a well-developed chorioallantoic placenta with a relatively thick syncytiotrophoblast through which the exchange must occur.

Although these studies have not revealed any common pharmacodynamic principles, they offer at least partial explanations for species differences.

CONSERVATION AND ITS EFFECT ON AVAILABILITY OF ANIMALS FOR TERATOLOGICAL INVESTIGATIONS

Since 1975, when India placed a temporary ban on the exportation of rhesus monkeys, there has been an increasing concern over the removal of primates from their natural habitat, regardless of origin. This concern was realized when India again placed a ban, effective April 1, 1978, on the exportation of all primates. Similar bans have also been enforced by most South American countries (with the exception of Bolivia). To offset this apparent decline in availability of some species, breeding programmes have been developed which will provide some of the animals required for medical research. Although this imposes some hardships in terms of the increased cost required for domestic breeding, there are some potential benefits. First of all, a more homogeneous population of animals will eventually be produced because the breeding will be confined to known groups of animals. Family lineage will be recorded within breeding populations, providing the possibility for both selected breeding and inbreeding in the production of lines or strains with characteristics suitable for the study of problems that occur in man.

The apparent reduction in natural primate populations available for research and the necessity for the increased use of alternative species, e.g. the crab-eating monkey for the rhesus monkey, will make it desirable if not crucial to exercise extreme care in the process of nonhuman primate species selection. Thus, it is important that investigators consider the factors discussed in this chapter, and others as they apply to their research objectives, in the selection of the most appropriate species for teratological studies.

References

1. Wilson, J. G. (1973). *Environment and Birth Defects.* (New York: Academic Press)
2. Wilson, J. G. and Fraser, F. C. (eds.). (1977). *Handbook of Teratology*, Vol. I, *General Principles and Etiology*, (New York and London: Plenum Press)
3. Butler, H. (1974). Evolutionary trends in primate sex cycles. *Contrib. Primat.*, 3, 2
4. Wright, E. M., Jr. and Bush, D. E. (1977). The reproductive cycle of the capuchin (*Cebus apella*). *Lab. Anim. Sci.*, **27**, 651
5. Hearn, J. P. and Lunn, S. F. (1975). The reproductive biology of the marmoset monkey, *Callithrix jacchus*. *Lab. Anim. Handb.*, **6**, 191
6. Srivastava, P. K., Cavazos, F. and Lucas, F. V. (1970). Biology of reproduction in the squirrel monkey (*Saimiri sciureus*). I. The estrus cycle. *Primates*, **11**, 125
7. Eaton, G. G., Slob, A. and Resko, J. A. (1973). Cycles of mating behaviour, oestrogen and progesterone in the thick-tailed bushbaby (*Galago crassicaudatus crassicaudatus*) under laboratory conditions. *Anim. Behav.*, **21**, 309
8. Bogart, M. H., Kumamoto, A. T. and Lasley, B. L. (1977). A comparison of the reproductive cycle in three species of lemur. *Folia Primat.*, **28**, 134
9. Stabenfeldt, G. H. and Hendrickx, A. G. (1973). Progesterone levels in the sooty mangabey (*Cercocebus atys*) during the menstrual cycle, pregnancy and parturition. *J. Med. Primat.*, 2, 1
10. Stabenfeldt, G. H. and Hendrickx, A. G. (1973). Progesterone studies in the *Macaca fascicularis*. *Endocrinology*, **92**, 1296
11. Atkinson, L. E., Hotchkiss, I., Fritz, G. R., Surve, A. H., Neill, J. D. and Knobil, E. (1975). Circulating levels of steroids and chorionic gonadotrophin during pregnancy in the rhesus monkey, with special attention to the rescue of the corpus luteum in early pregnancy. *Biol. Reprod.*, 12, 335
12. Stabenfeldt, G. H. and Hendrickx, A. G. (1972). Progesterone levels in the bonnet monkey (*Macaca radiata*) during the menstrual cycle and pregnancy. *Endocrinology*, **91**, 614
13. Kling, O. R. and Westfahl, P. K. (1978). Steroid changes during the menstrual cycle of the baboon (*Papio cynocephalus*) and human. *Biol. Reprod.*, **18**, 392
14. Graham, C. E., Collins, D. C., Robinson, H. and Preedy, J. R. K. (1972). Urinary levels of estrogen and pregnanediol and plasma levels of progesterone during the menstrual cycle of the chimpanzee: Relation to the sexual swelling. *Endocrinology*, **91**, 13
15. Shackleton, C. H. L. and Mitchell, F. L. (1975). The comparison of perinatal steroid endocrinology in simians with a view to finding a suitable animal model to study human problems. *Lab. Anim. Handb.*, **6**, 159
16. Tullner, W. W. (1974). Comparative aspects of primate chorionic gonadotropins. *Contrib. Primat.*, 3, 235
17. Luckett, W. P. (1974). Comparative development and evolution of the placenta in primates. *Contrib. Primat.*, 3, 142
18. Hendrickx, A. G. and Houston, M. L. (1971). Prenatal and postnatal development. In: E. S. E. Hafez (ed.). *Comparative Reproduction of Nonhuman Primates*, pp. 334–381. (Springfield, Ill.: Charles C. Thomas)
19. Butler, H. (1967). The giant cell trophoblast of the Senegal galago (*Galago senegalensis senegalensis*) and its bearing on the evolution of the primate placenta. *J. Zool., Lond.*, **152**, 195
20. Schwaier, A. and Kuhn, H. J. (1975). Chronology of the development of embryo and placenta of *Tupaia belangeri*. *Lab. Anim. Handb.*, **6**, 257
21. Hill, J. P. (1932). The developmental history of the primates. *Phil. Trans. Royal Soc. Lond. Ser. (B)*, 221, 45
22. Hendrickx, A. G. (1972). Early development of the embryo in non-human primates and man. *Acta Endocrinol., Kbh.*, **166 (Suppl.)**, 103

23. Butler, H. (1972). The chronology of embryogenesis in the lesser galago: A preliminary account. *Folia Primat.*, **18**, 368
24. Hendrickx, A. G., Houston, M. L., Kraemer, D. C., Gasser, R. F. and Bollert, J. A. (1971). *Embryology of the Baboon.* (Chicago and London: The University of Chicago Press)
25. Heuser, C. H. and Streeter, G. L. (1941). Development of the macaque embryo. *Contrib. Embryol., Carnegie Inst. Washington*, **29**, 17
26. Hendrickx, A. G. and Sawyer, R. H. (1975). Embryology of the rhesus monkey. In: G. Bourne (ed.). *The Rhesus Monkey.* Vol. II, *Management, Reproduction, and Pathology*, pp. 141–169. (New York: Academic Press)
27. Hendrickx, A. G., Sawyer, R. H., Lasley, B. L. and Barnes, R. D. (1975). Comparison of developmental stages in primates with a note on the detection of ovulation. *Lab. Anim. Handb.*, **6**, 305
28. Hendrickx, A. G. (1972). A comparison of temporal factors in the embryological development of man, Old World monkeys and galagos, and craniofacial malformations induced by thalidomide and triamcinolone. In: E. I. Goldsmith and J. Moor-Jankowski (eds.). *Medical Primatology 1972.* Part III, pp. 259–269. (Basel: Karger)
29. Streeter, G. L. (1942). Developmental horizons of human embryos: Description of age group XI, 13–20 somites, and age group XII, 21–29 somites. *Contrib. Embryol., Carnegie Inst., Washington*, **30**, 211
30. Streeter, G. L. (1945). Developmental horizons in human embryos: Description of age group XIII, embryos about 4 or 5 millimeters long, and age group XIV, period of indentation of the lens vesicle. *Contrib. Embryol., Carnegie Inst., Washington*, **31**, 27
31. Streeter, G. L. (1948). Developmental horizons in human embryos: Description of age groups XV, XVI, XVII, and XVIII, being the third issue of a survey of the Carnegie Collection. *Contrib. Embryol., Carnegie Inst., Washington*, **32**, 135
32. Streeter, G. L. (1951). Developmental horizons in human embryos: Description of age groups XIX, XX, XXI, XXII, and XXIII, being the fifth issue of a survey of the Carnegie Collection. *Contrib. Embryol., Carnegie Inst., Washington*, **34**, 165
33. Nishimura, H., Takano, K., Tanimura, T. and Yasuda, M. (1968). Normal and abnormal development of human embryos: First report of the analysis of 1213 intact embryos. *Teratology*, **1**, 281
34. Wilson, J. G. (1972). Abnormalities of intra-uterine development in non-human primates. *Acta Endocrinol., Kbh.*, **166 (Suppl.)**, 261
35. Hendrickx, A. G., Terrell, T. G., Andersen, A. C., Osburn, B. I., Sawyer, R. H. and Steffek, A. J. (1977). Induction of abnormal intra-uterine development with triamcinolone, vitamin A and X-irradiation in non-human primates. In: M. R. N. Prasad and T. C. Anand Kumar (eds.). *Use of Non-Human Primates in Biomedical Research*, pp. 149–169. (New Delhi: Indian National Science Academy)
36. Hendrickx, A. G. and Sawyer, R. H. (1978). Developmental staging and thalidomide teratogenicity in the green monkey (*Cercopithecus aethiops*). *Teratology*, **18**, 393
37. Butler, H. (1977). The effect of thalidomide on a prosimian: The greater galago (*Galago crassicaudatus*). *J. Med. Primat.*, **6**, 319
38. Esaki, K., Tanioka, Y., Ogata, T. and Koizumi, H. (1975). Effect of sodium dipropylacetate (Dpa) on rhesus monkey fetuses. *CIEA (Cent. Inst. Exp. Anim.) Preclin. Rep.*, **1**, 157
39. Wilson, J. G. (1974). Teratologic causation in man and its evaluation in non-human primates. In: A. G. Motulsky and W. Lenz (eds.). *Birth Defects*, pp. 191–203. (Amsterdam: Excerpta Medica)
40. Ulfelder, H. (1976). DES – transplacental teratogen – and possibly also carcinogen. *Teratology*, **13**, 101
41. Robboy, S. J., Scully, R. E. and Herbst, A. L. (1975). Pathology of vaginal and cervical abnormalities associated with prenatal exposure to diethylstilbestrol (DES). *J. Reprod. Med.*, **15**, 13
42. Bibbo, M., Gill, W. B., Azizi, F., Blough, R., Fang, V. S., Rosenfield, R. L., Schumacher, G. F. B., Sleeper, K., Sonek, M. G. and Wied, G. L. (1977). Follow-up study of male and female offspring of DES-exposed mothers. *Obstet. Gynecol.*, **49**, 1

43. Cosgrove, M. D., Benton, B. and Henderson, B. E. (1977). Male genitourinary abnormalities and maternal diethylstilbestrol. *J. Urology*, **117**, 220

44. Hendrickx, A. G., Benirschke, K., Thompson, R. S., Ahern, J., Lucas, W. E. and Oi, R. (1978). The effects of prenatal diethylstilbestrol (DES) exposure on the genitalia of pubertal *Macaca mulatta*. *Teratology*, **17**, 23A (Abstract)

45. Szabo, K. T., DiFebbo, M. E., Kang, Y. J., Palmer, A. K. and Brent, R. L. (1975). Comparative embryotoxicity and teratogenicity of various tranquilizing agents in mice, rats, rabbits, and rhesus monkeys. *Toxicol. Appl. Pharmacol.*, **33**, 124 (Abstract)

46. Jackson, B. A., Rodwell, D. E., Kanegis, L. A. and Noble, J. F. (1975). Effect of maternally administered minocycline on embryonic and fetal development in the rhesus monkey (*Macaca mulatta*). *Toxicol. Appl. Pharmacol.*, **33**, 156

47. Hendrickx, A. G. (1975). Teratologic evaluation of imipramine hydrochloride in bonnet (*Macaca radiata*) and rhesus monkeys (*Macaca mulatta*). *Teratology*, **11**, 219

48. Hendrickx, A. G., Sawyer, R. H., Terrell, T. G., Osburn, B. I., Henrickson, R. V. and Steffek, A. J. (1975). Teratogenic effects of triamcinolone on the skeletal and lymphoid systems in nonhuman primates. *Fed. Proc.*, **34**, 1661

49. Fantel, A. G., Shepard, T. H., Newell-Morris, L. L. and Moffett, B. C. (1977). Teratogenic effects of retinoic acid in pigtail monkeys (*Macaca nemestrina*). I. General features. *Teratology*, **15**, 65

50. Reynolds, W. A. and Pitkin, R. M. (1975). Methylmercury toxicity *in utero* in the macaque. *J. Med. Primat.*, **4**, 372 (Abstract)

51. Poswillo, D. E., Sopher, D. and Mitchell, S. J. (1972). Experimental induction of foetal malformation with 'blighted' potato: A preliminary report. *Nature (Lond.)*, **239**, 462

52. Poswillo, D. E., Sopher, D., Mitchell, S. J., Coxon, D. T., Curtis, R. F. and Price, K. R. (1973). Investigation into the teratogenic potential of imperfect potatoes. *Teratology*, **8**, 339

53. Allen, J. R., Marlar, R. J., Chesney, C. F., Helgeson, J. P., Kelman, A., Weckel, K. G., Traisman, E. and White, J. W., Jr. (1972). Teratogenicity studies on late blighted potatoes in nonhuman primates (*Macaca mulatta* and *Saguinus labiatus*). *Teratology*, **15**, 17

54. Vondruska, J. F., Fancher, O. E. and Calandra, J. C. (1971). An investigation into the teratogenic potential of captan, folpet, and difolatan in nonhuman primates. *Toxicol. Appl. Pharmacol.*, **18**, 619

55. Dougherty, W. J., Herbst, M. and Coulston, F. (1975). The non-teratogenicity of 2,4,5-trichlorophenoxyacetic acid in the rhesus monkey (*Macaca mulatta*). *Bull. Environ. Contam. Toxicol.*, **13**, 477

56. Kurent, J. E. and Sever, J. L. (1977). Infectious diseases. In: J. G. Wilson and F. C. Fraser (eds.). *Handbook of Teratology*, pp. 225–259. (New York and London: Plenum Press)

57. London, W. T., Fuccillo, D. A., Sever, J. L. and Kent, S. G. (1975). Influenza virus as a teratogen in rhesus monkeys. *Nature (Lond.)*, **255**, 483

58. London, W. T., Levitt, N. H., Kent, S. G., Wong, V. G. and Sever, J. L. (1977). Congenital cerebral and ocular malformations induced in rhesus monkeys by Venezuelan equine encephalitis virus. *Teratology*, **16**, 285

59. Andersen, A. C., Hendrickx, A. G. and Momeni, M. H. (1977). Fractionated X-radiation damage to developing ovaries in the bonnet monkey (*Macaca radiata*). *Radiat. Res.*, **71**, 398

60. Miller, P., Smith, D. W. and Shepard, T. H. (1978). Maternal hyperthermia as a possible cause of anencephaly. *Lancet*, i, 519

61. Hendrickx, A. G. and Stone, G. W. (1976). Preliminary studies on the embryotoxicity of hyperthermia in the bonnet monkey (*Macaca radiata*). *Teratology*, **13**, 24A (Abstract)

62. Poswillo, D., Nunnerly, H., Sopher, D. and Keith, J. (1974). Hyperthermia as a teratogenic agent. *Ann. Royal Coll. Surg. Engl.*, **55**, 171

63. Smith, C. C. (1969). Value of nonhuman primates in predicting disposition of drugs in man. *Ann. N.Y. Acad. Sci.*, **162**, 604

64. Smith, R. L. and Williams, R. T. (1974). Comparative metabolism of drugs in man and monkeys. *J. Med. Primat.*, **3**, 138

65. Smith, R. L. and Caldwell, J. (1977). Drug metabolism in non-human primates. In: D. V. Parke and R. L. Smith (eds.). *Drug Metabolism from Microbe to Man*, pp. 331–356. (London: Taylor and Francis, Ltd.)

66. Clifford, J. M. (1977). Drug disposition and effect in sub-human primates used in pharmacology. *Comp. Biochem. Physiol.*, 57C, 1
67. Krauer, B. and Krauer, F. (1977). Drug kinetics in pregnancy. *Clin. Pharmacokin.*, 2, 167
68. Nishimura, H. (1973). Comparative study on maternal–embryonic transfer of drugs in man and laboratory animals. In: L. O. Boréus (ed.). *Fetal Pharmacology*, pp. 47–53. (New York: Raven Press)
69. Wilson, J. G., Scott, W. J. and Ritter, E. J. (1975). Comparative distribution of teratogenic drugs in pregnant rats and rhesus monkeys. In: D. Neubert and H. J. Merker (eds.). *New Approaches to the Evaluation of Abnormal Embryonic Development*, pp. 311–325. (Stuttgart: Thieme Verlag)
70. Wilson, J. G., Scott, W. J., Ritter, E. J. and Fradkin, R. (1975). Comparative distribution and embryotoxicity of hydroxyurea in pregnant rats and rhesus monkeys. *Teratology*, 11, 169
71. Wilson, J. G., Ritter, E. J., Scott, W. J. and Fradkin, R. (1977). Comparative distribution and embryotoxicity of acetylsalicylic acid in pregnant rats and rhesus monkeys. *Toxicol. Appl. Pharmacol.*, 41, 67

2

Usefulness of Golden Syrian hamster in experimental teratology with particular reference to the induction of orofacial malformations

R. M. SHAH

> 'I shouldn't know you again if we did meet', Humpty Dumpty replied in a discontented tone, giving her one of his fingers to share. 'You're so exactly like other people'.
>
> 'The face is what one goes by, generally', Alice remarked in a thoughtful tone.
>
> 'That's just what I complain of', said Humpty Dumpty. 'Your face is the same as everybody has – the two eyes, so –' (marking their places in the air with his thumb) 'nose in the middle, mouth under. It is always the same. Now if you had two eyes on the same side of the nose, for instance – or the mouth at the top – that would be some help.
>
> 'It wouldn't look nice', Alice objected. But Humpty Dumpty only shut his eyes, and said, 'Wait till you've tried'.
>
> *Through the Looking-Glass*, Lewis Carroll

INTRODUCTION

There is something metaphysical about malformations in general, and orofacial malformations in particular. Some of them are easy to recognize with the naked eye; others require physiological, biochemical and/or microscopic analysis. Nevertheless, the extent and severity of these malformations can be immediately life-threatening or detrimental to later development.

Orofacial malformations are amongst the very serious health problems and remain as an unsolved evil up to the present time. Because of their congenital origin, they affect both the child and the parents soon after the baby is born. A child, with orofacial malformation presents a complex medical, dental, emotional, social, educational and vocational problem, and may require prolonged supervision for optimal rehabilitation. In order to prevent and treat these malformations, the knowledge of normal, and abnormal prenatal development of the orofacial region is crucial. Investigations of this region in humans are relatively scant, perhaps due to lack of accessibility to the appropriate embryonic material, and definitely due to moral and ethical considerations. Studies on laboratory animals have, however, contributed significantly towards understanding the mechanisms of these malformations.

Spatially and temporally, the organs and structures of the orofacial complex are related both developmentally and physiologically. They include palate, lip, mandible, cheek, tongue, teeth, nasal cavities, salivary glands, oral mucosa and temporomandibular joint. Coordinated development and function of these organs and structures during pre- and postnatal life is essential since alteration in any one of them could affect the other. Normal development of these structures can be found in any standard embryology textbooks, and will not be repeated here. This chapter rather discusses the usefulness of the hamster from a teratological viewpoint with special emphasis on the induction of orofacial malformations.

HAMSTER AS A MODEL IN THE STUDY OF MALFORMATIONS

Golden Syrian hamster, a desert animal, was introduced to extensive biological research during the 1940s. The hamster can be a very useful animal in experimental teratology, both for drug screening and for studying normal and abnormal developmental mechanisms. It compares well with the more commonly used laboratory animals, i.e. rat, mouse and rabbit. Unfortunately, with the exception of Canada[7], the teratological guidelines of many countries[1–5] and of the World Health Organization[6], do not include the hamster as one of the choice animals.

The gestation period of hamster is 15½–16 days which is less than that of the mouse (18–20 days), rat (21–22 days), chick (20–21 days), rabbit (31–34 days) and guinea pig (64–68 days). The oestrous cycle of the hamster is well-defined[8]. The animals are easy to breed and mating can be timed accurately. An average litter gives 11–12 offspring. The large size of the hamster fetus helps in gross observations and analytical procedures. The embryology of hamster is fairly well established[9–11], but this area still needs further work especially the ontogenesis of organs and structures from the molecular and functional viewpoint. Although major organogenesis occurs between days 8 and 10 of gestation, many organs such as palate, limb, eye, kidney, testes, heart, brain, etc. continue to develop thereafter (i.e. until day 14), and thus are susceptible to environmental insults. Data on spontaneous malformations is lacking. In our laboratory, during the past ten years, we have not observed any gross spontaneous malformations in hamster fetuses. Ferm (personal communication) observed that the incidence is much less than 1%. Large colonies of hamsters are easy to handle and the cost of maintenance is low. The animals are clean, domesticated, and friendly, and are generally resistant to disease. A stable heterogeneous stock and many inbred strains of hamsters are now available, though information on intraspecies differences is still lacking. Thus the hamster can be a very useful model for teratological drug testing. Already this animal has been used for a wide variety of biological research[12].

The hamster embryo responds to a variety of known teratogenic agents[11,13–75]. A comprehensive list of these teratogens and the organ systems affected are presented in Table 2.1. Many of these teratogens produce malformations of a particular organ, e.g. palate, or a malformation syndrome of developmentally, functionally and spatially related tissues, e.g. cleft palate, micrognathia and microglossia. An advantage of such a system, from the mechanistic viewpoint, is that it allows one to determine whether an organ or tissue is required for the normal development of other organs or tissues.

Table 2.1 Abnormalities produced by environmental agents in hamster fetuses

Agent	*Malformation(s)*	*References*
Heat	Exencephaly, encephalocele, cleft lip, cleft palate, limb anomalies, rib defects	13–15
Freezing	Hydrocephalus, anencephaly, exencephaly, anophthalmia, spina bifida, cleft lip, cleft palate, micrognathia, limb defects, oedema, growth retardation	16
Irradiation	Acrania, exencephaly, hydrocephalus, microphthalmia, anophthalmia, spina bifida, albinism, cranial blisters, atelocephalia, cleft lip, hernia, limb defects, oedema	17, 18
Hyperbaric oxygen	Exencephaly, spina bifida, cleft lip, limb defects, gut herniation	19
Vitamin A	Anencephaly, anopia, exophthalmosis, spina bifida, rib defects, cleft lip, cleft palate, limb defects, gut herniation	20
Retinoic acid	Encephalocele, omphalocele, microcephaly, exencephaly, spina bifida, exophthalmosis, microphthalmia, ear defects, facial clefts, cleft lip, cleft palate, micrognathia, maxillary hypoplasia, rib defects, tail anomalies, limb defects, anal atresia, genital malformation	11
Arsenic	Encephalocele, exencephaly	14, 21
Lead	Exencephaly, spina bifida, anophthalmia, rib defects, tail anomalies	22, 23
Zinc	Exencephaly, rib defects	24
Cadmium	Exencephaly, encephalocele, anophthalmia, exophthalmia, microphthalmia, cleft lip, cleft palate, micrognathia, tail anomalies, umbilical hernia, limb defects	23–26
Mercury	Exencephaly, encephalocele, microphthalmia, anophthalmia, oedema, growth retardation, ectopia cordis, cleft lip, cleft palate, rib defects, tail malformations, limb defects	27–29

(continued)

Table 2.1 (*continued*)

Agents	*Malformation(s)*	*References*
Indium	Limb defects	30
Ytterbium chloride	Exencephaly, encephalocele, hydrocephalus, meningocele, eye anomalies, spina bifida, oedema, limb defects, tail malformations, growth retardation, gut herniation, liver malformation	31
β-Aminopropionitril	Exencephaly, encephalocele, cleft palate, micrognathia, oedema, limb defects, growth retardation, rib defects, haemorrhage	32
Aminoacetonitrile	Encephalocele, ectocardia, umbilical hernia, limb defects, growth retardation	33
D-penicillamine	Exencephaly, encephalocele, limb defects, rib defects, growth retardation	33
Semicarbazide	Limb defects, growth retardation	33
Marijuana	Exencephaly, omphalocele, myelocele, spina bifida oedema, limb defects	34, 35
Morphine	Exencephaly, cranioschisis, anophthalmia, microphthalmia, myelocele, spina bifida, micrognathia, liver anomalies	36, 37
Thebaine	Myelocele, cranioschisis, anophthalmia, microphthalmia, spina bifida, micrognathia, liver anomalies	37
Heroin	Exencephaly, myelocele, cranioschisis, spina bifida, anophthalmia, microphthalmia, liver anomalies, micrognathia	36, 37
Codeine	Myelocele, spina bifida, cranioschisis, anophthalmia, microphthalmia, micrognathia, liver anomalies	37
Methadone	Exencephaly, cranioschisis	36, 37
Mescaline	Exencephaly, hydrocephalus, myelocele, omphalocele, meningocele, spina bifida, oedema, intracranial haemorrhage	38
Lysergic acid diethylamide	Exencephaly, hydrocephalus, myelocele, omphalocele, meningocele, spina bifida, oedema, intracranial haemorrhage	38
Bromolysergic acid	Exencephaly, hydrocephalus, myelocele omphalocele, meningocele, spina bifida, oedema, intracranial haemorrhage	38
Adrenocorticotropic hormone	Cleft palate, growth retardation	39
Hydrocortisone	Cleft palate, growth retardation	40–42
Triamcinolone	Cleft palate, growth retardation	43
Dexamethasone	Cleft palate, growth retardation	43
Prednisolone	Cleft palate, growth retardation	43
Corticosterone	Cleft palate, growth retardation	43
Vinblastine	Anophthalmia, microphthalmia, spina bifida, rib and vertebral defects	44

(*continued*)

Table 2.1 (*continued*)

Agents	*Malformation(s)*	*References*
Vincristine	Exencephaly, anophthalmia, microphthalmia, rib defects	44
Hydroxyurea	Cranioschisis, spina bifida, central axis deformities	45, 46
Urethane	Exencephaly, spina bifida, neural tube anomalies, cardiac defects	46
Cytosine arabinoside	Cerebellar hypoplasia	47
Pyrimethamine	Meningocele, craniorachischisis, haemorrhage, limb defects	48
Actinomycin D	Omphalocele, microcephaly, exencephaly, hydrocephaly, spina bifida, agnathia, micrognathia, limb defects	49
5-Fluoro-2-deoxy-cytidine	Cephalocele, open eye, microphthalmia, cleft palate, micrognathia, skeletal anomalies, limb defects, gut herniation, aplasia and hypoplasia of kidney	50
Hadacidin	Exencephaly, meningocele, acrania, anophthalmia, microphthalmia, open eye, cleft lip, cleft palate, micrognathia, limb defects, oedema, gut herniation, tail anomalies, tooth anomalies, aplasia and hypoplasia of salivary glands, dysmorphogenic changes in liver, lung and kidney	51, 52
5-Bromo-2-deoxyuridine	Encephalocele, spina bifida, microphthalmia, cleft lip, cleft palate, micrognathia, limb defects, tail anomalies, rib defects, growth retardation, gut herniation	53, 54
5-Fluorouracil	Exencephaly, open eye, cleft palate, micrognathia, limb defects, tail anomalies, gut herniation, growth retardation	54
6-Mercaptopurine	Cleft palate, micrognathia, microglossia, limb defects, gut herniation, growth retardation; dysmorphogenic changes in brain, eye, kidney, liver, salivary gland and lung; tooth anomalies	55
6-Aminonicotinamide	Exencephaly, hydrocephalus, microcephaly, spina bifida, anophthalmia, cleft lip, cleft palate, micrognathia, limb defects, rib defects	56
Colchicine	Exencephaly, anophthalmia, microphthalmia, umbilical hernia, rib defects	57
Thiram	Exencephaly, cranial pimple, spina bifida, cleft lip, cleft palate, micrognathia, rib defects, tail anomalies, umbilical hernia, kidney aplasia, cardiac abnormalities	58
Disulfiram	Exencephaly, cranial pimple, spina bifida, cleft palate, limb defects, rib defects, tail anomalies, kidney aplasia, cardiac abnormalities	59
Aldrin	Open eye, cleft lip, cleft palate, limb defects, rib defects, growth retardation	59

(*continued*)

Table 2.1 (*continued*)

Agents	*Malformation(s)*	*References*
Dieldrin	Platy crania, exencephaly, open eye, cleft lip, cleft palate, micrognathia, limb defects, rib defects, growth retardation	59
Endrin	Open eye, cleft lip, cleft palate, limb defects, rib defects, growth retardation	59
Adhesive spray	Exencephaly, oedema, limb defects, skeletal malformations, ectopic and cystic kidney	60
Thalidomide	Exencephaly, acrania, cleft lip, cleft palate, limb defects, tail anomalies	61
Acetazolamide	Exencephaly, microphthalmia, cleft lip, micrognathia, limb defects, gastroschisis	62, 63
Salicylamide	Cranial blisters, oedema, subdermal haemorrhage, umbilical hernia, albinism	64
Angiotensin	Hydrocephalus, omphalocele, myelocele, Meningocele, micrognathia, oedema	65
Trypan blue	Exencephaly, encephalocele, hydrocephalus, meningocele, omphalocele, tail anomalies, oedema, umbilical hernia	66, 67
Isoproterenol	Exencephaly, hydrocephalus, microcephalus, myelocele, anophthalmia, tail anomalies, oedema, umbilical hernia, ectopia corids	67
Diazepam	Exencephaly, cleft palate, limb anomalies, growth retardation	68
δ-Hydroxy-γ-oxo-L-norvaline	Hydrocephalus, limb defects	69
Ochratoxin A	Hydrocephalus, cleft lip, micrognathia, limb defects, tail anomalies	70
3,3′-dichloro-5,5′-dinitro-0,0′-biphenol	Skull anomalies, lumbosacral herniation, rib defects, skeletal malformations	71
Dimethyl sulphoxide	Exencephaly, neural tube anomalies, cleft lip, cleft palate, limb defects	72
Reovirus, type 1	Hydrocephalus	73
Reovirus, type 3	Infection of skin, oral mucosa, visual organs, muscles, nerve	74
H-1 virus	Exencephaly, spina bifida, microcephaly, facial clefts, micrognathia, umbilical hernias, cardiac anomalies, limb defects, liver anomalies	75

GLUCOCORTICOID HORMONES AND CLEFT PALATE

In 1949 Baxter and Fraser[76] experimentally induced cleft palate in the fetuses by treating pregnant mice with cortisone. In retrospect, the experiment was doubly significant. First, because glucocorticoid hormones are used extensively in humans, the observations caused some apprehension regarding their use during pregnancy. Several human cases of cleft palate and/or various other malformations were also linked to the glucocorticoid treatment during pregnancy. Second, the experiment provided researchers with an animal model to study the mechanism(s) of normal and abnormal palatal development. Since then, numerous attempts were made to explore the reaction of palatal tissues of different animal species to glucocorticoid treatment during pregnancy. These studies demonstrated that there are important intra- and interspecies differences in the potentiality of glucocorticoids to induce cleft palate. Table 2.2 shows that mouse, hamster and rabbit fetuses are more

Table 2.2 **Potency of glucocorticoid hormones to induce cleft palate in humans and laboratory animals***

Drug	*Human*	*Hamster*	*Mouse*	*Rabbit*	*Rat*
Cortisone	Suspected +ve	none	+ve	Mildly +ve	none
Hydrocortisone	†	+ve	+ve	+ve	none
Corticosterone	†	+ve	+ve	†	none
Triamcinolone	†	+ve	+ve	+ve	+ve
Dexamethasone	†	+ve	+ve	+ve	+ve
Prednisolone	Suspected +ve	+ve	+ve	+ve	none

* Data collected from numerous sources. For summary see Shah and Kilistoff (1976)[43]
† Not yet known

susceptible than rat to glucocorticoid assault during pregnancy. Recently Brent[79] suggested that if an agent is associated with a unique malformation, which when duplicated in various animal models resembles the one in humans, and if there is a dose–response relationship between the agent and the incidence of malformation, then it is likely that the particular agent may also be a human teratogen. In the light of proven teratogenicity of glucocorticoids in various laboratory animals and their suspected effect on the human embryo[80–93], it was suggested that prior to their therapeutic recommendation, all glucocorticoids should be tested for their teratogenic potency to induce cleft palate in at least two species, preferably mouse and hamster[43].

In Golden Syrian hamster palatal clefts can be induced with ease following treatment with all glucocorticoid hormones, with the exception of cortisone[40,42,43]. The most severe form of cleft involves the entire length of the secondary palate (Figure 2.1). The less severe forms show an array of incomplete cleft palate: cleft in the anterior and posterior part of the secondary palate with varying degrees of fusion in the middle (Figure 2.2); a cleft only of the anterior (Figure 2.3) or the posterior part of the secondary palate. Experimental induction of these morphological forms of cleft palate is dependent upon the time and dose of hormone treatment. For example, it was observed that day 11 of pregnancy is the most appropriate time to induce maximum frequency of complete cleft palate, and day $11\frac{1}{2}$ for incomplete cleft palate, with 40 mg hydrocortisone[42]. The relationship between the dose and the morphological form can readily be seen with prednisolone. Treatment of pregnant hamsters on day 11 with 10 mg prednisolone produces the maximum incidence of incomplete cleft palate, and 15 mg complete cleft palate[43]. On the basis of this 'dose–effect' relationship, one may suggest that a particular drug concentration in the maternal and/or fetal body needs to be maintained,

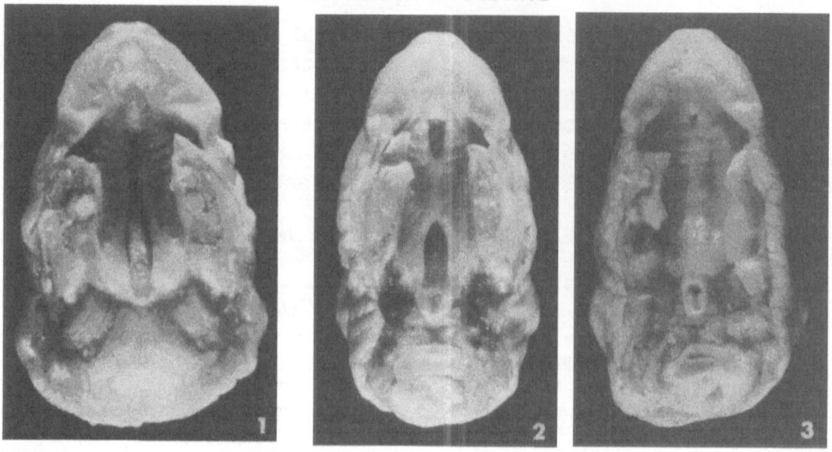

Figures 2.1–2.3 Hydrocortisone induced cleft palate in hamster fetuses at term 1, Complete cleft of the secondary palate following 40 mg hydrocortisone on day 11 of pregnancy. × 7 2, Incomplete cleft of the secondary palate following 20 mg hydrocortisone on day 11 of pregnancy. The cleft is present in both the anterior and the posterior thirds of the palate with fusion in the middle. × 7 3, Incomplete cleft of the secondary palate following 15 mg hydrocortisone on day 11 of pregnancy. The cleft is present only in the anterior thirds of the palate. × 7

during a precise interval, to produce a specific type of cleft palate. It may be that a larger dose produces a more severe form of cleft palate than a smaller dose. Studies on the pharmacokinetic aspects, i.e. absorption, metabolism, excretion etc., in the maternal and fetal system may, therefore, further our understanding of 'dose–effect' relationship during experimental clefting.

The aforementioned hamster model also allows one to study the mechanism of cleft formation in one part and fusion in the other part of the same palate following treatment with an environmental teratogen. Moreover, this system will facilitate the study of the suspected role of such neighbouring structures as the tongue, mandible, cranial base, nasal septum, etc. during normal and abnormal palatogenesis.

ANTINEOPLASTIC AGENTS AND OROFACIAL MALFORMATIONS

The antineoplastic chemicals as a group are probably the most potent teratogens known. In humans, the teratogenic risk following treatment with these chemicals will be great, because the drugs are usually administered in the range of maximum tolerated doses. This group of chemicals is distinguished from other groups of drugs by a very low therapeutic index, due to their relative inability to discriminate between normal and target tissues. Because of their broad spectrum of action the antineoplastic agents are also used as immunosuppressants to prevent rejection of tissue grafts, as an epilating agent, as a chemical sterilizer for insects, and in the treatment of autoimmune diseases and psoriasis. In recent years, the high degree of activity in the biological system of these drugs have also been exploited in studying mechanisms of mutagenesis, carcinogenesis and teratogenesis.

Although it is generally implied that the biological properties of antineoplastic agents are broadly similar, the chemicals vary widely in their final effects, depending upon the animal system, and the target tissue within the system. In general, however, antineoplastic agents act on proliferating cells primarily by altering nucleic acid formation and function, protein synthesis, and by inhibition and destruction of specific enzymes.

Several antitumour agents are now recognized to be teratogenic in different experimental animals. Only a few agents, however, have been tested in Golden Syrian hamster (Table 2.1)[44–56]. The general teratological effect of these antineoplastic agents in hamsters is similar to that in mice, rats and rabbits. They all produce numerous malformations of various organs and systems. However, the effect of a particular agent on an organ, a system or a tissue may differ among species. This species-specific target tissue effect has been successfully exploited in our laboratory in the production of orofacial malformations in hamsters. As already noted such a system allows one to study the role, if any, played by an individual organ, in the development of another organ or tissue, e.g. role of the tongue and mandible during palate development, etc.

Various orofacial structures of hamster react differently to different antineoplastic agents. For example, a syndrome of cleft palate, cleft lip and mandibular micrognathia is induced consistently with hadacidin[52] (Figures 2.4–2.11) and 5-bromo-2-deoxyuridine[54] but not with 5-fluorouracil[54] and 6-mercaptopurine[55]. These latter drugs do not appear to react with lip tissue. Similarly, a syndrome of cleft palate, mandibular micrognathia and microglossia is induced only by 6-mercaptopurine[55].

A dose and treatment time dependent correlation in the production of orofacial malformations (Table 2.3) suggests that a teratogenic agent may

Table 2.3 Antineoplastic agents and orofacial malformations in hamster

Drug	*Cleft palate*	*Cleft lip*	*Mandibular micrognathia*	*Microglossia*
Hadacidin*	+ve	+ve	+ve	none
5-Bromo-2-deoxyuridine*	+ve	+ve	+ve	none
5-Fluorouracil†	+ve	none	+ve	none
6-Mercaptopurine*	+ve	none	+ve	+ve

* A dose-dependent positive correlation for induction of malformation. Day 9 of treatment is most sensitive
† No correlation between induction of cleft palate and micrognathia

affect a particular site during organogenesis, resulting in the dysmorphogenesis of a particular organ(s). Johnston and associates[77] have suggested that the mesenchyme of the head region is primarily derived from the neural crest cells. The neural crest derived mesenchyme, also known as ectomesenchyme, takes part in the formation of various tissues and organs of the orofacial region. The teratogenic effect of hadacidin, 6-mercaptopurine, and 5-bromo-2-deoxyuridine, all of which produce a malformation syndrome, is the most prominent on day 9 of gestation[52, 54, 55], i.e. soon after the formation of the ectomesenchyme[78]. Thus it is quite possible that the common locus of action for these drugs may be the migration and/or subsequent differentiation

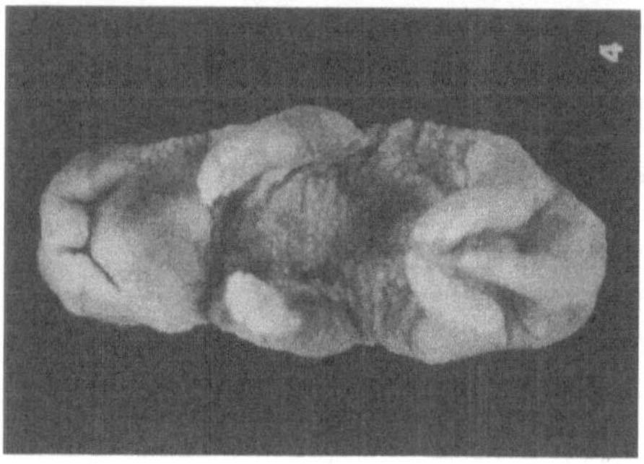
4

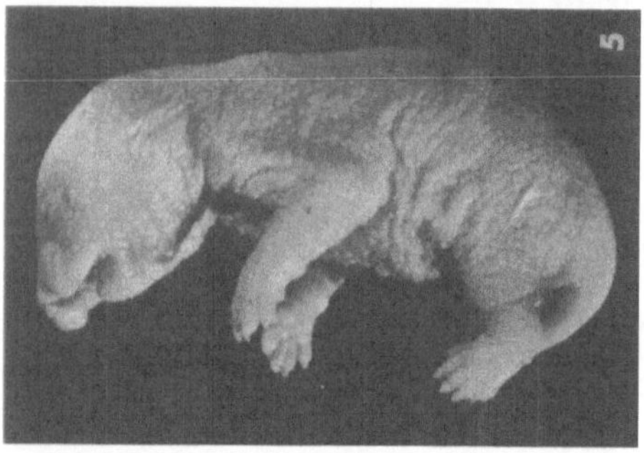
5

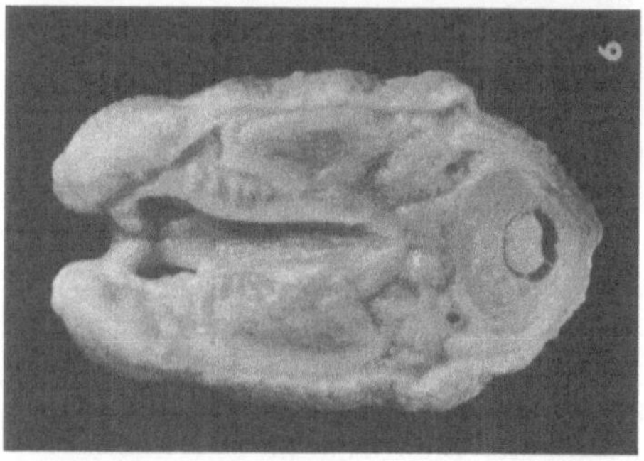
6

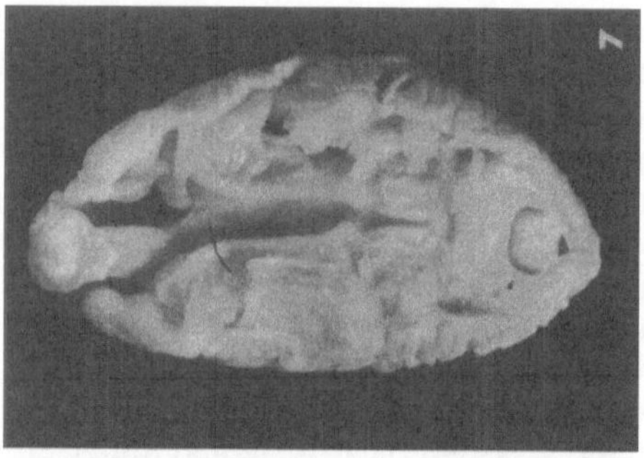
7

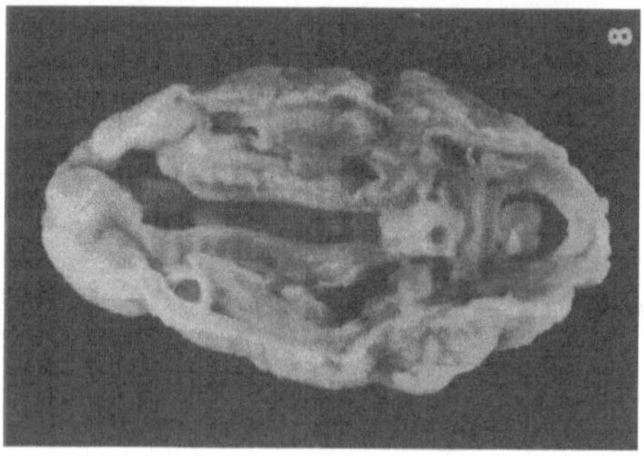
8

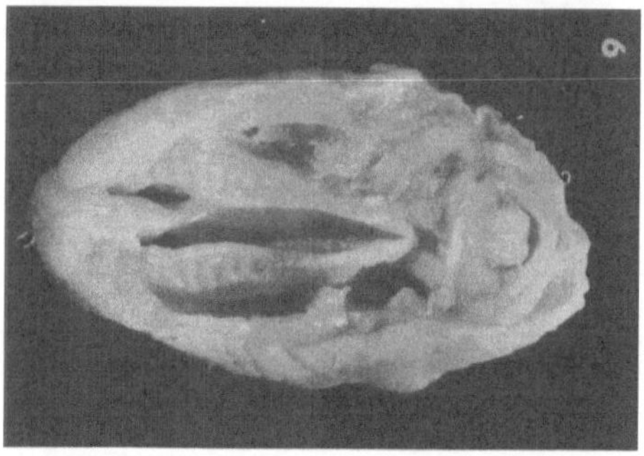
9

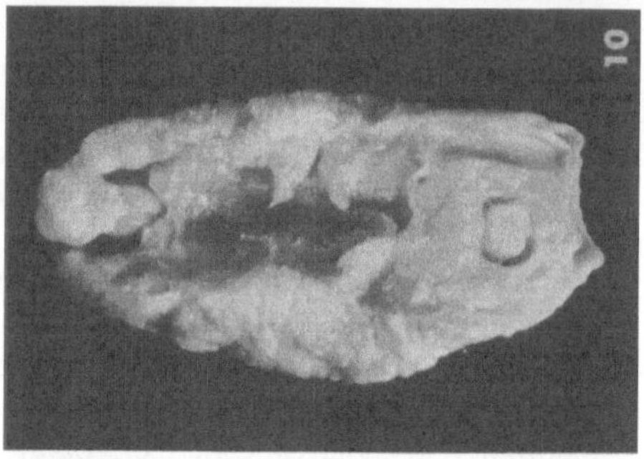
10

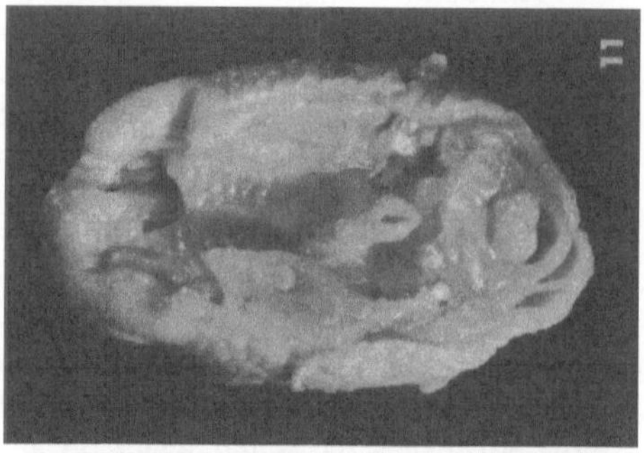
11

of the neural crest cells. Future studies should perhaps be directed towards behaviour of the ectomesenchyme and its role in the formation of various orofacial structures.

In addition to the aforementioned gross malformations, several histological defects are observed in the orofacial area, in term fetuses, following prenatal drug treatment during early pregnancy. Commonly observed microscopic defects are absent or flattened nasal septum, lack of stratification in the oral epithelia, aplasia and/or hypoplasia of various salivary glands, fusion between various intra-oral structures, e.g. palate, tongue, mandible, cheek, floor of the mouth; abnormalities of odontogenesis such as partial anodontia, separation of the dental papilla from the enamel organ, dysmorphogenic changes in the dental papillae, oronasal and nasofacial fistula, and ectopic cartilage in the tongue and cheek.

The dysmorphogenic alterations at the histological level were more pronounced following treatment with hadacidin[52] and 6-mercaptopurine[55] than 5-fluorouracil and 5-bromo-2-deoxyuridine[54]. In addition, many of these cellular and tissue level defects were observed in drug treated fetuses which did not show any gross abnormalities or general growth retardation. Functional abnormalities which may result from microscopic malformations may not become obvious until late in postnatal life. Clearly there is a risk factor involved in a fetus exposed to drugs during pregnancy. It is, therefore, of paramount importance that histological examination of different organs and tissues be made an essential requirement in the guidelines of drug safety evaluation procedures. Such examination, unfortunately is not required in the drug safety protocol of most nations, including the World Health Organization[1–7]. From the viewpoint of intra-uterine treatment, and ultimately prevention of congenital malformation, such information may prove useful if one is to rationally interfere with the maternal and/or fetal environment.

The foregoing discussion clearly suggests that the hamster is a useful animal for both teratological drug screening and for studying the mechanism(s) of normal and abnormal development. Although the animal is susceptible to several known teratogens many more groups of drugs should be evaluated to establish the versatility of the system. Orofacial malformations can be induced with great reliability by several teratogens. The time of treatment seems to be responsible for the type of malformation, whereas the dose for the severity of defect. The hamster is, therefore, well suited as an animal model for investigating the pathogenesis of orofacial malformations.

Figures 2.4–2.11 Orofacial malformations in hadacidin treated hamster fetuses at term 4, Bilateral cleft lip following 250 mg hadacidin on day 9 of pregnancy. × 1.875 5, Micrognathia and cleft lip in the fetus from Figure 2.4. × 1.875 6, Median cleft lip and complete cleft of the primary and the secondary palate following 200 mg hadacidin on day 8 of pregnancy. × 5.6 7, Bilateral cleft lip and complete cleft of the primary and the secondary palate in the fetus from Figure 2.4. × 5.6 8, Unilateral cleft lip and complete cleft of the secondary palate following 150 mg hadacidin on day 8 of pregnancy. × 5.6 9, Microform of unilateral cleft lip and complete cleft of the secondary palate following 125 mg hadacidin on day 8 of pregnancy. × 5.6 10, Bilateral cleft lip and small incomplete cleft in the anterior and posterior third of the secondary palate following 125 mg hadacidin on day 8 of pregnancy. × 5.6 11, Unilateral cleft lip and incomplete cleft in the anterior third of the palate following 150 mg hadacidin on day 10 of gestation. × 5.6

ACKNOWLEDGEMENTS

This work was supported by a grant from the Medical Research Council of Canada. The author remains grateful to V. Koulouris, V. Long, R. Paton and D. Burdett for various assistance.

References

1. Hebold, G. (1973). General and special toxicological testing of drugs. Recommendations of various countries. *Pharm. Ind.*, **35**, 205
2. Tuchmann-Duplessis, H. (1972). Teratogenic drug screening. Present procedures and requirements. *Teratology*, **5**, 271
3. Ferngren, H. and Forsberg, U. (1971). Evaluation in animals of teratogenic effects of drugs submitted to the Swedish Drug Control 1963–1968. *Proc. Eur. Soc. Study Drug Toxic.*, **12**, 347
4. Dunlop Committee for the Safety of New Drugs. Report. (1968)
5. Food and Drug Administration. (1966). Guidelines for reproduction studies for safety evaluation of drugs for human use
6. World Health Organization Scientific Group. (1967). Principles for testing of drugs for teratogenicity. *WHO Tech. Rep. Ser.*, **364**, 1
7. Health and Welfare Canada. (1973). The testing of chemicals for carcinogenicity, mutagenicity, teratogenicity
8. Orsini, M. W. (1961). The external vaginal phenomena characterizing the stages of the estrous cycle, pregnancy, pseudopregnancy, lactation and the anestrous hamster, *Mesocricetus auratus* Waterhouse. *Proc. Animal Care Panel*, **11**, 193
9. Graves, A. P. (1945). Development of the golden hamster, *Cricetus auratus* Waterhouse, during the first nine days. *Am. J. Anat.*, **77**, 219
10. Boyer, C. (1968). Embryology. In: R. A. Hoffman, P. F. Robinson and H. Magalhaes (eds.). *The Golden Hamster: Its Biology and Use in Medical Research*, pp. 73–90. (Ames: The Iowa State University Press)
11. Shenefelt, R. E. (1972). Morphogenesis of malformations in hamsters caused by retinoic acid: relation to dose and stage of treatment. *Teratology*, **5**, 103
12. Homburger, F. (1976). Potential contribution of inbred syrian hamsters to future toxicology. *Adv. Modern Toxicol.*, **1**, 35
13. Kilham, L. and Ferm, V. H. (1976). Exencephaly in fetal hamsters following exposure to hyperthermia. *Teratology*, **14**, 323
14. Ferm, V. H. and Kilham, L. (1977). Synergistic teratogenic effects of arsenic and hyperthermia in hamsters. *Envir. Res.*, **14**, 483
15. Umpierre, C. C. and Dukelow, W. R. (1977). Environmental heat stress effects in the hamster. *Teratology*, **16**, 155
16. Smith, A. U. (1957). The effects on foetal development of freezing pregnant hamsters (*Mesocricetus auratus*). *J. Embryol. Exp. Morphol.*, **5**, 311
17. Harvey, E. B. and Chang, M. C. (1962). Effects of radiocobalt irradiation of pregnant hamsters on the development of embryos. *J. Cell. Comp. Physiol.*, **59**, 293
18. Jensh, R. P. and Magalhaes, H. (1962). The effect of whole body X-irradiation on the central nervous system of golden hamster embryos. *Penn. Acad. Sci.*, **36**, 194
19. Ferm, V. H. (1964). Teratogenic effects of hyperbaric oxygen. *Proc. Soc. Exp. Biol. Med.*, **116**, 975
20. Marin-Padilla, M. and Ferm, V. H. (1965). Somite necrosis and developmental malformations induced by vitamin A in the golden hamster. *J. Embryol. Exp. Morphol.*, **13**, 1
21. Ferm, V. H. and Carpenter, S. J. (1968). Malformations induced by sodium arsenate. *J. Reprod. Fertil.*, **17**, 199
22. Ferm, V. H. and Carpenter, S. J. (1967). Developmental malformations resulting from the administration of lead salts. *Exp. Mol. Pathol.*, **7**, 208
23. Ferm, V. H. (1969). The synteratogenic effect of lead and cadmium. *Experientia*, **25**, 56
24. Ferm, V. H. and Carpenter, S. J. (1968). The relationship of cadmium and zinc in experimental mammalian teratogenesis. *Lab. Invest.*, **18**, 429

25. Ferm, V. H. (1971). Developmental malformations induced by cadmium. A study of timed injections during embryogenesis. *Biol. Neonate*, **19**, 101
26. Gale, T. F. and Ferm, V. H. (1973). Skeletal malformations resulting from cadmium treatment in the hamster. *Biol. Neonate*, **23**, 149
27. Gale, T. F. and Ferm, V. H. (1973). Embryopathic effects of mercuric salts. *Life Sci.*, **10**, 1341
28. Gale, T. F. (1973). The interaction of mercury with cadmium and zinc in mammalian embryonic development. *Envir. Res.*, **6**, 95
29. Gale, T. F. (1974). Embryopathic effects of different routes of administration of mercuric acetate in the hamster. *Envir. Res.*, **8**, 207
30. Ferm, V. H. and Carpenter, S. J. (1970). Teratogenic and embryopathic effects of indium, gallium and germanium. *Toxicol. Appl. Pharmacol.*, **16**, 166
31. Gale, T. F. (1975). The embryotoxicity of ytterbium chloride in golden hamsters. *Teratology*, **11**, 289
32. Wiley, M. J. and Joneja, M. G. (1976). The teratogenic effects of β-amino propionitrile in hamsters. *Teratology*, **14**, 43
33. Wiley, M. J. and Joneja, M. G. (1978). Neural tube lesions in the offspring of hamsters given single oral doses of lathyrogens early in gestation. *Acta Anat.*, **100**, 347
34. Geber, W. F. and Schramm, L. C. (1969). Teratogenicity of marihuana extract as influenced by plant origin and seasonal variation. *Arch. Int. Pharmacol. Therap.*, **177**, 224
35. Geber, W. F. and Schramm, L. C. (1969). Effects of marihuana extract on fetal hamsters and rabbits. *Toxicol. Appl. Pharmacol.*, **14**, 276
36. Geber, W. F. and Schramm, L. C. (1969). Comparative teratogenicity of morphine, heroin and methadone in the hamster. *Pharmacologist*, **11**, 248
37. Geber, W. F. and Schramm, L. C. (1975). Congenital malformations of the central nervous system produced by narcotic analgesics in the hamster. *Am. J. Obstet. Gynecol.*, **123**, 705
38. Geber, W. F. (1967). Congenital malformations induced by mescaline, lysergic acid diethylamide, and bromolysergic acid in the hamster. *Science*, **158**, 265
39. Shah, R. M. (1977). Induction of cleft palate in hamster fetus following prenatal treatment with ACTH. *Toxicol. Appl. Pharmacol.*, **42**, 229
40. Shah, R. M. and Chaudhry, A. P. (1973). Hydrocortisone induced cleft palate in hamsters. *Teratology*, **7**, 191
41. Chaudhry, A. P. and Shah, R. M. (1973). Estimation of hydrocortisone dose and optimal gestation period for cleft palate induction in golden hamster. *Teratology*, **8**, 139
42. Shah, R. M. and Travill, A. A. (1976). The teratogenic effects of hydrocortisone on palatal development in hamster. *J. Embryol. Exp. Morphol.*, **35**, 213
43. Shah, R. M. and Kilistoff, A. (1976). Cleft palate induction in hamster fetuses by glucocorticoid hormones and their synthetic analogues. *J. Embryol. Exp. Morphol.*, **36**, 101
44. Ferm, V. H. (1963). Congenital malformations in hamster embryos after treatment with vinblastine and vincristine. *Science*, **141**, 426
45. Ferm, V. H. (1965). Teratogenic activity of hydroxyurea. *Lancet*, i, 1338
46. Ferm, V. H. (1966). Severe developmental malformations. Malformations induced by urethane and hydroxyurea in the hamster. *Arch. Pathol.*, **81**, 174
47. Fischer, D. S. and Jonas, A. M. (1965). Cerebellar hypoplasia resulting from cytosine arabinoside treatment in the neonatal hamster. *Clin. Res.*, **13**, 540
48. Sullivan, G. E. and Takacs, E. (1971). Comparative teratogenicity of pyrimethamine in rats and hamsters. *Teratology*, **4**, 205
49. Elis, J. and DiPaolo, J. A. (1970). The alteration of actinomycin D teratogenicity by hormones and nucleic acid. *Teratology*, **3**, 33
50. Degenhardt, K. H., Yamamura, H., Franz, J. and Kleinebrecht, J. (1971). Dose response to 5-fluoro-2-deoxyuridine in organogenesis of the golden hamster. *Cong. Anom.*, **11**, 41
51. Roux, C. and Horvath, C. (1970). Effet teratogene de l'hadacidine chez la souris et le hamster. *C. R. Soc. Biol.*, **164**, 2171
52. Shah, R. M. (1977). Effects of prenatal administration of hadacidin, a cancer chemotherapeutic agent, on the development of hamster fetuses. *J. Embryol. Exp. Morphol.*, **39**, 203
53. Ruffolo, P. R. and Ferm, V. H. (1965). The embryocidal and teratogenic effects of 5-bromodeoxyuridine in the pregnant hamster. *Lab. Invest.*, **14**, 1547

54. Shah, R. M. and MacKay, R. A. (1978). Teratological evaluation of 5-fluorouracil and 5-bromo-2-deoxyuridine on hamster fetuses. *J. Embryol. Exp. Morphol.*, **43**, 47
55. Shah, R. M. and Burdett, D. N. (1979). Developmental abnormalities induced by 6-mercaptopurine in the hamster. *Canad. J. Physiol. Pharmacol.*, **57**, 53
56. Turbow, M. M., Clark, W. H. and DiPaolo, J. A. (1971). Embryonic abnormalities in hamsters following intra-uterine injection of 6-aminonicotinamide. *Teratology*, **4**, 427
57. Ferm, V. H. (1963). Colchicine teratogenesis in hamster embryos. *Proc. Soc. Exp. Biol. Med.*, **112**, 775
58. Robens, J. F. (1969). Teratological studies of carbaryl, diazinon, norea, disulfiram and thiram in small laboratory animals. *Toxicol. Appl. Pharmacol.*, **15**, 152
59. Ottolenghi, A. D., Haseman, J. K. and Suggs, F. (1974). Teratogenic effects of aldrin, dieldrin and endrin in hamsters and mice. *Teratology*, **9**, 11
60. Murphy, J. C., Collins, T. F. X., Black, T. N. and Osterberg, R. E. (1975). Evaluation of the teratogenic potential of a spray adhesive in hamsters. *Teratology*, **11**, 243
61. Homburger, F., Chaube, S., Eppenberger, M., Bogdonoff, P. D. and Nixon, C. W. (1965). Susceptibility of certain inbred strains of hamsters to teratogenic effects of thalidomide. *Toxicol. Appl. Pharmacol.*, **7**, 686
62. Layton, W. M. (1971). Teratogenic action of acetazolamide in golden hamsters. *Teratology*, **4**, 95
63. Storch, T. G. and Layton, W. M. (1973). Teratogenic effects of intra-uterine injection of acetazolamide and amiloride in hamsters. *Teratology*, **7**, 209
64. Lapointe, R. and Harvey, E. B. (1964). Salicylamide induced anomalies in hamster embryos. *J. Exp. Zool.*, **156**, 197
65. Geber, W. F. (1969). Angiotens in teratogenicity in the fetal hamster. *Life Sci.*, **8**, 525
66. Ferm, V. H. (1958). Teratogenic effects of trypan blue on hamster embryos. *J. Embryol. Exp. Morphol.*, **6**, 284
67. Geber, W. F. (1969). Comparative teratogenicity of isoproterenol and trypan blue in the fetal hamster. *Proc. Soc. Exp. Biol. Med.*, **130**, 1168
68. Shah, R. M., Donaldson, D. and Burdett, D. (1979). Teratological evaluation of diazepam in hamster. *Canad. J. Physiol. Pharmacol.* (In press)
69. Mizutani, M. and Ihara, T. (1973). Teratogenicity of δ-hydroxy-γ-oxo-L-norvaline. 1. Studies in mice, rats, hamsters and rabbits. *Teratology*, **8**, 99
70. Hood, R. D., Naughton, M. J. and Hayes, A. W. (1976). Prenatal effects of ochratoxin A in hamsters. Teratology, **13**, 11
71. Juszkiewicz, T., Rakalska, Z. and Dzierzawski, A. (1971). Effect embryopathique du 3, 3-dichloro-5, 5-dinitro 0, 0-biphenol (Bayer 9015) chez le hamster dore. *J. Eur. Toxic.*, **4**, 525
72. Ferm, V. H. (1966). Congenital malformations induced by dimethyl sulphoxide in the golden hamster. *J. Embryol. Exp. Morphol.*, **16**, 49
73. Kilham, L. and Margolis, G. (1969). Hydrocephalus in hamsters, ferrets, rats and mice following inocculation with reovirus type I. *Lab. Invest.*, **21**, 183
74. Kilham, L. and Margolis, G. (1974). Congenital infections due to reovirus type 3 in hamsters. *Teratology*, **9**, 51
75. Ferm, V. H. and Kilham, L. (1964). Congenital anomalies induced in hamster embryos with H-1 virus. *Science*, **145**, 510
76. Baxter, H. and Fraser, F. C. (1949). Production of congenital defects of offspring of female mice treated with cortisone. *McGill Med. J.*, **19**, 245
77. Johnston, M. C., Bhakdinaronk, A. and Reid, Y. C. (1973). An expanded role of the neural crest in oral and pharyngeal development. In: J. F. Bosma (ed.). *Oral Sensation and Perception: Development in the Fetus and Infant*, pp. 37–52. (Washington: U.S. Government Printing Office)
78. Nishimura, H. and Shiota, K. (1977). Summary of comparative embryology and teratology. In: J. G. Wilson and F. C. Fraser (eds.). *Handbook of Teratology*, **Vol. 3**, pp. 119–154. (New York: Plenum Press)
79. Brent, R. L. (1978). Editor's note. *Teratology*, **17**, 183
80. Guilbeau, J. A. (1953). Effects of cortisone on the fetus. *Am. J. Obstet. Gynecol.*, **65**, 227
81. Wells, C. N. (1953). Treatment of hyperemesis gravidarum with cortisone. I. Fetal results. *Am. J. Obstet. Gynecol.*, **66**, 598

82. Doig, R. K. and Coltman, O. M. (1956). Cleft palate following cortisone therapy in early pregnancy. *Lancet*, ii, 730
83. Harris, J. W. and Ross, I. P. (1956). Cortisone therapy in early pregnancy – relation to cleft palate. *Lancet*, i, 1045
84. Reilly, W. A. (1958). Hormone therapy during pregnancy: effects on the fetus and newborn. *Qu. Rev. Pediatr.*, **13**, 198
85. Bongiovanni, A. M. and McFadden, A. J. (1960). Steroid during pregnancy and possible fetal consequences. *Fert. Steril.*, **11**, 181
86. Volpato, S. and Scarpa, P. (1961). Bilateral complete agenesis of the radius in a newborn (eventual teratogenic activity of cortisone). *Acta Pediatr. Scand.*, **14**, 154
87. Popert, A. J. (1962). Pregnancy and adrenocortical hormones. Some aspects of their action in rheumatic disease. *Br. Med. J.*, **1**, 967
88. Noda, T., Ueda, K. and Satoyama, M. (1963). A case of malformed infant born to a mother treated with adrenocorticoids during pregnancy. *Sanfujinka No Shimpo*, **15**, 189
89. Malpas, P. (1965). Foetal malformation and cortisone therapy. *Br. Med. J.*, **1**, 795
90. Warrell, D. W. and Taylor, R. (1968). Outcome for the fetus of mothers receiving prednisolone during pregnancy. *Lancet*, i, 117
91. Serment, H. and Ruf, H. (1968). Les dangers pour le produit de conception de medicaments administres a la femme enceinte. *Bull. Fed. Soc. Gynecol. Obstet. Lang. Fr.*, **20**, 69
92. Khudr, G. and Olding, L. (1973). Cyclopia. *Am. J. Dis. Child.*, **125**, 120
93. Schatz, M., Patterson, R., Zeitz, S., O'Rourke, J. and Melam, H. (1975). Corticosteroid therapy for the pregnant asthmatic patient. *J. Am. Med. Ass.*, **234**, 804

3
Morphogenetic systems and the central phenomena of teratology

R. JELÍNEK AND Z. RYCHTER

DEDICATION

In memory of the late Walter Landauer

INTRODUCTION

Any monograph on experimental teratology appearing at present brings the evidence that experimental teratology, nowadays, more than being an exact science, dwells on an empirical basis. The more voluminous the monograph, the more apparent is the editor's endeavour for managing the facts into some kind of an organized entity, into some system. At the same time, however, it becomes obvious that experimental teratology lacks the theory that would provide this branch of science with reliable foundations granting some destination and sense. It is evident that the ultimate aim of any teratological investigation is to prevent, as soon as possible and ever so effectively, the occurrence of inborn defects. The total elaborate empiricism of experimental teratology does not allow more, from the practical point of view, than synthesis and generalization of data according to the prescriptive rules. This character is inherent even in the relevant technical reports and recommendations[1] of the World Health Organization where indefinite terms are often used as, for instance, low, medium and high dose of a substance, the teratogenic properties of which are to be investigated. It is only the semantic forms expressing simple guides how to progress in order to attain our aim – to demonstrate whether, and to what extent is a substance tested teratogenic. From the bulk of facts examined by trial and error, some facts capable of being generalized are gradually assorted, that conduct all activities aimed to the satisfaction of our immediate need – prevention of inborn defects. It is the process that, it must be admitted, allows one to react upon new events, but seems to be quite incompetent with regard to prediction.

Before making predictions in any sphere of reality, we must set down the laws for assorting and fixing the actual invariant objective relations. Not till then will it be possible to create any programme of rational activity, which in our case is the screening of environmental factors, in particular of all drugs, for embryotoxicity. The identified laws together with a certain conceptive idea often lead to the formation of a good theory relevant to that sphere of the reality. The theory helps to elucidate the facts, to develop on their basis a predictive system, and to influence the choice of future problems as well as of methods for their solution. Such an organizing function of theory is badly needed in present teratological research.

This contribution aims to demonstrate that the conceptive idea of the theory of experimental teratology does, perhaps, already exist. May it allow us to identify soon the laws required to predict and cope effectively with the man-made risk of inborn defects.

MORPHOGENESIS

Morphogenesis, as it concerns development of a multicellular organism, can be objectively studied at three principal levels of biological organization: cell, organ and organism. In addition to these, teratology, as a branch of science concerned with the causes, mechanisms and manifestations of developmental errors[2], involves the population, at least in the form of epidemiological studies (population teratology[3]). Intermingling of the four levels of bio-organization represents one of the characteristic features of the present empirical approach to teratology phenomena though any scientific investigation cannot avoid some forms of epistemological reduction. Bearing in mind the actual ontological pluralism of the matter, let us consider which level of biological organization is the most profitable and comprehensive to start with when trying to elucidate the basic features of maldevelopment. At first, we are dealing with development of multicellular organisms – that means the communities where individual cells lose their autonomy determined by their genomes being, under normal conditions, hierarchically subordinated to the higher levels of organization. At this point the development ceases to be ruled singly by inherent genetic information, although possessed by any cell of the whole organism, and results from interaction between genetic endowment and the environment, to start with the environment self-assembled by the presence of neighbouring cells. The environment forces the individual cells to act, to adjust their gene expression to the continuously or discretely changing conditions along the invariant chain of developmental steps characterizing normal ontogenesis. Thus, during development, any type of cell organization is preceded in time by its components and precursors[4], commitment and differentiation of any individual cell is determined by its 'phylogeny' i.e. by the pathway of its precursors on the structural and time pattern of the past morphogenesis. The phenomena of adjusting gene expression to changing microenvironment are highly intricate and at present not well understood in molecular terms[5]. So, in the light of the current view, morphogenesis is a self-programming process that uses genome as a 'master library'. One point of crucial importance is to be stressed. *The phenomenon of differentiation seems to be a collective event rather than a private affair of the individual cell.*

According to Solter[6], who studied differentiation in teratomas from embryonic cells – a model in which differentiation could not be totally unlike normal – the stem cells produced a small nest of cells which then collectively differentiated into one tissue cell type. The above mentioned arguments imply that the first level of biological organization at which deviations from normal morphogenesis are to be effectively studied, is the level of *local cell populations*. This means the groups of cells capable of maintaining their internal milieu, homeostasis, and, therefore, their specific trend of differentiation (quantitative data see[7]). Moreover, as it will be demonstrated later, to this level any phenomena usually described at the organismal level can be reduced for analytical purposes. Before dealing with the problem of the basic mechanisms responsible for establishment and formation of micromilieu of the local cell populations, we shall make several remarks on the role of the genome in morphogenesis and maldevelopment.

Several decades ago the idea prevailed that, for any organ system or even an organ, a universe of special genes existed whose function was simply to build up the structure in question. Such an idea seems to be held by many people even today. The answer of Hans Grüneberg, one of the great personalities involved in studying effects of mutant genes, to the question of how do genes affect the skeleton seems to be in the apparent controversy: 'The answer is – largely indirectly. This is not peculiar to the skeleton. It applies quite generally to the effects of mutant genes on any system studied on the morphological level . . . Some forty years ago, we all hoped that the study of the development of mutant genes would lead to deeper insights into causalities of embryology. After all this time and effort, we can no longer disguise the fact that that hope has, by and large, been disappointed'[8]. This statement, based upon many years of precise and exhaustive investigations, does, by no means, detract from the role of genome in morphogenesis, but supports the conception of a 'master library'. The proper programme is chosen according to the microenvironmental conditions that, in their sum, may be called the *morphogenetic situation*. It is the morphogenetic situation that forces the genome of individual cells to enter the next step of genetic expression, i.e. either differentiation or commitment. The process naturally proceeds without any change in DNA sequence and may be referred to as cellular reprogramming. In case of genetic damage, as it takes place in hereditary diseases, the defect transmitted by gametes occurs in any of the cell clones derived from the zygote. This means that the 'master library' of *any* individual cell is defineably defective, but the expression of the erroneous DNA sequence, that practically equals the possibility of recognizing the defect, depends upon the occurrence of situation when the proper action of mutant gene is needed. Either in morphogenesis, or in the function of the fully differentiated cell of the adult organism, or in both. In the first case, structural defects of predominantly genetic origin develop, in the second case, inborn defects of metabolism, and in the third case, the hereditary syndromes involving both types of developmental errors. In the apparently non-affected cell populations the defect is masked simply by the fact that the mutant gene need not have been expressed, either in the ontogeny or in the adult functional state. The same applies for somatic mutations which may occur, no doubt rarely, during any period of individual life.

Evidently, the above mentioned concept is at variance with the opinion that the primary and determining phenomenon of morphogenesis is the progressive differentiation of individual cells ruled by genetic pacemakers. On the contrary, it relegates differentiation to the submissive role of a secondary phenomenon dependent upon both the initial contents of a genetic 'master library' and on the position of a cell (including its precursors) within the net of space and time coordinates of ontogenesis. This view may be supported by the data resulting from the numerous experiments carried, for instance, on early mouse embryo (see review[5]) or on interspecific chimaeras. If differentiation seems not to be the primary phenomenon of morphogenesis, then what kind of processes determine, in fact, the fate and the individual cell patterns of local cell populations? It is a difficult task, at present, to give any decisive and exhaustive answer. Nevertheless, it is possible to enumerate and describe those morphogenetic processes that take part in the formation of any embryonic component and which evidently participate in determining what we have called the morphogenetic situation.

BASIC MORPHOGENETIC PROCESSES

The determined effort to identify the main morphogenetic events affected by the teratogen action, can already be traced in the work of Stockard[9]. Experimenting on embryos of *Fundulus heteroclitus*, he was able to recognize that 'the structural quality may be affected by many things, but always depends directly upon the rate of development'. On slowing development (e.g. by lowering the temperature), the embryonic components (that is embryonic organs and their parts) lower their rates in a corresponding fashion. The faster growing components are still progressing at a faster rate than the slower-growing parts. The immediate consequence of the experimental intervention inducing an inborn defect is always lowering of the developmental rate of the affected part of the embryo, whereas at the moment of intervention the intensively growing areas are more sensitive to teratogenic action than the others growing more slowly. This statement which formed the firm grounds for formulating the theory of critical developmental periods (critical moments[9]) has been valid up to now, at least for cytotoxic agents. Stockard's conclusions have stressed the importance of *proliferation* as one of the basic morphogenetic processes that, being adversely affected by a teratogen, directly relates to the origin of inborn defects.

As the volume of teratological knowledge has been increasing, attempts were made, though in part on an empirical or hypothetical basis, to define some other fundamental morphogenetic events, the disturbance of which might always initiate the development of structural defects. So, for instance, Knorre[10] proposed several morphogenetic processes as a basis for systematic reclassification of malformations. Gustafson and Toneby[11] requested morphologists to describe the morphogenetic events in terms of changes in cellular contact, changes in cell motility, changes in the rate of proliferation or cell death, and Saxen[12] introduced under the term 'stages of action' determination, proliferation, cellular organization, migration, and morphogenetic cell death as sensitive stages of development of the target tissue. The concept of morphogenetic processes of a fundamental nature percolates even through the pages of textbooks and monographs[13–15] involving, besides proliferation, migration and cell death, also the phenomena of secondary character (induction, determination and differentiation).

The first attempt, as far as we know, to propose the theory of basic morphogenetic processes was made at the 2nd Conference of the European Teratology Society[16,17] in a concise formalized form by Rychter and Jelínek[18]. The reasons for this were to choose the reference level at which, on the one hand, all disturbances of embryonic metabolism and physiological functions converge into a remarkable morphogenetic effect and, on the other hand, all the varying final states of the obvious morphogenetic deviations can be reduced to a simple form. According to this theory, within local cell populations four basic morphogenetic events occur, namely, cellular *proliferation*, cell *distribution*, *integration* of cells into the higher-order entities by means of cell contacts and specialized junctions, and finally *reduction* of cell numbers by varying mechanisms of cell death[16,19,20]. Development of primordia of organ systems, organs and organ constituents (which all are included under the term *embryonic components*) represent invariant sequences of combinations of the basic morphogenetic processes. Certain combinations of basic morphogenetic processes occur in embryos of all vertebrate species determining in the development of their embryonic components the *isofactorial phases*. Action of teratogenic agents is reflected in the disturbance of a definite group of combinations of basic morphogenetic processes. So far the main propositions of the theory is elaborated in detail in the quoted monograph[18], together with a survey of data relating to particular basic morphogenetic processes. It is necessary to point out that the processes involved are not only elucidated by relatively simple techniques, but can also be quantitatively evaluated. In this way the development of any embryonic component can be recorded in a quantitative manner, in the form of curves giving the picture of changes in proliferation rate or incidence of cell death, and in the form of diagrams depicting the course of cell migration and, eventually, the patterning of cell contacts. Course of events during development of inborn defect can be expressed in the similar way. Several studies performed in accordance with this approach can be mentioned concerning, for instance, cell cycle in the neural tube of mutant mouse[21], proliferation in palate shelves of corticoid-treated mice[22], morphogenesis of the heart in normal chicken embryos[23,24], and the dynamics of cell death in bulbar and atrioventricular cushions after cyclophosphamide treatment on the same experimental model[25].

The basic morphogenetic processes operate at the level of local cell populations creating the 'critical cell masses' necessary for the onset of differentiation, bring together the 'inductors' and 'reactors', determine the pattern of proliferative structures, link up the individual cells by communicating networks enabling the transfer of morphogenetic signals and, finally, deplete the surplus material having not been included into the species-specific invariant patterns of embryogenesis. The last statement seems to be implied too generally. What is in principle the species specific pattern of embryogenesis? And at this point it is necessary to introduce the new term – the *morphogenetic system* (MGS).

THE DEFINITION AND PROPERTIES OF MORPHOGENETIC SYSTEMS

According to Waddington[26], individual development comprises differentiation in time (histogenesis), differentiation in space (regionalization) and differentiation in shape (morphogenesis) occurring simultaneously in intimate relationship with one another. Let us pause for a while and study the term regionalization which implies that during early development the originally totipotent cell mass is distributed into regions with different prospective potency. These regions can be progressively recognized as the anlages of organ systems, organs and organ constituents, that are related to concrete structures composing the embryonic body. Hence, the primordia of organ systems (e.g. the neural tube), organs (e.g. the diencephalic vesicles), and organ constituents (e.g. layers of the optic cup as the retinal anlage) can be generally called the *embryonic components*. From the examples introduced, it is evident that the embryonic components, characterizing any defined developmental horizon, are preceded in time by precursors of the broader developmental potency. Embryonic components include the compartment of morphogenetically active cells, that is the cells that proliferate, migrate, integrate by establishing cell contacts, or die. These cells constitute the *embryonic morphogenetic system* which is defined as the *group of cell populations carrying, creating and executing the morphogenetic programme for development of an embryonic component*. Impairment of function of the embryonic MGSs results in gross structural abnormalities – the malformations. The period of function of MGSs is by no means limited to the embryonic phase of development. After the anlages of the main organs are laid down in their conventional adult shape, that is in the fetal period of development, the MGSs are withdrawn from the organ level to that of the organ constituents. Induction of gross abnormalities is no longer possible. Interference with function of the *fetal MGSs* may be reflected in the development of minor defects being more and more restricted to the tissue level, manifesting themselves, if ever, by impaired postnatal function. The functional differentiation of tissues generally culminates around birth. At that time, a number of MGSs operate that gradually complete the preceding morphological development. They can be defined as groups of cell populations carrying, creating and executing the morphogenetic programme for developing the morphological substrate for the specific integrated life functions. Deviant activity of the *perinatal MGSs* may be followed by the impaired pattern of complex life manifestations, such as behaviour. Even after birth the morphogenetic systems do not cease to exist. Although their occurrence is limited to tissues with the renewing and expanding cell populations, and regenerates, the proper functioning of the *postnatal MGSs* represents a matter of vital importance. Consequences of their impairment or imbalanced functions are within the area of pathology. Table 3.1 summarizes the results of an attempt to classify the MGSs according to periods of their prevailing existence and activity. By introducing the concept of MGSs, the basic morphogenetic processes have attained the concise spatial and chronological framework, being related to particular embryonic and fetal components and, perhaps, to any morphogenetically active

Table 3.1 Taxonomy of morphogenetic systems

Type	*Level of organization*	*Example*	*Dysfunction consequences*
Embryonic	Organ systems, organs	Caudal morphogenetic system, neural tube, limbs	Monstrosities, malformations
Fetal	Organ constituents	Coronary vessels	Minor structural abnormalities
Perinatal	Developing tissues	Fine structure of the CNS	Functional and behavioural defects
Postnatal	Tissues	Bone marrow, epidermis	Functional disturbances, carcinogenesis

structure of the newborn, adolescent, and the adult.

The function of any MGS has two distinct points (Figure 3.1). The first point is *qualitative* and concerns differentiation in shape (Figure 3.1A). This remained for a long time the domain of classical embryology. A thorough study of shape differentiation is always directed to overemphasizing the structural diversity and to the loss of cognizance of the common features of development. The second aspect of morphogenetic function is *quantitative*, governed by the general laws of growth[27] (Figure 3.1B). Investigation of this aspect is similarly legitimate as the study of shape differentiation, representing the other side of the same coin. Exact rendering of growth processes enables one, moreover, to compare the function of MGSs of two entirely different embryonic components, and what is even more important from the point of view of teratogenesis, to discern deviations from normality by comparing growth of an embryonic component under the influence of teratogen with that of a control one. It might be useful to draw attention to the convention of considering growth retardation as one of the manifestations of embryotoxic action. Growth and shape differentiation seem to be dialectically linked, both being subjected to the common denominator of the basic morphogenetic processes. Hence, when searching for the major events during maldevelopment, it is not recommended to start directly with analysis of the basic morphogenetic processes; however, it is more useful to choose any gross quantitative parameter relating to the sum of products of the morphogenetic function followed. The parameter used as an indicator of morphogenetic function enables one to examine a sufficiently large experimental group, i.e. the 'population sample' for any developmental stage examined. This seems to be unavoidable for the reason that experimental teratology faces the phenomena that are multifactorially conditioned even in a highly standardized experimental material. Also the detailed analysis of basic morphogenetic processes, if conclusive results are needed, must follow this general rule. Such a type of an analysis is apparently more laborious and time-consuming. Therefore, it is recommended to start with only after study using some gross quantitative parameter. This procedure allows one to concentrate on the critical moments of development and/or maldevelopment, when more detailed analysis finds its proper place.

The curve of a gross developmental parameter attains, in most cases, the asymmetrically sigmoid character (Figure 3.2) which is often modified either

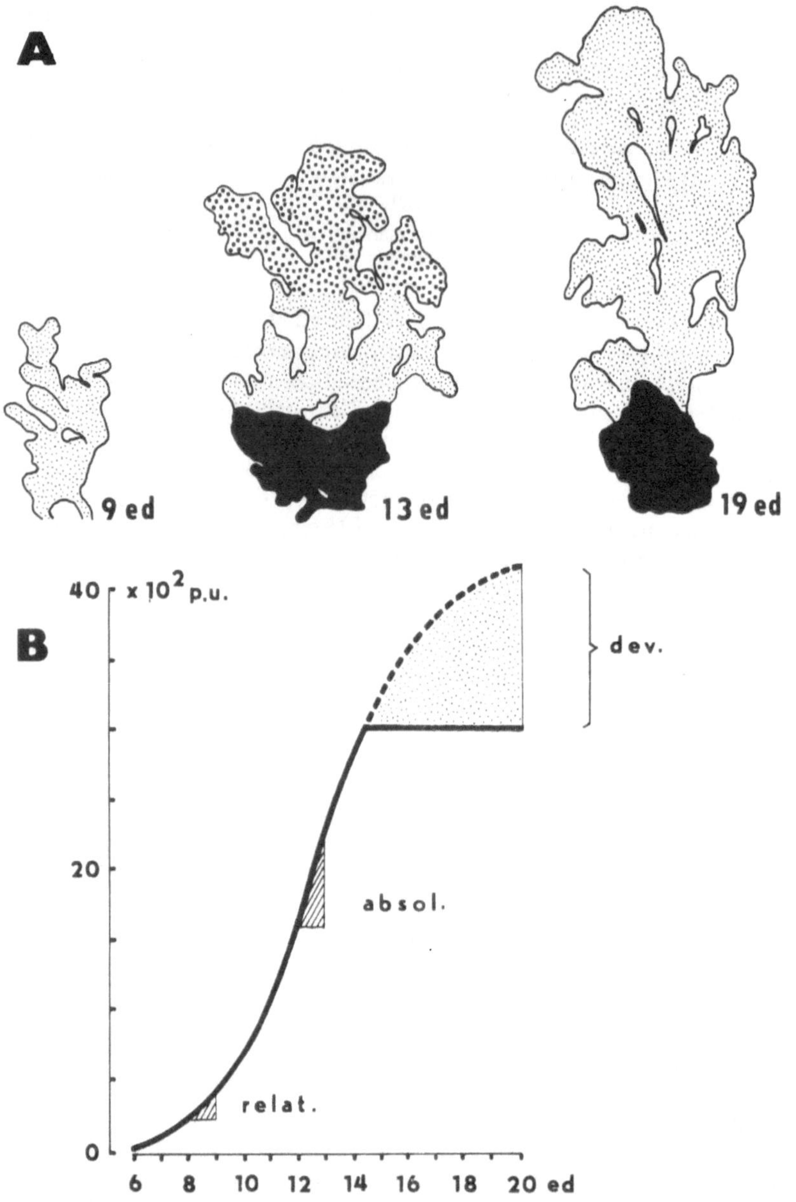

Figure 3.1 A, Development in shape of chick embryonic telencephalic choroid plexus. B, Growth curve of chick embryonic telencephalic choroid plexus. Greatest relative increments occur in its initial exponential phase while maximum absolute increments are seen in linear phase. Abscissa: embryonic days (ed); ordinate: area of choroid plexus in planimetric units (p.u.); dev., significant deviation from theoretical sigmoid growth curve

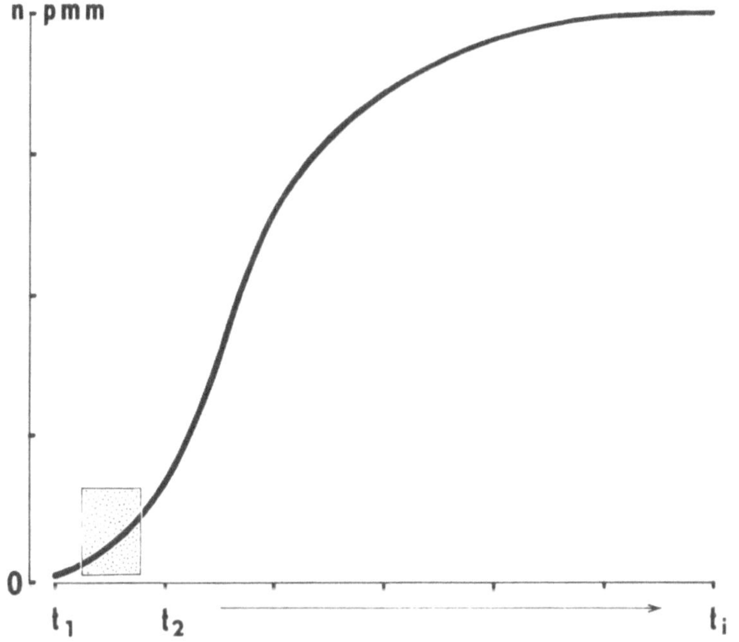

Figure 3.2 Typical asymmetrical sigmoid growth curve with prolongated terminal phase. Experimental intervention is most effective when performed in initial exponential phase (stippled area)

in its initial exponential (Figure 3.3) or in its final, decaying exponential phase (Figure 3.1B). These irregularities can be explained both by the mode of origin of the initial anlage, as in the case of caudal morphogenetic system, and by the onset of sensitivity to regulatory signals of corticoids, as in the case of the choroid plexus[28, 29]. The middle, linear phase of the sigmoid seems to occur regularly in most of the embryonic components. According to our experience, the teratogenic insult is most effective when administered in the first, exponential phase of the growth curve characterized by immense relative increments (Figures 3.2 and 3.1B). This applies at least for the group of 'classical' teratogens with remarkable cytotoxic activity.

Let us return to the question posed at the end of the previous section. What is, in principle, the species-specific pattern of embryogenesis? In terms of the advanced theory, it is the pattern of succession of morphogenetic systems, whose functions can be described quantitatively in terms of the basic morphogenetic processes: proliferation, distribution, integration and reduction of the defined cell populations.

MORPHOGENETIC SYSTEMS AND THE DOSE–RESPONSE PHENOMENA

The question of demonstrating dose–response relationships in teratology has been one of the most discussed topics. There is no room for those who require short and categorical answers. First, it is useful to avoid or to exclude from

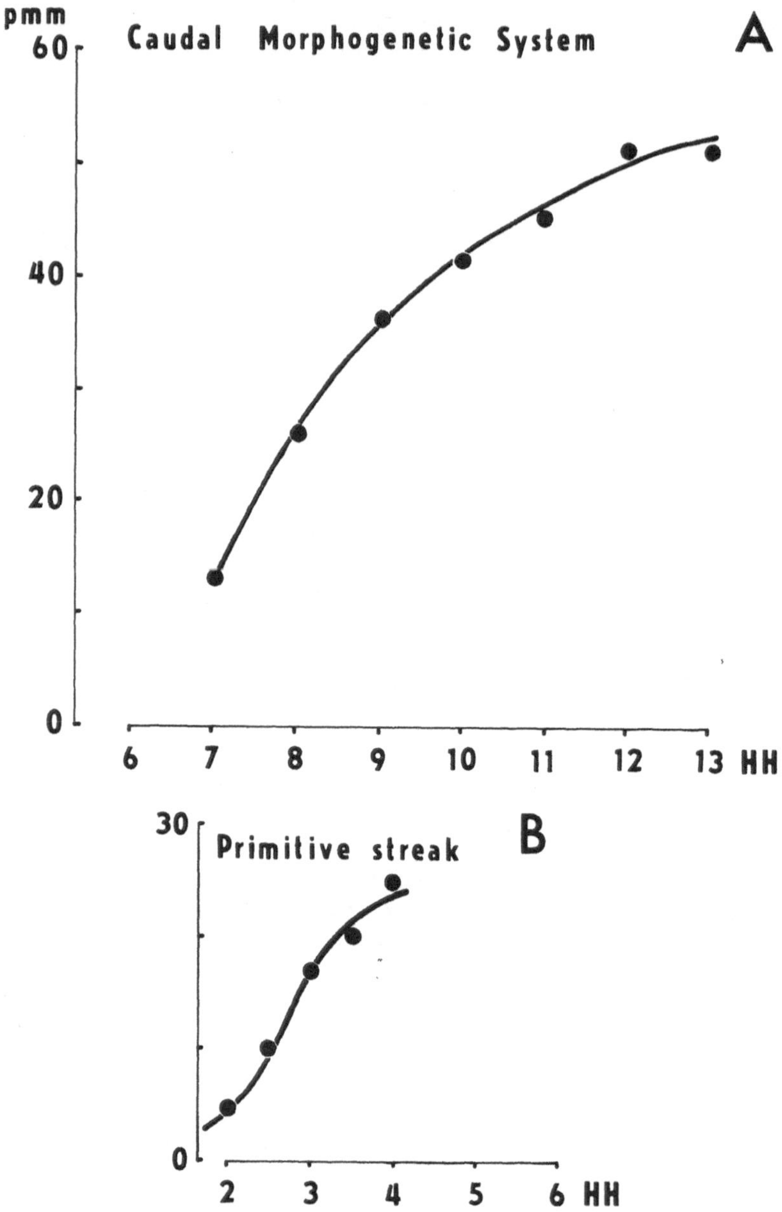

Figure 3.3 A, Function of chick caudal morphogenetic system as depicted by gross developmental parameter, length of trunk. The curve lacks its initial phase being dependent upon preceding formation of primitive streak (B). Abscissae: developmental stages according to Hamburger, Hamilton; ordinates: length in projection millimetres (pmm) at magnification × 13

our considerations the extra-embryonic part of the maternal–fetal complex, because it is rarely possible to increase the dosage sufficiently without poisoning the maternal organism – a situation which is beyond the scope of pure teratology. The target of teratogenic action, the embryo, finds itself in a substantially different position than at the beginning of experimentation. It is not surprising that the classical concept of dose–response relationships, as developed in pharmacology and toxicology, falls into ruins. The second difficulty stems from the well-known transformation 'malformed – dead' that led to adopting the phenomenon of embryonic decay among the three manifestations of embryotoxicity. In exact evaluation, however, this proposition seems to be of limited use. The reason is that the embryolethality need not be causally associated with the morphogenetic effect but rather with embryophysiological disturbances (e.g. cardiotoxicity[15]). In brief, *the most convenient object for studying the dose-response phenomena in teratology is the morphogenetic system exposed to increasing doses of a substance administered directly into the immediate environment.* Accepting this recommendation we are able to grasp the whole situation in terms of three variables: the *target structure*, or more precisely the MGS of this structure (embryonic component), *the factor* investigated (at several dosage levels), and the *time of exposure.* The system 'target structure – dose – time of exposure' itself represents a great simplification. It marks out, in the development of the embryo, the definite time interval to be investigated, and makes redundant dealing with developmental stages that precede or follow the phase when the selected MGS operates. The effectiveness of a dose can be evaluated either *discretely*, by the incidence of some morphological change (malformation), or *continuously*, using a quantitative parameter. *In the latter case, information is provided by any member of the set.* As an example of both possible modes of evaluation, the final record of testing of actinomycin D using the Chick Embryotoxicity Screening Test (CHEST)[30,31], may be introduced. This elementary screening technique employs the caudal morphogenetic system for estimating the beginning of the embryotoxicity dose range. The substance tested is injected in mounting doses (ten multiples) subgerminally to embryos at HH stages 10–11. The effects of the doses are evaluated 24 h after by measuring the newly-formed part of trunk (Figure 3.4A). The first effective dose is found to reduce significantly the mean length of the trunk, indicating in this way interference with function of the caudal morphogenetic system. This statement serves as a starting point for the second step of the test based upon administration of three doses around the estimated beginning of the embryotoxicity range to embryos on days 2, 3, and 4. The results are, in this case, evaluated on day 8 when the incidence of the malformed and dead fetuses are summed up to estimate roughly the dose–response relationships and the stage effect in the selected morphogenetic systems (Figure 3.4B). In this test, both the continuous (measuring the caudal part of trunk), and the discrete parameters (expressing the incidence of the abnormal fetuses) are employed leading, in the case of actinomycin D, to the same conclusion. *Using CHEST, the positive dose–response relations can be easily demonstrated for any effective substance tested.* This is fixed in the case of the discrete mode of evaluation by the fact that several MGSs are employed as indicators of embryotoxic effect. The

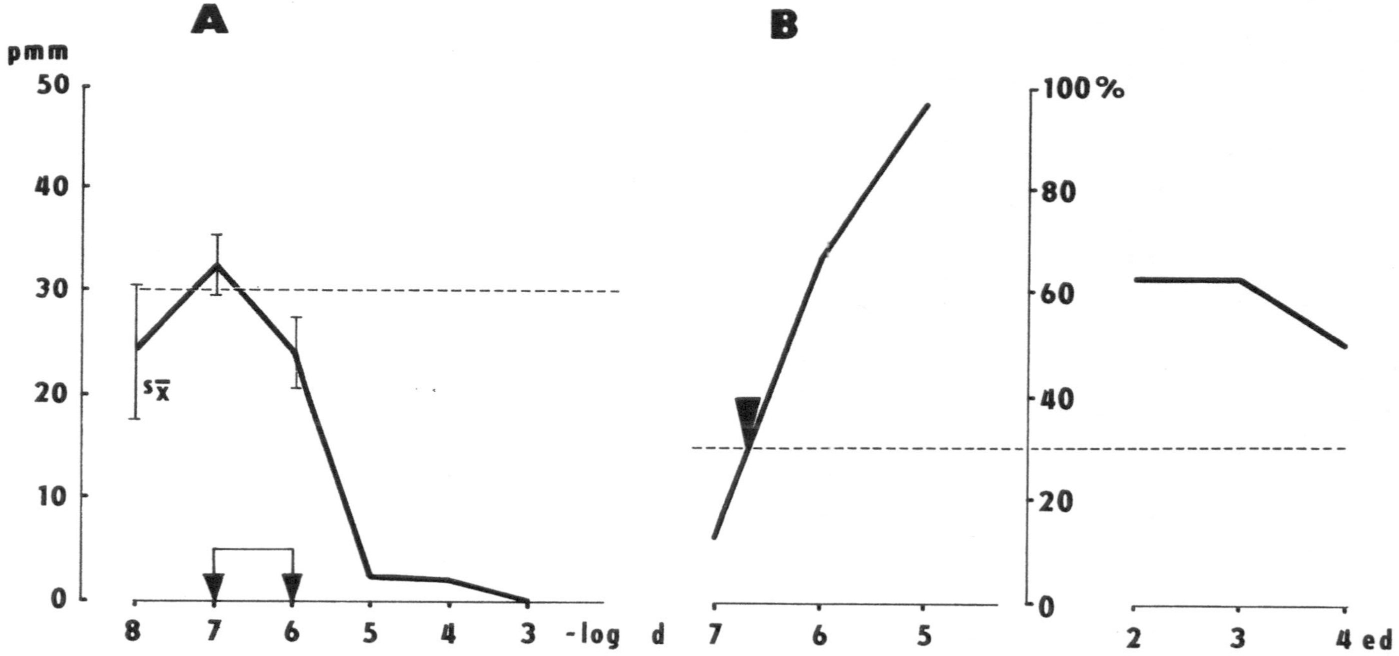

Figure 3.4 Final record of testing actinomycin D with CHEST (Chick Embryotoxicity Screening Test[30,31]). A, dependence of length of newly formed part of trunk upon dosage of drug. Abscissa: logarithm of dose (mg), ordinate: length of trunk in pmm (projection millimetres at magnification × 13). B, Left: Dose effect of actinomycin D as revealed by incidence of dead and malformed fetuses following administration on days 2, 3 and 4. Abscissa: logarithm of dose (mg). Right: Stage effect of actinomycin D as revealed by mean incidence of dead and malformed fetuses following application of the three above-mentioned doses. Arrows indicate beginning of embryotoxicity range as revealed either by continuous (A) or discrete (B) mode of evaluation

following example is mentioned as a warning against the use of only one index malformation. The thoraco-lumbo-sacral myelodysplasias are caused by a moderate disturbance of function of the caudal morphogenetic system[32]. When the disturbing factor reaches a certain level of intensity, the myelodysplasias are transformed to the syndrome of caudal regression which, influenced by the serious damage to the caudal morphogenetic system, is lethal in most cases. The incidence of the malformation at birth therefore decreases. So, for this reason the dysraphic malformations of the central nervous system cannot be employed as marker malformations in population teratology because their dose–response relationships are evidently biphasic reaching, approximately in the middle of the course, the negative trend (Figure 3.5)[32]. Following this trend of reasoning, the well-known decrease in neural dysraphias observed in the highly developed countries could hardly be

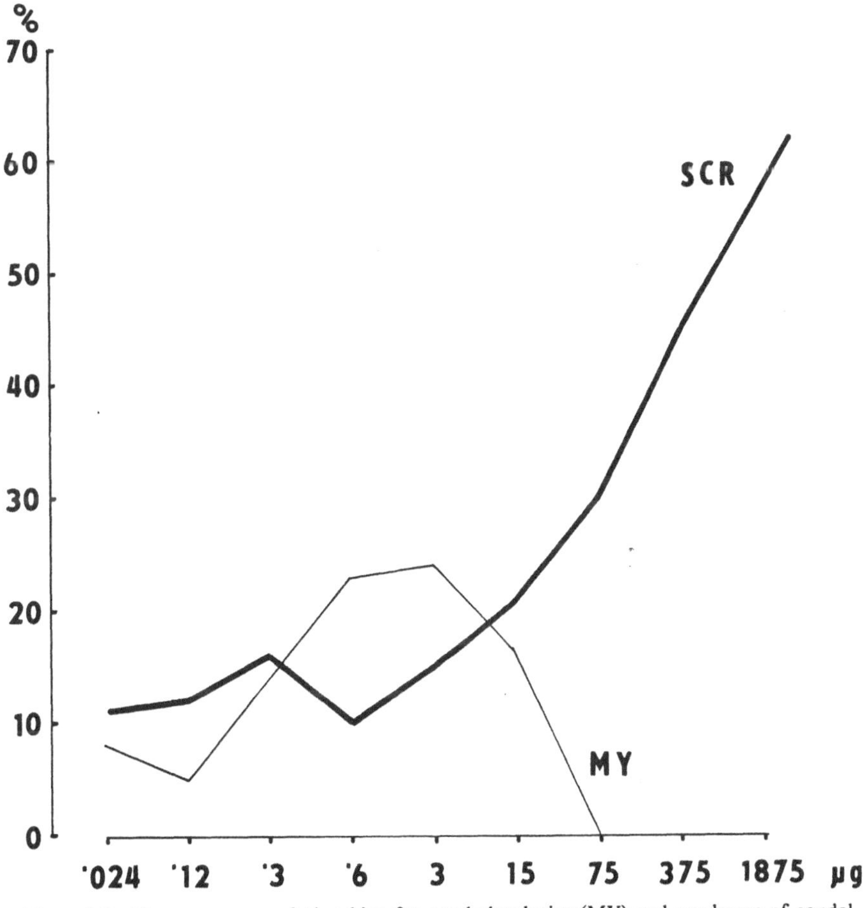

Figure 3.5 Dose–response relationships for myelodysplasias (MY) and syndrome of caudal regression (SCR) as evaluated by their incidence in chick embryos after administering mounting doses of 6-azauridine on days 1.5, 2, 3 and 4. Abscissa: dose of 6-azauridine, ordinate: mean incidence of the malformations

put into causal relationships with the improvement of hygienic standards, as is often done, but rather with increasing environmental stress. The same applies for cleft lip and palate, demonstrating that only those inborn defects that are properly identified with regard to their behaviour on the dose scale can be employed as marker malformations in population teratology.

The methods using the continuous mode of evaluation of morphogenetic functions (i.e. quantitative parameters) conceal another type of deficiency inherited in the regulative capacity of the morphogenetic systems. Teratogenesis is almost always concerned, as a rule, with moderate and transient disturbing phenomena that can be traced in the altered, still not abolished, morphogenetic functions. The teratogenic impulse, insulting the dynamic equilibrium of a morphogenetic system, induces disturbances in the further course of its morphogenetic function, capable of confusing the incidental observer studying the dose–response phenomena. The event can be easily illustrated in the case of the mandible representing, at least in the mouse, the

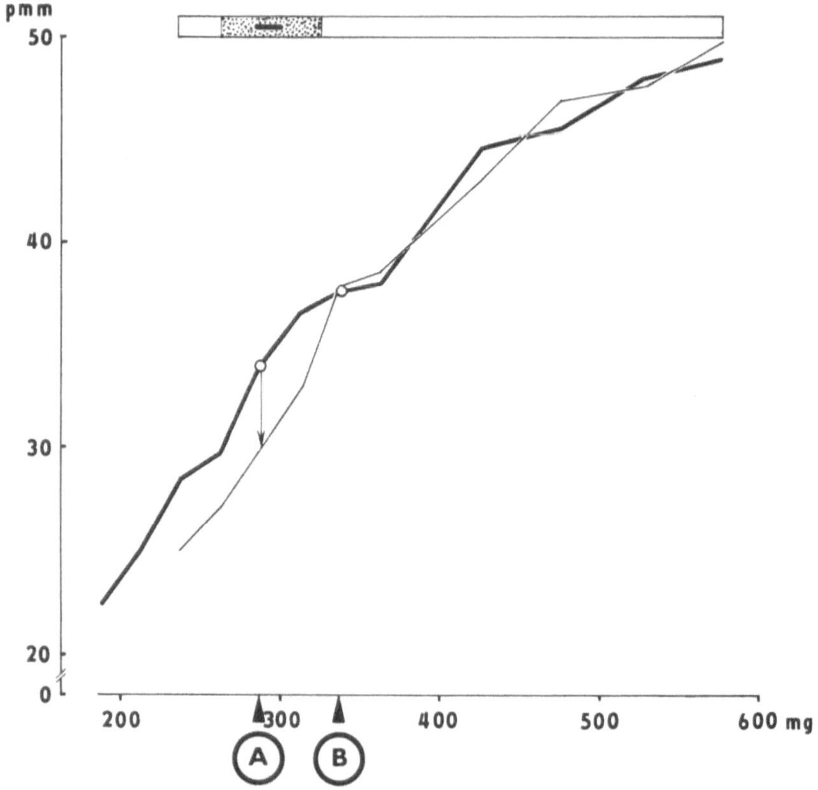

Figure 3.6 Growth of mandible in mouse fetuses after corticoid treatment on day 12 (thin line) compared with control untreated group (bold line). Abscissa: weight of embryos of comparable chronological age; ordinate: values of index of mandibular growth in projection millimetres (pmm) at linear magnification × 13. References of two observers situated at points A and B, with respect to the effect of cortisone acetate on the growth of the mandible, would substantially differ

important morphogenetic subsystem engaged in horizontalization of the palatal shelves[33]. When cortisone acetate is applied on day 12, the initial growth retardation encountered reaches its maximum just at the time when palatal shelves normally elevate. This phase is, nevertheless, soon followed by complete restitution. In this way, the two observers planted at the interval as depicted on Figure 3.6 will fall into an absurd but irreconcilable disagreement when considering the effect of one and the same dose. One can hardly imagine what would happen if they decided to settle their dispute in a gentlemanly fashion and re-evaluate their conflicting observations by administering several doses with the aim of identifying the dose–response relationships. A similar effect was observed in the polydactylous mutant strain of rats[34] (Figure 3.7). A further example should illustrate the most intricate situation introduced up to this time – the situation when the dose–response is investigated in the different phases of development. Studying the effects of hydrocortisone on the developing chick choroid plexus Šťastný and Rychter[29] were able to demonstrate that at the beginning of development hydrocortisone exerted no influence. Then the effect appears and culminates between days 13–15, disappearing again till the end of incubation. These three distinct phases, charac-

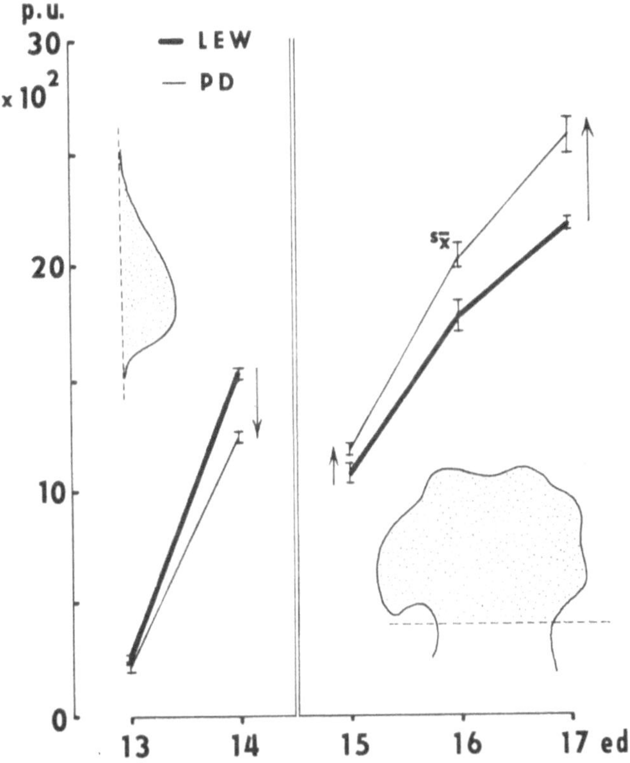

Figure 3.7 Growth of polydactylous leg primordia in rat. Abscissa: embryonic days (ed), ordinate: area of leg primordium in planimetric units (p.u.). LEW, control specimens; PD, embryos with polydactyly

terized by the different response of the choroid plexus MGS, were further studied using a battery of doses ranging from those, differing slightly from the natural level of corticoids in the blood plasma of the relevant developmental stage, to the irrational ones, limited only by the solubility of the substance used in a given amount of saline[35,36]. The concept of morphogenetic systems made it possible to select the time intervals representative of the definite developmental phases. Administration on day 7 representing the exponential phase of growth, unexpectedly caused the stimulation of development when moderate dosage was used (Figure 3.8A). Inhibition was observed only after the highest doses. In the linear growth phase, represented by days 11–13, the maximum inhibitory effect was observed when the intermediate doses were applied (Figure 3.8B). The low, and even the maximal ones exerted no effects at all. In the second series of experiments hydrocortisone was applied on day 15, when the growth of the choroid plexus culminates, and the results were evaluated on day 17. At that stage the stimulating effect was seen with both the low and the high dosage (Figure 3.9A). On day 17 hydrocortisone induced no effect at all (Figure 3.9B). The detailed analysis of these complex phenomena led to the hypothesis of the transient presence of hydrocortisone receptors within the cell populations constituting the MGS of the choroid plexus on day 13, that increase the susceptibility to the inhibitory effect of hydrocortisone 40–50 times. The hypothesis was further supported by the demonstration of the antagonistic effect of progesterone[37]. This clearly illustrates that the dose–response relationships may be influenced by the development of specific receptor systems – a situation which is to be presumed in case of any substance occurring naturally in the organism, or its analogue.

Finally, we wish to mention that, provided some more complicated MGS as the MGS of an organ constituent is used for studying the dose–response phenomena, it is necessary to use the repeated dosage. As the example, blocking the development of the chick coronary arteries with 6-azauridine[38] may be introduced (Figure 3.10). The doses repeated daily from days 7 to 9 were the most effective. This 'interval' (in contrast to the 'point' administration) was selected on the basis of knowledge of the developmental mechanisms contributing to the formation of the coronary bed[39,40]. Influencing in this way the initial exponential phase, including its transition into the linear phase of development of the coronary bed, the authors were able to study the consequences of this invariant intervention on the development of the myoarchitectonics of the heart wall[41].

Concluding this section we wish to emphasize one point of crucial importance. As we have mentioned above, the experimental procedure involving the direct application of a substance into the immediate environment of an embryonic MGS enables one to demonstrate the positive dose–response relationships for *any* substance tested. In this way, the obsolete empirical concept of a teratogen as 'an agent that may interfere with development in a detectable way'[1], is overcome and a new definition, based upon the quantitative way of reasoning, is to be established. *Only those agents that interfere with functions of the embryonic, fetal or perinatal morphogenetic systems, in doses that must be considered in actual exposure of human population, can be termed teratogens.*

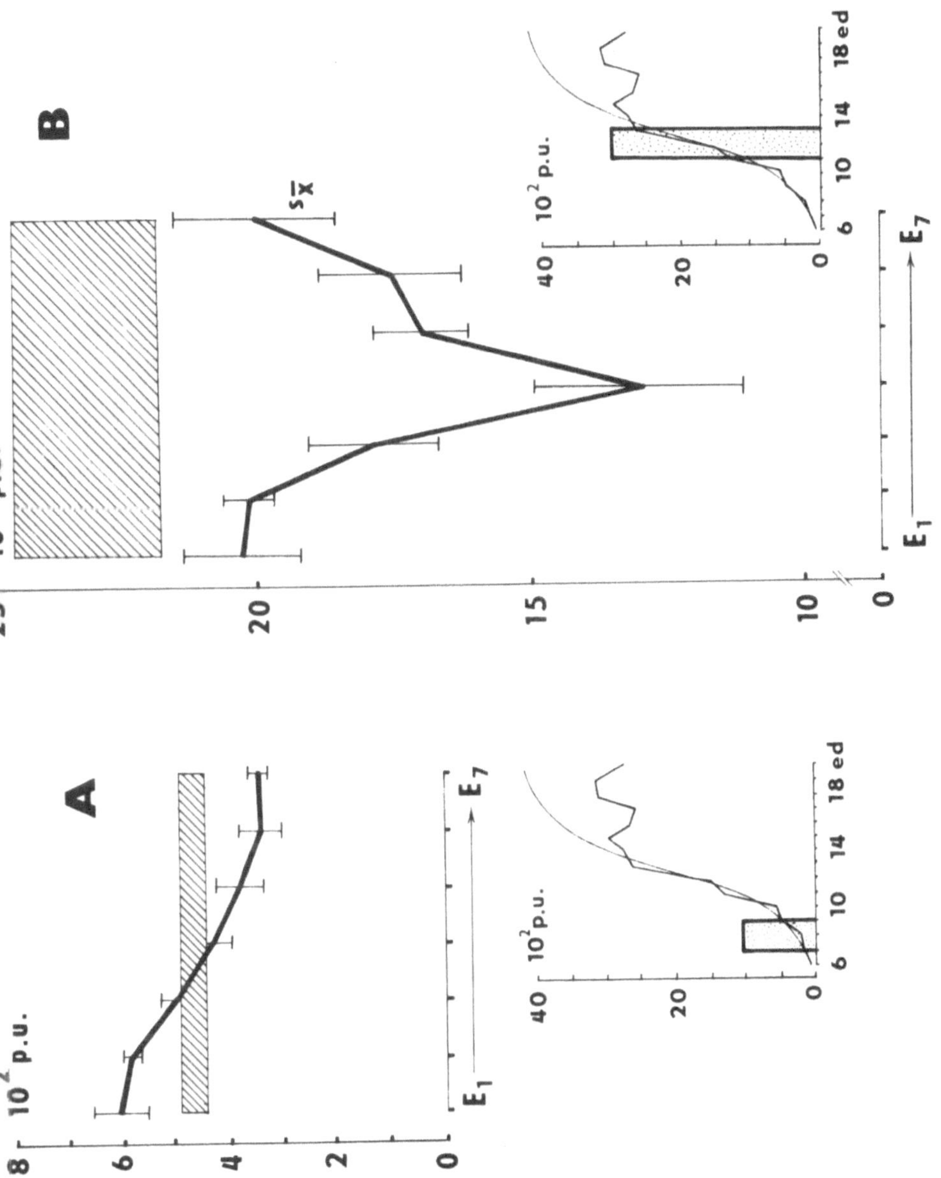

Figure 3.8 Dose–response curve for chick telencephalic choroid plexus and hydrocortisone. E_1, 5^2 multiple of natural plasma level of corticoids; E_7, 5^8 multiple of the same; p.u., planimetric unit; ed, embryonic day

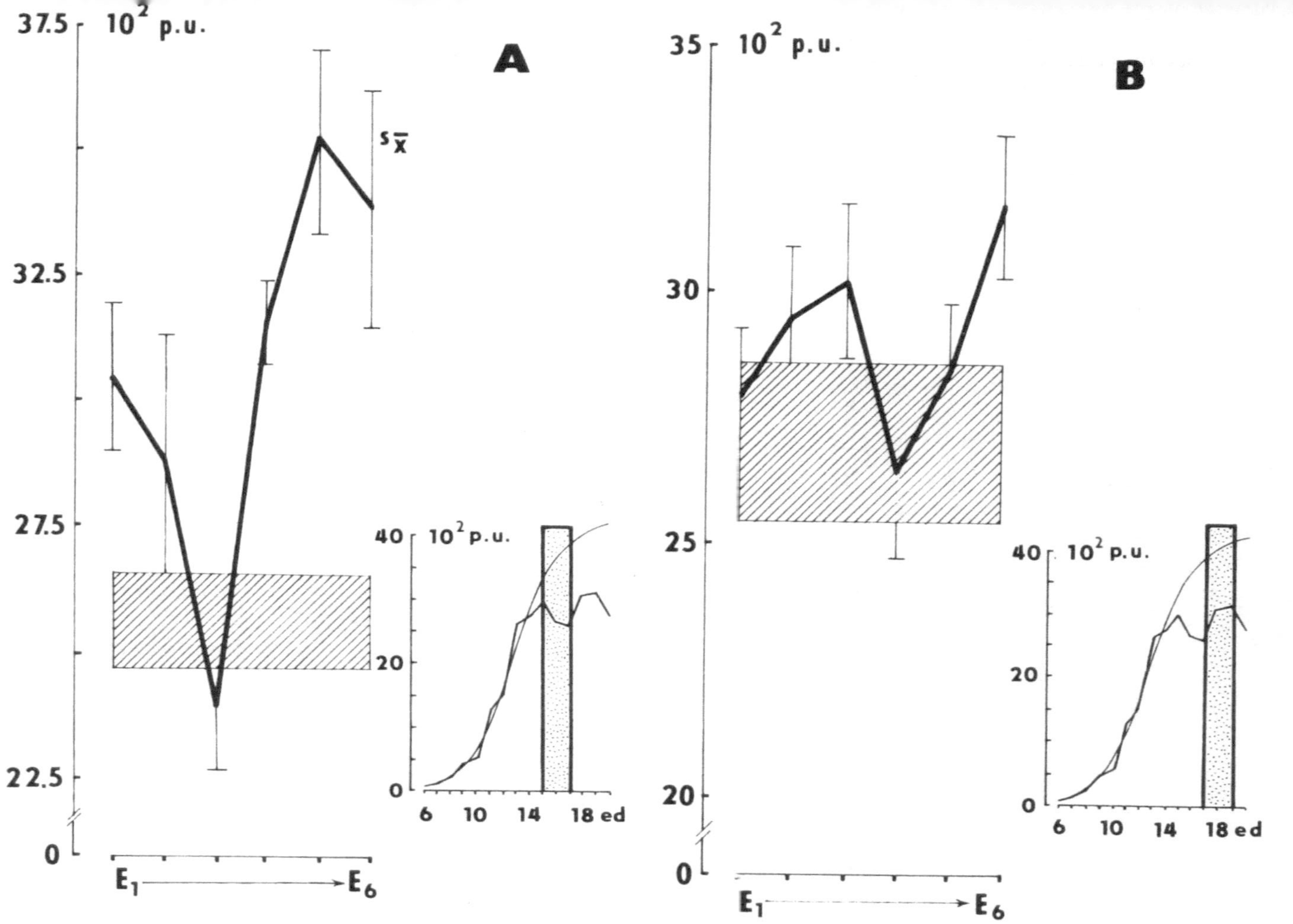

Figure 3.9 Dose–response curve for chick telencephalic choroid plexus and hydrocortisone. E_1, 5^2 multiple of natural plasma level of corticoids, E_6, 5^7 multiple of the same; p.u., planimetric unit; ed, embryonic day

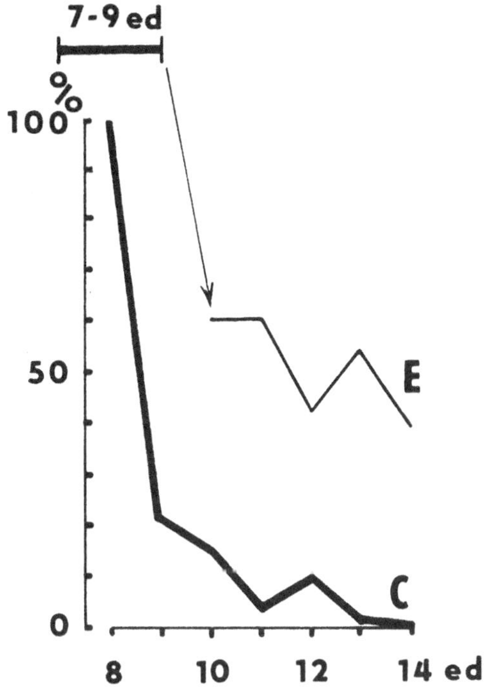

Figure 3.10 Blocking the vascularization of chick heart after 6-azauridine treatment repeated on days 7–9. Abscissa: embryonic days, ordinate: percentage of non-vascularized area at anterior heart wall. E, curve for experimental embryos; C, curve for control specimens

MORPHOGENETIC SYSTEMS AND THE CRITICAL DEVELOPMENTAL PERIODS

In the preceding section an attempt was made to define the teratogenic situation as a function of the three variables: target structure, dose and time of exposure, and the role of dose was investigated mainly on the background of the remaining two fixed variables. Now, being aware of the general properties of the morphogenetic system, and knowing something about its response to various doses of a substance, we may concentrate on the role of the third variable, the developmental stage during which the substance is applied, from the point of view of the term critical period. This concept was introduced in experimental teratology by Stockard[9], designating such developmental phases in which intensive developmental processes make the embryo more susceptible to the action of environmental factors, leading to structural defects. What is valid for the whole embryo, holds also good for any of its morphogenetic systems. It appears that the morphogenetic function may be most easily impaired when the disturbing agent is administered somewhere in the initial exponential phase. This applies at least for a large group of cytotoxic agents. It does not mean, however, that the phase when it is possible to disturb the morphogenetic function, as reflected in the course of a quantitative parameter, may be readily identified with the critical period regarding, in any case, the *qualitative* phenomena caused by entering the abnormal developmental pathway at the defined critical moments of development. Therefore, in this respect we can hardly succeed in our search for critical periods in the adult morphogenetic systems; on the other hand, their susceptibility is most probably related to the existing functional cycles. For the above mentioned reasons we are forced to use such methods of evaluation labelled in the preceding section as discrete. This means that we are forced to express the critical periods by a frequency curve depicting the incidence of abnormalities of the embryonic component in question, as plotted to the time axis of the experimental intervention (further the curve of critical period). In most cases, it attains the character of normal distribution (Figure 3.11B). Otherwise, the critical period is inherent in the invariant pattern of morphogenetic events accomplished by the basic morphogenetic processes and may be identified precisely only when a number of substances of various nature is administered in a broad spectrum of doses. The conventional use of two or three dosage levels of a single substance results in the trivial conclusion that the critical period is, as a rule, slightly shorter than the function of the relevant MGS, and that the critical period may be different for any substance employed. This conclusion reflects, of course, the differential affinity of a substance with respect to various combinations of the basic morphogenetic processes, as conceived by Rychter and Jelínek[18], the rapid changes in susceptibility, influenced by the development of specific receptors, as well as the sequence of morphogenetic systems during individual development. For each step, the curve of the critical period attains its specific character (Figure 3.11). The curve of the critical period of an organ system (Figure 3.11A) differs from those of an organ (Figure 3.11B) and of an organ constituent which, due to the complex dynamics and interaction of the relevant cell populations, presents usually two or even more peaks (Figure 3.11C). The peculiar form of the curve of an organ system differs from the normal distribution in that it lacks the initial ascending phase; the anlage of an organ system forms rapidly and, as usual, in a way that substantially differs from the mode of the development following. It is, therefore, extremely vulnerable, at least for cytotoxic agents; and the embryos, provided still alive, exhibit severe changes. The extent and even the nature of the effects changes considerably in accordance with the changed pattern of morphogenetic systems (Table 3.1). The critical periods were analysed in detail in the papers of Rychter[17], Rychter and Jelínek[42] and in their monograph on experimental teratology[18]. In the present section, we have restricted our discussion to fundamental concepts in order to illustrate the approach to critical periods from the standpoint of morphogenetic systems.

MORPHOGENETIC SYSTEMS AND SPECIES SPECIFICITY

Species specificity in teratology represents a problem of outstanding practical significance. Anyone working in the field of toxicology is aware of the fact that no test performed on laboratory animals warrants complete safety for man. The same applies in screening for embryotoxicity, but in the most extreme form. Besides the fact that the experimental object is, as in other

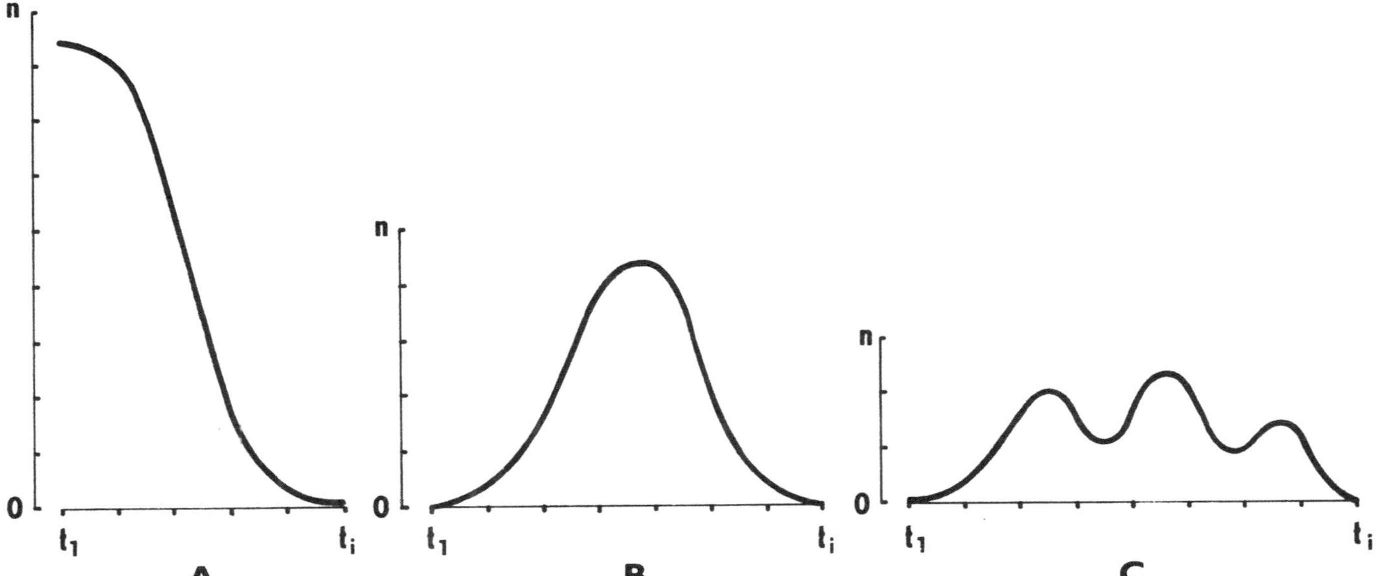

Figure 3.11 Various types of curves for critical periods. A, organ-system type; B, organ type; C, multiphasic type characteristic for organ constituents

branches of basic medical sciences, never identical with the object of investigation, the human being, there even exists one substantial complication: the action of the teratogen is mediated by the maternal organism that in itself represents a strong and probably major source of variability in the response of the embryo[43,44]. Hence, the first and maybe the most important cause of species specificity in teratology is the *different concentrations of teratogens and/or of their species-specific metabolites within the embryonic tissues.* The significant role of concentration of a parent substance can be easily demonstrated on the case of cytembena. Cytembena (Spofa) (natrium bromebricum) a cytostatic agent that is highly effective in gynaecological carcinomas, exerts no adverse effects on the progeny when administered to pregnant mice. When injected, at a dose comparable to the therapeutic one, intra-amniotically, it is embryotoxic. The reason for this is simply that it does not pass, in considerable amounts, the placental barrier. The same applies for the rat (Marhan, personal communication). With this fact in mind, it is hardly surprising that cytembena is highly toxic to the embryonic chick. An example from the second category provides the case of another cytostatic agent, butocine. This substance exerts only a slight embryotoxic influence when applied in extremely high doses to the developing chick. In contrast, if administered to pregnant rats or rabbits, its embryotoxic properties are pronounced at relatively low dosages. We attempted to investigate whether some of the metabolites might not be responsible for the embryotoxicity in mammals. The Chick Embryotoxicity Screening Test (CHEST[30,31]) disclosed that at least one of the metabolites, 3-p methoxybenzoylacrylic acid, exerted a highly deleterious effect upon function of the caudal morphogenetic system (Jelínek, Kocna, Peterka, unpublished) (Figure 3.12).

In neither case was the difference betwen the chick and mammalian embryos response caused by the difference in sensitivity of their morphogenetic systems proper. These, and many other facts[45] may speak in favour of the idea that at least some MGSs (i.e. MGSs taking part in development of the different embryonic components) are similarly sensitive within all vertebrates including man, when directly exposed to similar teratogenic factors (Figure 3.13). We believe that this is sufficiently true to serve as the basis for a rapid embryotoxicity screening procedure. However, in this section we wish to consider these phenomena more deeply. If we administer a substance directly into the embryo's immediate environment, and if the concentration used is effective, the interspecific difference caused by the maternal organism is, to a large extent, obliterated and the embryos are malformed. We still need an explanation of why the malformation spectrum need not be exactly the same in different species, however, even if the moment of administration roughly corresponds to similar developmental periods. We shall leave, for a while, the question of development of specific receptors as it has been discussed in a preceding section, and concentrate on the fact that development of one embryonic component may be, and often actually is, achieved by more than a single MGS. These *morphogenetic subsystems* are either related as precursors (as, for instance the primitive streak in relation to the caudal morphogenetic system), or act and exist relatively independently (as, for example, Meckel's cartilage or the anterior skull base with respect to palatal shelves in

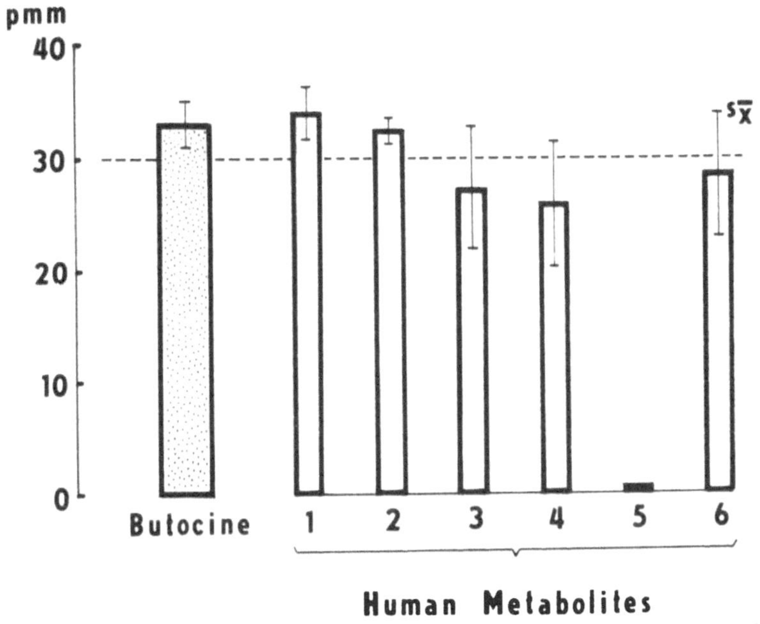

Figure 3.12 Effect of butocine and of its major human metabolites upon chick caudal morphogenetic system. While parent substance exerts no influence at all, metabolite No. 5 produces complete block of morphogenetic function. Ordinate: length of newly-formed part of trunk 24 h after administration of 30 μg doses

the development of secondary palate). During normal development, the action of morphogenetic subsystems is highly integrated and malformation is caused by increasing the disproportion resulting from a distorted spatial pattern at the critical moments of development, for instance at the moment of closure of the neural tube[46] or horizontalization of palatal shelves[47]. It is evident, from the preceding considerations, that the impairment of function of a MGS need not be always followed by the occurrence of an inborn defect. The dose of a teratogen may cause, in one species, a symmetrical and reversible growth retardation; meanwhile in the other species a remarkable growth disproportion initiating maldevelopment. Moreover, it is known that one and the same dosage of a cytotoxic agent may simultaneously cause the growth inhibition in one MGS and acceleration of development in the other, leaving the rest of the subsystems involved possibly unaffected[47]. Agents of this type are known to damage at first the rapidly proliferating tissues[48] and the absolute degree of mitotic activity varies not only among the MGSs and subsystems of the same species, but evidently among similar MGSs of different species. So it is, at any rate, more effective, when testing the effect of a substance on morphogenesis, to evaluate the quantitative, i.e. the growth parameters, than the qualitative ones, i.e. the incidence of malformations. Moreover, it is useful to be aware of the fact that any significant variation, that is even a *positive* deviation from the normal course, is to be suspected as being the first sign of an adverse effect. To be certain, it is recommended that

6-Azauridine

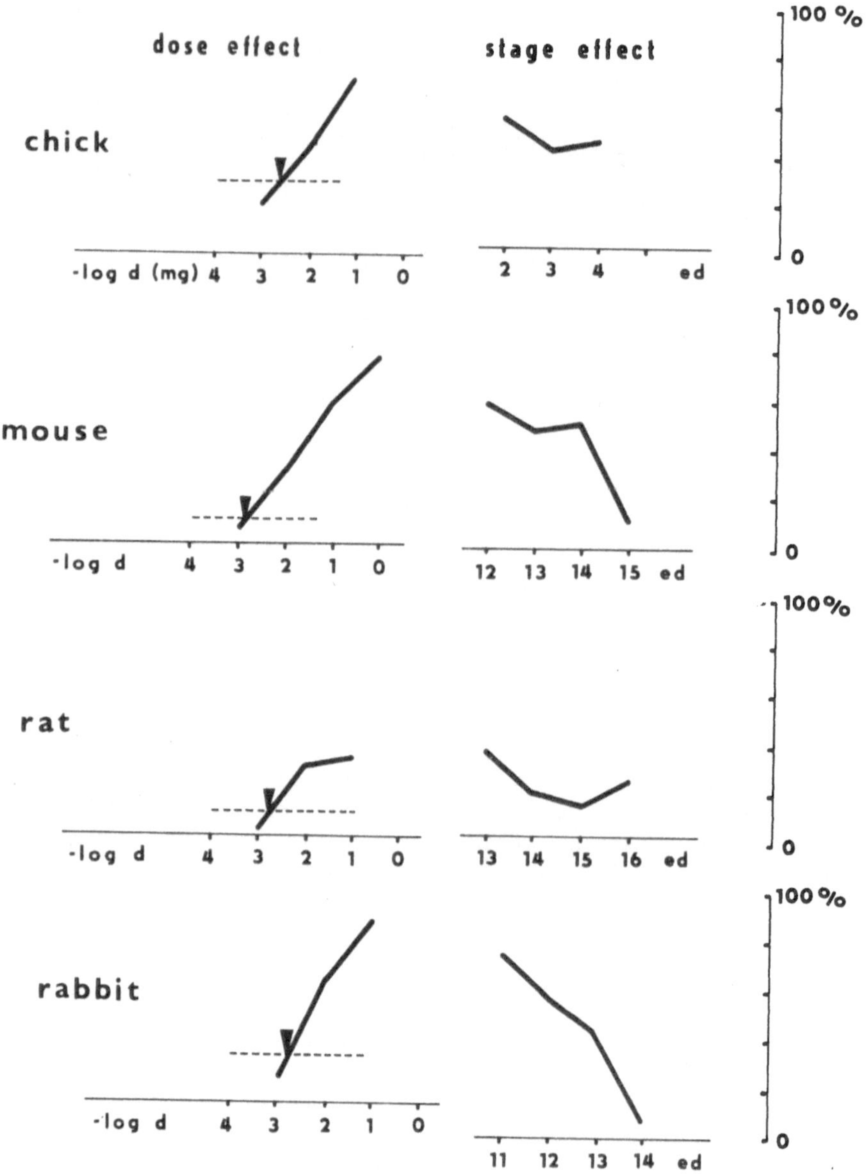

Figure 3.13 Effects of intra-amniotic 6-azauridine in embryos of four animal species. Abscissae: logarithm of dose, days of administration (ed); dashed horizontals, levels of non-specific factors effects. Ordinates: percentages of dead and malformed specimens. Arrows indicate beginning of embryotoxicity dose range

the experiment be repeated using the higher dosage (most conveniently by one order). Hence, the second factor of species specificity in teratology appears to be the *varying intensity of the basic morphogenetic processes within the similar MGSs at the time* of exposure, *as well as the varying contribution of morphogenetic subsystems to the developmental processes at the critical moments of morphogenesis.*

Finally, we wish to deal with the third complex of factors that seem to be intimately bound to cell differentiation. The morphogenetically active compartment of an embryonic component, i.e. the MGS, comprises, as a rule, the less differentiated cells. As development proceeds, the embryonic MGSs are replaced by fetal ones, those by the perinatal MGSs and, eventually, by the postnatal ones. Representatives of this hierarchy appear not to be similarly sensitive to equal doses of a teratogen. Two general types of response could be derived from the results obtained when testing more than 100 different kinds of substances[49]. The first type, comprising the classical cytotoxic agents, is characterized by decreasing sensitivity along the time axis (Figure 3.4). The younger the embryo, the more susceptible the MGSs. In the second type the relations are reversed at least within some limited period of development. The typical representative of the second group is hydrocortisone, and all substances acting on the embryophysiological sphere, especially on blood circulation[50]. The amplification of the effect in the more advanced developmental stages amounts, though infrequently, to even several orders. The first type of response may be explained by the progressive increment in embryonic mass, lowering of the relative amount of a substance arriving at the responding cell populations and/or the increasing redundancy and regulatory potential of MGSs, and/or the progressive slowing down of developmental rate causing the cells to be less susceptible to the cytotoxic action. Response of the second type, characterized by an increasing sensitivity, is suggested to be linked with the development of specific receptors. Their presence on cells make the cells extremely responsive, so that the sensitivity of the responding MGSs dramatically increases. The results of studies of Goldman *et al.*[51], and Bonner and Slavkin[52], correlating palatal cortisol receptors with the susceptibility to cleft palate teratogenesis, are in agreement with such a concept. It is necessary to point out that the development of receptors within MGSs may be interpreted as specific differentiation towards the adult, hence species specific cell lines. It considers, above all, substances of endogenous origin and their analogues. Due to this possibility, it seems useful, when evaluating the risk of an embryotoxic substance, that increase its effect with increasing embryonic age, to consider also its effect on the adult organism to which the results are to be extrapolated. In any case, this type of a response needs to be intensively studied with special reference to its significance in comparative teratology and embryotoxicity testing.

The last of possible sources for species specificity in teratology is the remote possibility that the *biochemical background for the basic morphogenetic processes may be, at least in some animal species, different in principle.* Before refuting this, much work needs to be done in both basic and in applied teratological research. For the time being, no more efficient method can be recommended than to investigate quantitatively the direct interaction between a parent substance and, perhaps, its metabolites occurring in man, and the model representatives of the hierarchy of morphogenetic systems.

References

1. W.H.O. (1977). Non-Mendelian developmental defects: animal models and implications for research into human disease. *Bull. Wld. Hlth. Org.*, **55**, 475
2. Wilson, J. G. (1975). Reproduction and teratogenesis: current methods and suggested improvements. *J. Ass. Off. Anal. Chem.*, **58**, 657
3. Kučera, J. (1976). Population teratology – method and new discipline (in Czech). *Čas. Lék. Čes.*, 115, 1473
4. Bunge, M. (1977). Levels and reduction. *Am. J. Physiol.*, **233**, R 75
5. Maclean, N. (1976). *Control of Gene Expression*. 348. (London: Academic Press)
6. Solter, D. (1975). Embryo-derived teratoma: a model system in developmental and tumor biology. *Symp. Soc. Develop. Biol.*, **2**, 243
7. Faber, J. (1971). Vertebrate limb ontogeny and limb regeneration: morphogenetic parallels. *Adv. in Morphogenesis*, **9**, 127
8. Grüneberg, H. (1975). How do genes affect the skeleton? In: D. Neubert and H.-J. Merker (eds.). *New Approaches to the Evaluation of Abnormal Embryonic Development*, pp. 354–362. (Stuttgart: Thieme)
9. Stockard, Ch. R. (1921). Developmental rate and structural expression: an experimental study of twins, 'double monsters' and single deformities, and the interaction among embryonic organs during their origin and development. *Amer. J. Anat.*, **28**, 115
10. Knorre, A. G. (1968). Congenital malformations, their forms and causes. In: G. A. Bairov (ed). *Surgery of Congenital Malformations in Children* (in Russian), pp. 5–24. (Moscow: Medgiz)
11. Gustafson, T. and Toneby, M. I. (1971). How genes control morphogenesis. The role of serotonine and acetylcholine in morphogenesis. *Am. Scientist*, **59**, 452
12. Saxén, L. (1976). Review article. Mechanisms of teratogenesis. *J. Embryol. Exp. Morph.*, **36**, 1
13. Langman, J. (1975). *Medical Embryology*. 3rd Ed. 421 p. (Baltimore: Williams & Wilkins)
14. Saxén, L. and Rapola, J. (1969). *Congenital Defects*. 247. (New York: Holt, Rinehart and Winston Inc.)
15. Berry, C. L. and Poswillo, D. E. (eds.). (1975). *Teratology. Trends and Applications*. (Berlin: Springer)
16. Jelínek, R. (1973). The contribution of embryological principles to teratology. *Acta Univ. Carolinae, Monograph.* **56–57**, 17
17. Rychter, Z. (1973). Concept of critical periods in teratology. *Acta Univ. Carolinae, Monograph.* **56–57**, 45
18. Rychter, Z. and Jelínek, R. (1978). *Foundations of Experimental Teratology* (in Czech). 159. (Praha: Avicenum)
19. Jelínek, R. and Rychter, Z. (1970). Present problems with the testing of teratogenic effects of drugs (in Czech). *Cs Pediatrie*, **25**, 521
20. Jelínek, R. (1976). The view of an academic scientist on existing requirements for testing of prenatal toxicity. Presented at the *5th Conference of European Teratology Society*, September 23–26, Gargnano, Italy
21. Wilson, D. B. (1974). Proliferation in the neural tube of the splotch (Sp) mutant mouse. *J. Comp. Neur.*, **154**, 249
22. Jelínek, R. and Dostál, M. (1975). Inhibitory effect of corticoids on the proliferative pattern in the mouse palatal processes. *Teratology*, **11**, 193
23. Pexieder, T. (1973). The tissue dynamics of heart morphogenesis. II. Quantitative investigations. A. Method and values from areas without cell foci. *Ann. Embryol. Morph.*, **6**, 325
24. Pexieder, T. (1973). The tissue dynamics of heart morphogenesis. II. Quantitative investigations. B. Cell death foci. *Ann. Embryol. Morph.*, **6**, 335
25. Pexieder, T. (1974). Der Einfluss von Cyclophosphamid auf die physiologischen Zelltodten im Herz des Hühnerembryos. *Vehr. Anat. Ges.*, **68**, 841
26. Waddington, C. H. (1966). *Principles of Development and Differentiation*. 115. (New York: Macmillan)

27. Bertalanffy, L. von (1960). Principles and theory of growth. In: W. Nowinski (ed.). *Fundamental Aspects of Normal and Malignant Growth*, pp. 137–259. (Amsterdam: Elsevier)
28. Rychter, Z. and Šťastný, F. (1976). Morphological development of the choroid plexus in chick embryo. *Folia Morph. (Praha)*, **24**, 317
29. Šťastný, F. and Rychter, Z. (1976). Quantitative development of choroid plexuses in chick embryo cerebral ventricles. *Acta Neurol. Scandinav.*, **53**, 251
30. Jelínek, R. (1977). The Chick Embryotoxicity Screening Test (CHEST). In: D. Neubert, H. J. Merker and T. E. Kwasigroch (eds.). *Methods in Prenatal Toxicology*, pp. 381–386 (Stuttgart: Thieme Verlag)
31. Jelínek, R., Rychter, Z. and Peterka, M. (1976). Cs. Authors' Certificate No. 2170
32. Jelínek, R., Rychter, Z. and Klika, E. (1971). Syndrome of caudal regression and dysraphic malformations of the spinal cord. *Folia Morph. (Praha)*, **19**, 58
33. Jelínek, R. and Peterka, M. (1977). The role of the mandible in palatal development revisited. *Cleft Palate J.*, **14**, 211
34. Rychter, Z., Seichert, V. and Křen, V. (1978). Morphogenetic study of PLS in three different genetic backgrounds. *Folia Biol. (Praha)*, **24**, 365
35. Rychter, Z. and Šťastný, F. (1977). Analysis of dose-response effect of hydrocortisone in the chicken telencephalic choroid plexus. *Physiol. Bohemoslov.*, **26**, 465
36. Šťastný, F. and Rychter, Z. (1977). Developing choroid plexus as a target structure for cortisol in chick embryos. *Proc. Int. Union Physiol. Sci.*, **13**, 714
37. Šťastný, F. and Rychter, Z. (1977). Suppression of the growth effect of hydrocortisone on the chick embryo choroid plexus by progesterone. Time and dose dependence. *Physiol. Bohemoslov.*, **26**, 473
38. Rychter, Z. and Jelínek, R. (1973). Change in shape and location of non-vascularized area of ventricular myocardium in chick embryo during terminal phase of heart vascularization after administration of 6-aza-uridine in different doses. *Folia Morph. (Praha)*, **21**, 1
39. Rychter, Z. and Ošťádal, B. (1971). Mechanism of the development of coronary arteries in chick embryo. *Folia Morph. (Praha)*, **19**, 113
40. Rychterová, V. (1977). Formation of the terminal vascular bed in the chick embryo heart. *Folia Morph. (Praha)* **25**, 7
41. Rychterová, V., Rychter, Z. and Jelínek, R. (1973). The effect of 6-azauridine on the ventricular myocardium of chick embryos (contribution to the teratogenic action of drugs). *Acta Univ. Carolinae, Monograph.* **56–57**, 191
42. Rychter, Z. and Jelínek, R. (1973b). Systematics of the term critical period in experimental teratology (in Czech). *Cs Fysiol.*, **22**, 301
43. Dostál, M. and Jelínek, R. (1973). Corticoid-induced cleft palate as a model system for the distinction of maternal and fetal genomes interacting with exogenous teratogen. *Folia Biol. (Praha)*, **19**, 153
44. Jelínek, R. and Dostál, M. (1973). Species specificity in teratology in the light of analysing the intraspecies differences in mice. *Folia Morph. (Praha)*, **21**, 94
45. Marhan, O. and Jelínek, R. (19—). Efficiency of embryotoxicity testing procedures. III. A comparison between results of the official, basic, and compromising methods. (In preparation).
46. Jelínek, R. (1968). Experimental Dysraphia of the Central Nervous System in Chick Embryo. (Prague: unpublished thesis)
47. Peterka, M. and Jelínek, R. (1978). Growth disproportion as the cause of cleft palate. *XIX. Congr. Morph. Symp.*, pp. 277–282 (Praha: Univerzita Kerlova)
48. Connors, T. A. (1975). Cytotoxic agents in teratogenic research. In: C. L. Berry and D. E. Poswillo (eds.). *Teratology. Trends and Applications*, pp. 49–79. (Berlin: Springer)
49. Jelínek, R. and Peterka, M. (1978). One hundred substances tested with CHEST. Presented at the *6th Conference of European Teratology Society*, September 4–7, Budapest
50. Jelínek, R. Benešová, O., Horák, J. and Souček, K. (1978). Cardiotoxicity of tricyclic antidepressants and Maprotiline in the chick embryo. A comparative study. *Activ. Nerv. Sup. (Praha)*, **20**, 52
51. Goldman, A. S., Katsumata, M., Yaffe, S. and Shapiro, B. H. (1976). Correlation of palatal cortisol receptor levels with susceptibility to cleft palate teratogenesis. *Teratology*, **13**, 22A
52. Bonner, J. J. and Slavkin, H. C. (1976). Cortisone-induced cleft palate susceptibility linked to corticosteroid receptor affinity. *J. Dent. Res.*, **55**. B 201

4
Some problems of chemical teratogenesis

S. SANDOR AND MARIA CHECIU

The purpose of these notes and comments is to draw attention to some facts and problems revealed by our work in experimental teratogenesis. In fact, the conclusions reached from the evidence presented below are concerned also with other than chemical noxious factors and should therefore be considered as tentative contributions to the 'principles of teratogenesis'. The following topics will be discussed:

1. The 'early malformative syndrome'.
2. The laterality of structural anomalies.
3. Repair in teratogenesis.
4. Some remarks concerning the extrapolation of experimental results to human.

THE 'EARLY MALFORMATIVE SYNDROME'

The first few post-implantation days in commonly used laboratory rodents are crucial for the further fate of the conceptus. Basic developmental events, such as the differentiation of ecto-, meso- and endoderm, the establishment of axial and other organ primordia, of some essential morphogenetic patterns, and the formation of extra-embryonic appendages (ectoplacental cone, yolk sac placenta, amnion, allantois), are all concentrated within a period of about 4–5 days. The slightest disturbance of these complex and integrated processes, if not regulated, may have various deleterious effects upon subsequent embryo- and fetogenesis. General or even partial retardation of growth and (or) differentiation, the abnormal morphogenesis of some anlagen or more or less marked destructive phenomena may lead to abnormal relations between the parts of the developing whole, between the embryo and its environment, and may modify the inductive interactions or other processes which determine normal development. Undoubtedly, early post-implantation stages have a 'key' position in prenatal development. If so, a systematic control of morphologically detectable pathological changes appearing during this period must be considered, first of all in pathogenetic investigations. By such an early control the primary target of the presumed early noxious influence, the first steps of a possible pathogenetic pathway (at least at microscopical level) may be more accurately determined.

At least three important methodological aspects are to be mentioned here:

1. In order to detect possible early changes, the usual at term control* must be completed by macro- and microscopical control at these stages. The correlation of the data thus obtained, with late at term observations may reveal some new pathogenetic 'linkages'. One or two (or even more) intermediary controls offer, of course, supplementary possibilities for such dynamical insights. Such repeated controls, if any, are usually made by killing the animals at different stages of pregnancy, i.e. on separate litters. In 1971 we proposed an experimental design, which allows one to control the induced teratogenic events within the same litter, at two or three successive stages of development[1,2]. Repeated partial hysterectomies are carried out on the same pregnant animal and provide embryos and fetuses of increasing developmental age.† This dynamic intralitter investigation of possible teratogenic changes has in our opinion and experience a real advantage, as compared with the same 'multiphase' control on separate litters. It diminishes, first of all, the normal variability of specimens. A detailed analysis made by other authors[4] in rats and also our own determinations made in albino mice (unpublished data) firmly suggest that the intralitter variability in these two species is significantly lower than the inter-litter variability. As variability is not related to size and weight, it can be presumed that this statement is valid also for the reactivity toward exogenous factors. On the other hand, the so called 'litter-effect'[4], the frequently observed different reaction of each materno-embryonic complex toward environmental agents can be avoided.

From the various aspects of this still unexplained 'litter effect', one was recently observed in our laboratory in experiments on rats with 6-aminonicotinamide[5]. The localized disorganization of the retinal structure with aberrant optic fibres, one of the typical dysmorphogenetic effects upon eye development, obtained by the administration of the antivitamin on day 11 of pregnancy, showed a striking 'litter effect'. While other pathological changes following injection on days 9, 10 or 15 could be detected with a variable frequency, in practically every litter aberrant fibres were found in some litters in a relatively high number, but in others were totally absent. It follows that the 'litter effect' may be also time-dependent and may exhibit a clear-cut phase specificity.

Returning to the methodology proposed, as repeated surgical intervention does not disturb subsequent development of the remaining conceptuses (according to the control series with two or three operations and mortality control), it offers the possibility for examining within the same litter the pathogenesis of structural anomalies or of other changes induced by an environmental agent.

2. In both human and in experimental prenatal pathology, the changes in

* Survey of the literature reveals that the overwhelming majority of investigations in experimental teratogenesis resumes too late, at term control of structural malformations

† At the time of our publication, we were not aware of an essentially similar methodology applied in a study on the rat yolk sac[3]

development are not exhausted by microscopical or macroscopical structural anomalies or other qualitative changes. One of the most frequent deviations from normal development is undoubtedly the modification of the developmental rate. It appears not only during late fetal stages or at term (when it is usually controlled by the ossification centres and of the general development of the skeleton, by weighing and size measurements), but also during the early postimplantation stages. An attempt to determine the rate of retardation assumes the application of statistically valid biometrical methods. As to early developmental stages, both in mammals and in other experimental animals, this problem is in our opinion still unsolved. The existing, valuable Normentafels involve only descriptive data with respect to some selected morphogenetic indices, without taking into account the well known variability and the necessity for its more or less exact statistical analysis. Some years ago, in our laboratory, an attempt was made to partially fill this gap by a detailed biostatistical analysis of the early development of the chick embryo[6]. The application of this complex biometrical methodology in a study of the effect of ethanol on the developing chick embryo, enhanced the detection of several minute early changes[1]. In continuation a similar but less complex methodology for early mammalian embryos has been proposed. The statistically assured registration of the presence or absence of the main morphogenetic characteristics during the first days of postimplantation development allows an exact comparison between the developmental rates, between control and experimental series. Using this methodology, focused upon day 10 of pregnancy, a stage most adequate for morphological analysis, with a sufficient number of intra- and extra-embryonic structures early, almost general retardation under the influence of ethanol intoxication could be detected[7] (see Table 4.1).

3. It seems perhaps obvious, but nevertheless it is worth emphasizing, that the early embryo must always be controlled together with its direct environment, with the extra-embryonic structures, surrounding membranes and adjacent maternal tissues. In fact, we are dealing at these early preplacental stages (the term is not completely correct, as there exists the yolk sac placenta)

Table 4.1 Effects of ethanol on early rat development

Morphogenetic characteristics on day 10	*Control (present %)*	*Ethanol 2 g/kg on days 6 and 7 (present %)*	*Difference from control group*
Proamniotic cavity	0	23.6	+23.6*
Amniotic folds	0	23.6	+23.6*
Amniotic cavity	100	76.0	−24.0
Exocoelom	94	47.0	−47.0*
Ectoplacental cavity	100	76.4	−23.6
Allantoic stalk	60.6	23.6	−37.0
Head process	24.2	70.6	+46.6*
Chorda	66.7	23.6	−43.1*
Neural plate	75.8	29.4	−46.4*
Somites	18.1	0	−18.1*
Head fold	54.3	12.0	−42.3*
Heart anlage	39.4	5.9	−33.5

* $p < 3\delta$

with a kind of embryo–maternal unit, analogous to the current notion of feto-placental unit, used for fetal stages of development. Data on the early changes of the decidua, of the uterine vessels, of the ectoplacental cone and of other extra-embryonic appendages may reveal unexpected targets and pathogenetic pathways.

With the continuous improvement of tissue culture techniques, the early postimplantation stages are studied also *in vitro*. *In vitro* studies may furnish valuable data concerning the direct or indirect mode of action of environmental agents. As the embryos are generally cultured without surrounding membranes, for a complete assessment of the possible early changes, *in vitro* examination must be supplemented by *in utero* control.

An attempt was made to obtain a general survey of a number of up to date available data reported with respect to primary 'target' effects and early pathological changes induced by various experimental procedures.

A. Even such an obviously incomplete review reveals some rather well established variants of the 'target' effect (at microscopic level):

(a) *Oedematous changes of the intra-embryonic mesenchyme* (*mesoderm*), involving the rarefaction of the cell population, blisters and blebs, the dilation of blood vessels. Factors inducing this effect are: excess of vitamin A (*in utero*[8,9], *in vitro*[10]), bisazo dyes (*in utero*[11,12], *in vitro*[13]) (Figures 4.1, 4.2, 4.3 and 4.4), panthotenic acid deficiency (*in utero*, cited by[14]), linoleic acid deficiency (*in utero*, cited by[14]).

(b) *Changes in somite size and pattern, necrosis in the somites.* Causal factors: rastinon (a blood sugar lowering drug) (*in utero*[15]) and the antivitamin 6-aminonicotinamide (*in vitro*[14]).

(c) *Inhibition of allantois development.* Causal factors involved: myelosan (a cytostatic drug) (*in utero*[16]) and tetracycline (*in vitro*[17]).

(d) *Regressive changes of the ectoplacental cone.* Causal factors: X-rays (irradiation was carried out on the gonads of males one week before mating) (*in utero*[18]), aminopterin (a folic acid antagonist) (*in utero*[19]), delagile (chloroquine) (an antirheumatic and antimalarial drug) (*in utero*, unpublished data) (Figures 4.5, 4.6 and 4.7).

(e) *Regressive changes of the decidua.* Causal factors cited are: actinomycin, methotrexate, protein-free diet (*in utero*[20]), daraprim (chloridine) (an antimalarial drug) (*in utero*[21]), 6-aminonicotinamide (*in utero*[5]) (Figures 4.8 and 4.9).

(f) *Selective destruction of embryonic, and persistence of extra-embryonic structures.* Causal factor: 6-mercaptopurine (*in utero*[22]) (Figures 4.10, 4.11 and 4.12).

(g) *Various disturbances of early embryogenesis.* Causal factor: ethanol (*in utero*[7]) (Figures 4.13, 4.14 and 4.15).

B. In a number of cases it is rather difficult to separate the 'target' effect from the early intra-embryonic changes. With regard to *in vitro* results, which exclude the detection of extra-embryonic 'targets', the suspected primary effect is always obtained by direct contact with the agent; thus, all but one of the *in utero* existing conditions is reproduced. Nevertheless, as mentioned above, the *in vitro* induced pathological changes can furnish additional evidence to *in utero* obtained previous results. In the case of bisazo dyes e.g. the

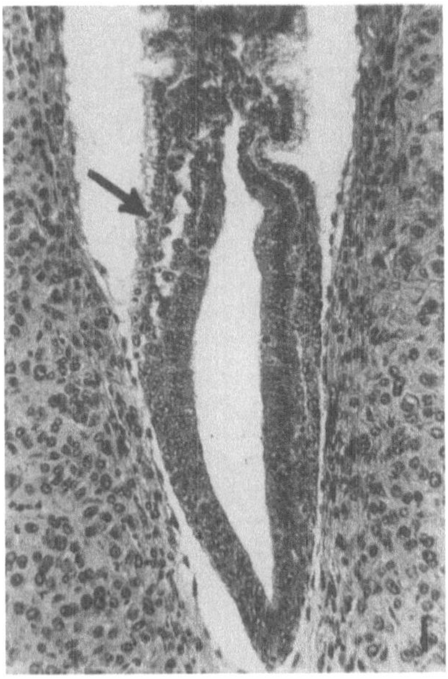

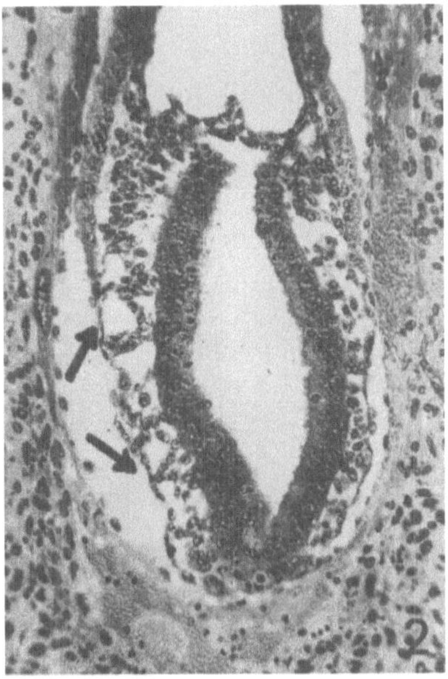

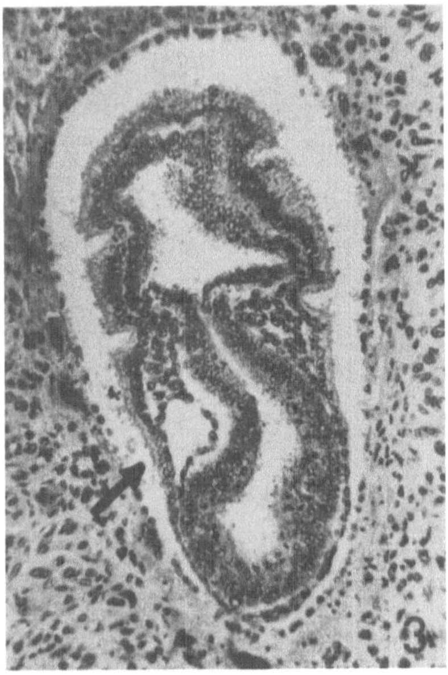

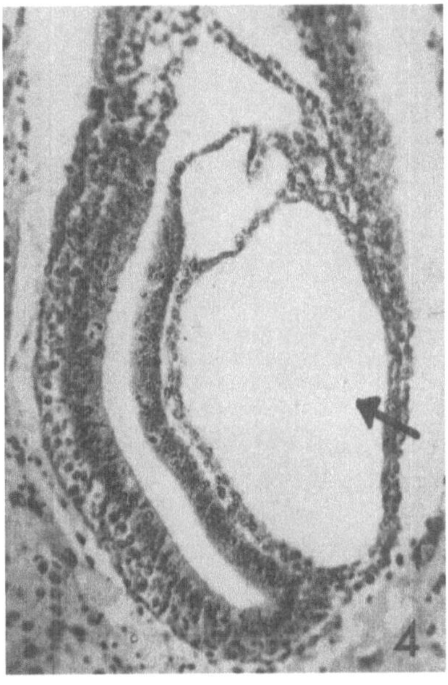

Figure 4.1 Rat embryo – day 10 of pregnancy. Trypan blue, 10 mg/kg, i.p. on day 9. Incipient rarefaction of the intra-embryonic mesoderm (see arrow)

Figure 4.2 Rat embryo – day 10 of pregnancy. Niagara sky blue 6B, 10 mg/kg, i.p. on day 9. Generalized rarefaction of the intra-embryonic mesoderm (see arrows)

Figure 4.3 Rat embryo – day 10 of pregnancy. Niagara sky blue 6B, 20 mg/kg, i.p., on day 9. Well formed 'empty' space within the intra-embryonic mesoderm (see arrow)

Figure 4.4 Rat embryo – day 10 of pregnancy. Trypan blue, 15 mg/kg, i.p. on day 9. Huge 'empty' bleb within the intra-embryonic mesoderm (see arrow)

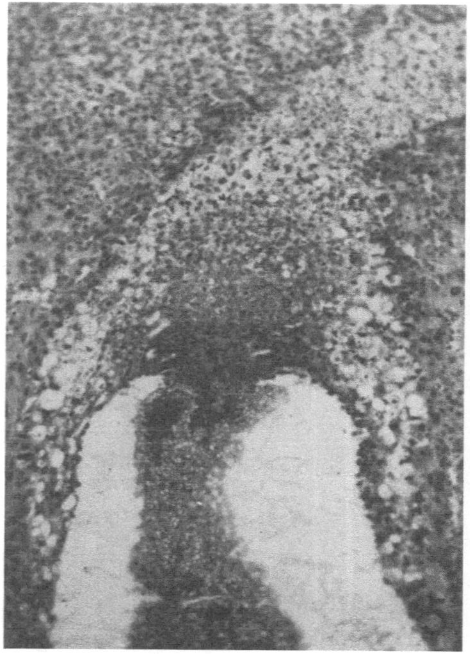

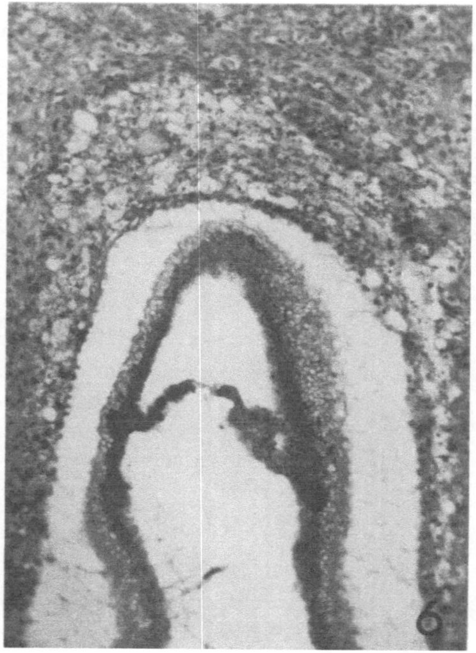

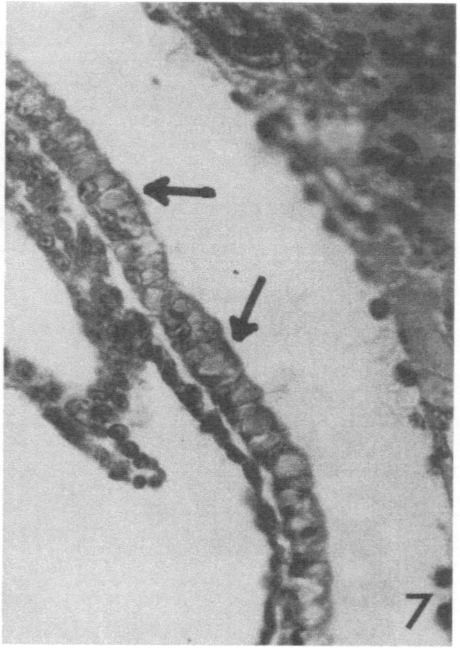

Figure 4.5 Rat embryo – day 10 of pregnancy. Delagile, 1 g/kg per os, on day 9. Regressive changes of the ectoplacental cone and of the adjacent decidua

Figure 4.6 Rat embryo – day 10 of pregnancy. Delagile, 1 g/kg per os, on day 9. Regressive changes of the ectoplacental cone and of the adjacent decidua

Figure 4.7 Rat embryo – day 10 of pregnancy. Delagile, 1 g/kg, per os, on day 9. Marked vacuolation of the extra-embryonic endoderm. Acidophyl substance in the vacuoles (see arrows)

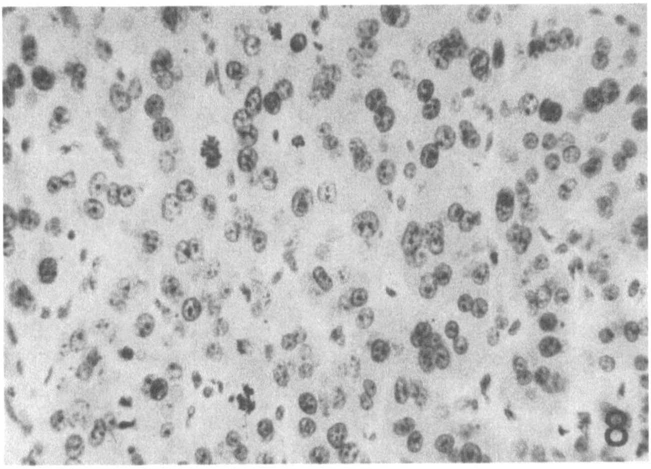

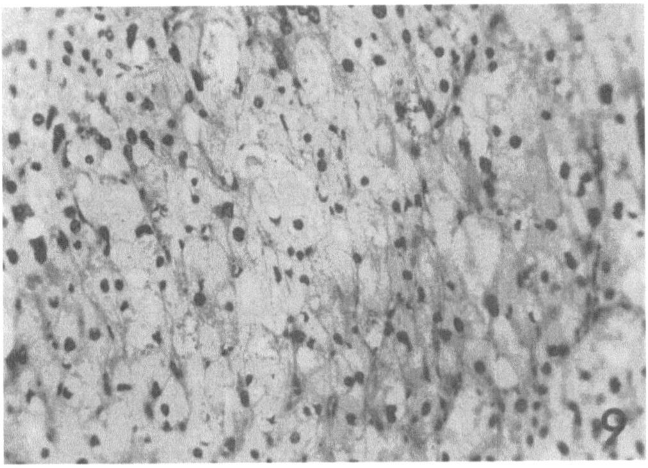

Figure 4.8 Rat embryo – day 10 of pregnancy. Normal antimesometrial decidua

Figure 4.9 Rat embryo – day 10 of pregnancy. 6-Aminonicotinamide, 4 mg/kg, i.p., on day 9. Marked degenerative changes (pycnosis, vacuolation) in the antimesometrial decidua

in vitro effects reported[13] are obviously in support of the 'direct action' hypothesis[23], in contrast to the 'altered yolk sac endoderm function' theory[24–27]. The *in vitro* effect of vitamin A, very similar to the bisazo dye-induced changes, also suggests a direct action upon the embryo. On the other hand, our own observations obtained *in utero* with trypan blue and Niagara blue and the results reported as to *in utero* experiments with vitamin A excess may be pertinent with the second indirect alternative. Moreover, in unpublished, preliminary experiments conducted in our laboratory involving prolonged hypobaric hypoxia of pregnant female rats (during early postimplantation stages) early mesodermal changes, very similar to those found after bisazo dye or vitamin A treatment, have been detected (Figure 4.16), confirming the possibility of early trophic disturbances via the yolk sac placenta. These

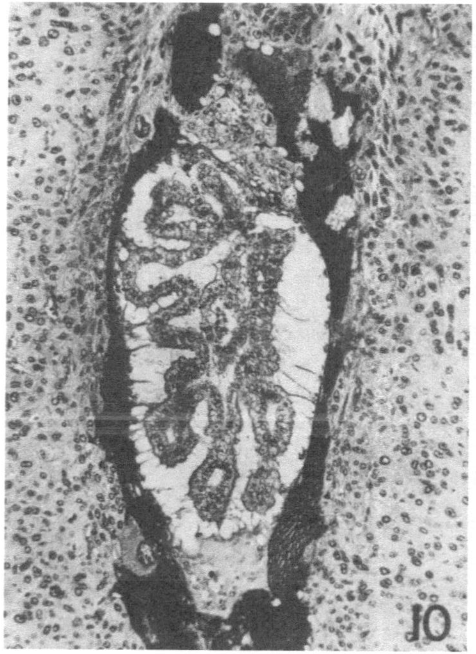

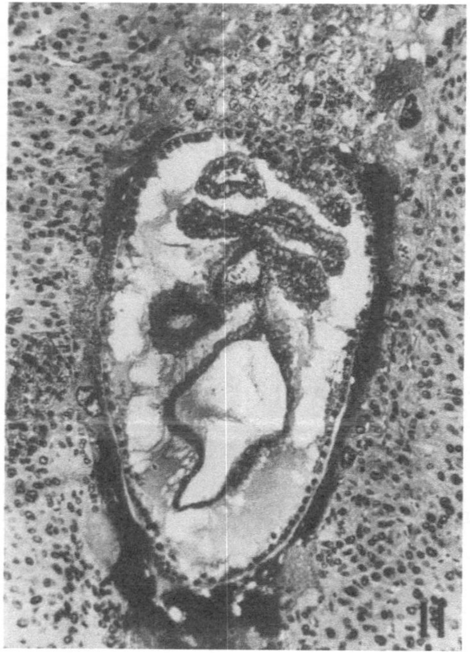

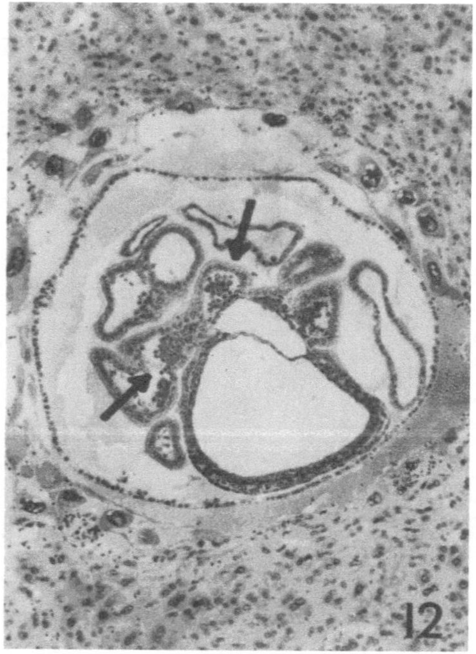

Figure 4.10 Rat embryo – day 11 of pregnancy. 6-Mercaptopurine, 10 mg/kg, i.p. on days 8 and 9. Absence of embryonic structures, abnormally proliferated extra-embryonic endoderm

Figure 4.11 See Figure 4.10

Figure 4.12 See Figure 4.10. Hyperplasia of blood islands (see arrows)

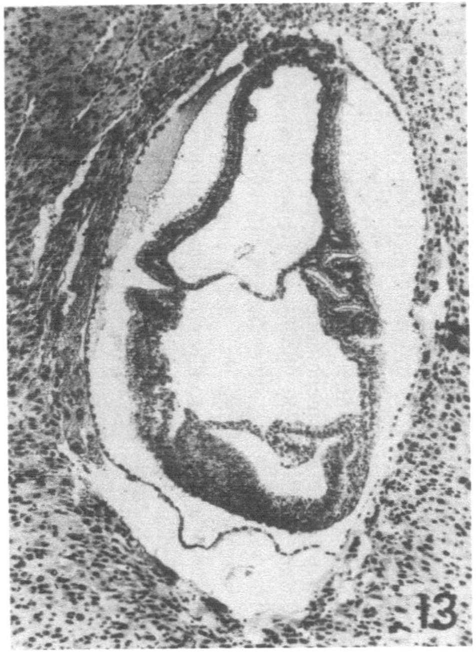

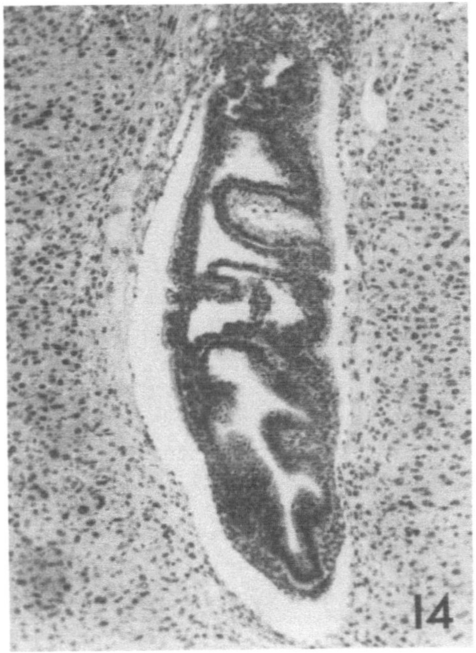

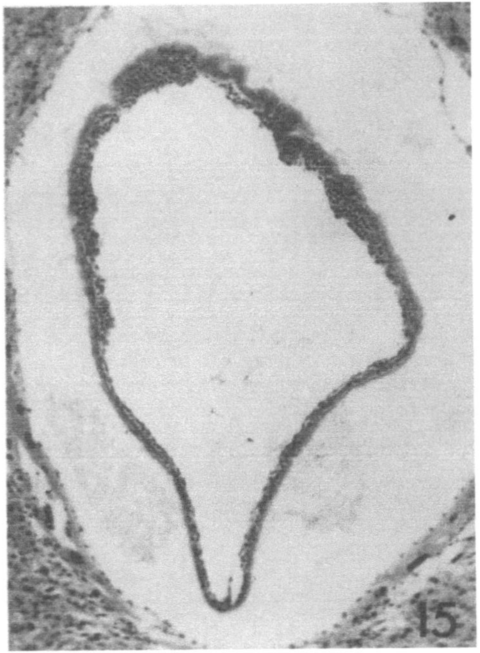

Figure 4.13 Rat embryo – day 10 of pregnancy. Ethanol, 2 g/kg, i.v. on days 6 and 7. Slight deformation of the neural plate

Figure 4.14 Rat embryo, day 10 of pregnancy. Ethanol, 2 g/kg, i.v. on days 6 and 7. Marked deformation of the embryonic and extra-embryonic structures

Figure 4.15 Rat embryo, day 10 of pregnancy. Ethanol, 2 g/kg, i.v. on days 6 and 7. Absence of embryonic structures. Endodermal vesicle with several blood islands

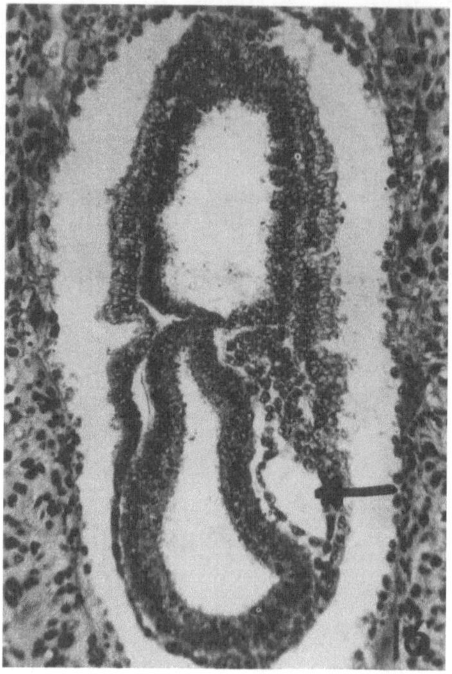

Figure 4.16 Rat embryo – day 10 of pregnancy. Hypobaric hypoxia, 20 hours, on day 9. Well formed 'empty' space within the intra-embryonic mesoderm (see arrow)

seemingly contradictory evidence allows, for the moment, no definite statements on matter of the primary 'target' effect.* It is quite possible that the environmental agents mentioned (and perhaps also other factors) may lead to similar dysmorphogenetic effects by both direct and indirect action.

C. If the allantois or the ectoplacental cone is involved in the primary 'target' effect, disturbances in the development of the chorioallantoic placenta and of the developing embryo itself are almost surely secondary, both extra-embryonic appendages being essential for the establishment of normal materno-embryonic and materno-fetal morpho-functional relations. *Mutatis mutandis*, a similar situation arises if the decidua is the site of the first detectable pathological changes. The decidual tissue is supposed to have several functions: control of placentation, mechanical protection of the developing embryo, nutrition and blood supply, hormone synthesis, immunological barrier, etc.[29–31]. The marked destructive phenomena which appear in the decidua under the influence of daraprim[21] and 6-aminonicotinamide[5] must supposedly alter, or even partially or totally stop these multiple functions. The secondary intra-embryonic changes may appear very early, almost syn-

* The yolk sac endoderm (extra-embryonic endoderm) is not mentioned as a site of primary 'target' effect. Besides the still unsettled discussion on this problem, it should be mentioned that of the three agents, surely acting with certainty upon the yolk sac endoderm and causing maldevelopment (bisazo dyes, triton-1339 and antisera)[24–28] intra-embryonic changes were also investigated only in the case of bisazo dyes early

chronously with the decidual defect[21] or, as in our experiments with 6-aminonicotinamide somewhat later as detectable dysmorphogenetic features.

D. The early pathological features reported after ethanol[7] and 6-mercaptopurine treatment[22] strongly suggest a direct, primary effect on the developing embryo (in the first case a large spectrum of more or less severe disturbances, in the second case a striking uniformity of the pathological lesions).

E. As already mentioned, one of the purposes of the early postimplantation control is the correlation of early and late pathological changes. This point has a special importance. If the early detectable changes present some regular, necessary correlations with late developmental defects they may have a certain predictive value. That means that the efficiency of current 'industrial' screening may be increased by shortening the time of control. If the early changes detected after the administration of a test substance could be involved in one of the groups of 'target' effect and of primary intra-embryonic changes just outlined, the late effects to be expected could possibly be predicted (first of all concerning mortality, retardation etc.). Moreover, such a correlation, if stated and strengthened also by intermediate, multiphase control (see above) may offer a valuable insight into some pathogenetic pathways. Unfortunately, from the *in utero* studies only a part produced the evidence necessary. Some of the studies were able to establish a plausible relationship between early and late findings, thereby indicating the pathways of certain dysmorphogenetic events (exencephaly after *in utero* excess of vitamin A[8], eye anomalies after *in utero* treatment with bisazo dyes and 6-aminonicotinamide[5,11], vertebral defects after *in utero* treatment with rastinon[32]).

When, at the early postimplantation stages, the allantois or the ectoplacental cone are damaged or even completely prevented from development, late control reveals a high mortality rate combined with the persistence of retarded specimens or almost total fetal wastage[16,19]. When, inversely, following the treatment, only the extra-embryonic structures persisted[22], later control revealed 100% mortality. A striking fact derived from the results of several studies is the obvious difference between the susceptibility of embryonic and extra-embryonic structures to various teratogens. One is reminded of the situation with respect to the chick embryo complex, where extra-embryonic appendages show another type of reactivity than the embryo itself[1,33–36].

F. Of course, the data mentioned are still fragmentary. But even this limited evidence reveals some characteristic features of early effects induced by environmental agents. In 1975 the term 'early malformative syndrome' was tentatively coined[12]. It includes both the primary 'target' effect, and the first, microscopically detectable changes in embryonic structures. Since then, further findings from our laboratory and several reports are in agreement with the first, preliminary statements. In our opinion, the establishment of the 'early malformative syndrome' and its systematic investigation, may contribute to at least three directions of teratological research:

(i) Systematic investigation of intra- and extra-embryonic early changes, if added to the usual teratological screening methodology, may in a number of cases shorten the time of investigation, by 'predicting' mortality and retardation. If, after early treatment, no early changes are detected, then there is little possibility of late structural anomalies occurring.

(ii) The detection of the primary 'target' effect and of the consecutive, first micro- or macroscopical changes of the embryonic structures may, if completed by multiphase control of later effects, reveal pathogenetic pathways leading to certain structural anomalies.

(iii) It seems that the localization and the intensity of the first lesions depends both on the quality (chemical structure etc.) of the agent acting and on the different susceptibility of the embryonic and extra-embryonic structures to various deleterious influences. Thus, the effects detected may serve as a sensible test of this double-dependent noxious action. Perhaps, it will be possible to group the noxious factors according to their primary 'target' effect (a first attempt in this direction has been made above). Moreover, by comparison as to their early, primary effect of related, even slightly differing agents, the significance of structural differences in determining the prenatal noxious action[37] can be studied.

THE LATERALITY OF ANOMALIES

Structural anomalies of paired organs are uni- or bilateral. This seemingly gratuitous routine statement is in fact one of the most intriguing and still unsolved problems of theoretical and practical teratology. On one hand, the obvious importance of uni- or bilaterality of malformations appears when the individual, familiar and social inference of maldevelopment is envisaged. It is enough to recall the difference in functional handicap between uni- or bilateral anophthalmia, uni- or bilateral phocomelia or even uni- or bilateral syndactyly. On the other hand, these anomalies are not to be separated from the general 'principle' of bilateral symmetry in development. Determined by still unknown genetic and epigenetic factors bilateral symmetry is manifest in all vertebrates and in many invertebrates. A modification must at once be introduced: within the general bilaterally symmetrical organization of the (embryonic, fetal or adult) organism several, more or less expressed asymmetries are present, from slight hardly detectable differences between right and left to the almost or totally asymmetric organs or organ systems. Extrapolating the well-known slight differences of paired adult organs, of right and left body-half, paired organ anlages and developing primordia are also supposed to be endowed with a certain degree of inequality (at a certain moment of development, due to differences in developmental rate, etc.).

It would lead too far giving a detailed account of problems and facts concerning bilateral symmetry. It is however worth emphasizing that in spite of its practical and theoretical importance, the uni- and bilaterality of dysmorphogenesis, the mechanisms and causal factors which determine this recognizable phenomenon, have up to now, scarcely entered the sphere of teratological research.

Indeed, only a strikingly low number of descriptive or experimental contributions have been directed at all to the problem discussed. A part of this limited information is not exactly concerned with uni- or bilaterality, but these topics are closely related. In particular, a pioneering study[38] investigated the non-random laterality of malformations in human paired structures. As pointed out by the authors, a statistical evaluation of more than 20000 cases with various separate unilateral defects revealed a 'non-random significant tendency to affect a particular side'. In the opinion of the authors, this very obvious situation, which has its equivalent also in mammals 'implies subtle developmental differences between the two sides for each tissue, allowing one of the sides a greater liability for a particular type of defect in morphogenesis'. Let us defer discussions concerning these presumed developmental differences and note the rather astonishing fact that even during the studies with such a great amount of data, no attention was given to the basic fact: the *unilaterality* of anomalies.

As quoted in the paper mentioned, anomalies induced by experimental teratogens also show a typical one-sided predominance (acetazolamide in rats leads to more severe right forepaw polydactyly[39]; thalidomide in the rabbit leads to 73% left-sided forelimb reduction[31]). These examples are surely not exhaustive. Even a summary survey of the literature shows that the overwhelming majority of the authors do not take into account (or do not mention) the uni- or bilateral localization of the anomalies induced (including also our own, previous studies!).

From the available data of the recent literature, there exists one ingenious experimental approach aimed at obtaining an insight into the factors determining the laterality of maldevelopment[40]. Assuming that the peculiar right-sided localization of acetazolamide-induced forelimb malformations in mice is due to normal asymmetry of the mouse embryo (coiling into a right-handed helix) the substance was administered to mice with embryos homozygous for a gene causing *situs inversus viscerum*. Confirming the above-mentioned hypothesis, the limb malformations obtained were both left and right-sided[39,40].

Recently, in our laboratory some observations were made as to the laterality of experimental eye anomalies induced by bisazo dyes and 6-aminonicotinamide[5,11] (see Table 4.2).

From the data presented two main conclusions may be drawn. Firstly that anomalies induced by the administration of teratogens during early post-implantation stages (9–11 days) are both uni- and bilateral. The percentage of bilateral anomalies seems to increase with increasing age at treatment. Secondly, that anomalies induced by the administration of teratogens after the 12th day of pregnancy are always bilateral.

Some reported data are consistent with these findings. After early administration (on day 8 of pregnancy) of trypan blue in rats frequent unilateral eye defects were reported[41]. X-ray irradiation of mouse embryos on day 8 of pregnancy induced uni- and bilateral eye anomalies[42].

Table 4.2 Laterality of experimentally-induced ocular defects

Substance	*Day of administration*	*Ocular defects*	
		Unilateral	*Bilateral*
Trypan blue	9	75	17
Niagara blue	9	12	13
6-AN	9–11	16	28
6-AN	13–17	—	195

Based upon these data, a working hypothesis has been tentatively outlined. *Scheme 1* presents its main points:

Scheme 1

As seen, the period of administration (in laboratory rodents, in our scheme, in rats) determines mainly three closely correlated variables, essential for the outcome of teratogenesis: treatment route, the mode of action and the 'developmental situation' encountered by the teratogen. Whereas the first two variables are at least partially known or may be approached experimentally without excessive difficulties, the third awaits new experimental models. In our opinion the working hypothesis presented is at any rate worth confirming.

REPAIR IN TERATOGENESIS

Almost every handbook or treatise of embryology contains one – usually the final – chapter dealing with problems of regeneration. Several common features of regenerative phenomena and embryonic development, a considerable amount of experimental evidence brought on regeneration, merely in embryonic stages of lower vertebrates, explain this usual association. On the other hand, surprisingly little work has been carried out on regenerative events during the development of higher vertebrates on possible repair of injuries affecting the developing embryo or fetus. As is already known, various environmental factors, including those originating from the direct maternal environment, may cause a more or less extensive loss of embryonic or fetal cells and tissues. Degenerative and necrotic changes are not only constitutive parts of normal development but also a rather frequent initial component of dysmorphogenesis[43,44]. It would be without doubt of considerable interest to know, to understand and to 'handle' the possible repair of these regressive phenomena.

Theoretically, the outcome of an affected pregnancy is at least partially the result of an interplay between destruction and repair. If the reparative phenomena are weak, destruction may induce further disturbances of development. If, however, the embryonic or fetal organism has sufficient regenerative capacities it may counteract the destructive action and achieve even a *restitutio ad integrum*. It is obviously not enough to put forward general statements about the possible role of repair in teratogenesis. Schematizations concerning the possibilities of regeneration and regulation at successive periods of prenatal development (blasto-, embryo- and fetogenesis)[45] are useful, giving a conceptual framework for further investigations. But, without a sufficient number of concrete observations and experimental findings, even the best scheme is unsatisfactory. As a matter of fact, our knowledge as to regeneration and regulation potencies of injuries suffered during the embryonic and fetal period in higher vertebrates is almost exclusively hypothetical (for the few existing data see below).

Regulation during blastogenesis, i.e. during preimplantation stages in laboratory mammals, is somewhat better known. In experiments with various environmental agents (radiations, chemicals, viruses, etc.) acting during this period, further development has either totally stopped (lethal effect) or showed no essential deviation from normal development (total regulation). Thus, during this developmental period, no true malformative effect could be obtained.* Such *a posteriori* statements, obtained mainly by late control of *in utero* or *in vitro* treated ova, with subsequent development in their own or in a 'foster' mother, do not explain the mechanism of early regulation.

Observations concerning intra-uterine regenerative processes in man are, of course, even less frequent. Perhaps, the particular body of such (indirect) observations and of derived ideas concerns vestigial signs of partially or totally regenerated (healed) intra-uterine malformations (hare lip, cleft palate, myelocoele etc.)[46,47]. It is further possible, in our opinion, that at least some of the discrepancies between very early and postnatal human structural anomalies[49–52] are due to regenerative (healing) processes during intra-uterine development. But, unless experimentally verified, both these important assumptions remain hypothetical, *a posteriori* deductions.

The evidence derived from experimental investigations on regulative and reparative potencies in higher vertebrate embryos are but indirectly concerned with our problem. Localized microinterventions (microsurgery, focused irradiations, microinjections) applied in these studies can be but conditionally homologous to the usually generalized action of noxious environmental agents. Nevertheless, they contribute essentially to our conceptual framework, to the understanding of reparative phenomena in pathological development. It would go beyond the purposes of these notes to present the rather great number of regulative and reparative events observed in various experimental models worked out mainly on early chick and duck blastoderms. (The reader is advised to consult, if interested, among others, the valuable, systematic reviews concerning this evidence[53,54].) In our laboratory, within the larger framework of studies on embryonic axis development, marked regenerative and regulative potencies of the early chick blastoderm have been identified[35,55–58]. Healing of experimentally induced defects, reorganization and even *restitutio ad integrum* of the mechanically destroyed tissue continuity, postgeneration phenomena, etc. revealed at least in this species and during the developmental stages studied impressive possibilities of defect reparation and pattern reconstitution. In early postimplantation mammalian embryos, various microsurgical interventions on the developing axial organs also revealed important reparative potencies[59,60].

Reparative potencies were also approached by investigating the wound healing process in chick embryos and fetuses[61,62]. As to this special potency studied in the skin and in the cornea, a main component, the epithelial migration begins at relatively late stages while proliferation is present beginning with late embryogenesis. The stepwise regeneration of fetal rat skin, following an injury by iodine solution has also been studied in detail[63].

There exists one more source of evidence which indirectly suggests and confirms the remarkable regenerative potencies of embryonic tissues: the

* Some evidence exists confirming this almost generally accepted rule. Thus, from pregnant rabbits treated with actinomycin D during early preimplantation stages of pregnancy, embryos and fetuses with structural anomalies of the central nervous system and with other anomalies were obtained[48]

spontaneous and experimental reaggregation of dissociated embryonic cells. The Herbst–Holtfreter–Weiss–Moscona line of investigation, this constantly developed experimental model, strongly suggests, that within the whole developing embryo or fetus similar reorganization is possible. Furthermore, *in vivo* investigations have to gather the facts necessary to support this belief.

The systematic or even incidental experimental investigation of reparative phenomena involved in developmental pathology is still at its onset.

Regeneration (repair) during teratogenesis may be approached in several ways. It may be suggested indirectly, by correlating the statistical data of experiments. Thus, for example, in experiments using X-ray irradiation of rat embryos[64] the time dependent decrease of the percentage of anomalies, which bypassed by far the increase of mortality suggested the presence of normalizing and reparative events. Other experimental observations are based upon successive morphological control during a part of or during the whole prenatal development. Almost all the findings reported concern regenerative processes in the central nervous system.

In a multiphase microscopical study of the spinal cord in X-ray irradiated mouse embryos[30] the initially observed heavy necrotic changes disappeared stepwise, the normal structure being reconstituted beginning with 2–3 days after the experimental injury. In some cases, however, the architecture of the spinal cord was severely disturbed. Experiments concerning DNA synthesis inhibition and cell death induced in rat embryos by hydroxyurea[32] showed that '... although an appreciable amount of cell death was also evident in the ependymal layer of the neural tube, gross malformations of the nervous system were quite rare'. Besides the possible remaining functional anomalies studied in continuation by the authors, the possibility of a true total absence of malformations, due to efficient reparative phenomena was envisaged. In our own investigations[22] on 6-mercaptopurine induced anomalies in developing rat embryos, the multiphase control applied also revealed, among others, the absence at term of morphologically detectable malformations of the brain, although in earlier stages extensive cell death appeared in various regions of the central nervous system (Figure 4.17). On the other hand, besides practically normal features the spinal cord showed, in some cases, obvious, disorganizing reparative phenomena (supplementary lumina, rosettes, etc.) (Figure 4.18). They were considered 'inefficient' variants of the reparative events, supposed to be consecutive, in the central nervous system, to the previously observed necrotic changes. In these notes on repair in teratogenesis we focused our attention on the true reparative phenomena which are able to counteract, more or less, the destructive effect of various teratogens. It must be pointed out, however, that 'inefficient', disorganizing proliferative phenomena, even malformation-producing 'regeneration' was observed in birds and mammals following various experimental interventions, mainly in the central nervous system[65–75].

A more recent study of the effects of 6-mercaptopurine on the fetal rat central nervous system[76] demonstrated a remarkable repair by phagocytic and regenerative processes of the marked destruction induced. It was stated, that '... necroses ... disappeared almost entirely shortly before birth with partial normalization of the cytoarchitecture' and only minimal remaining defects, as

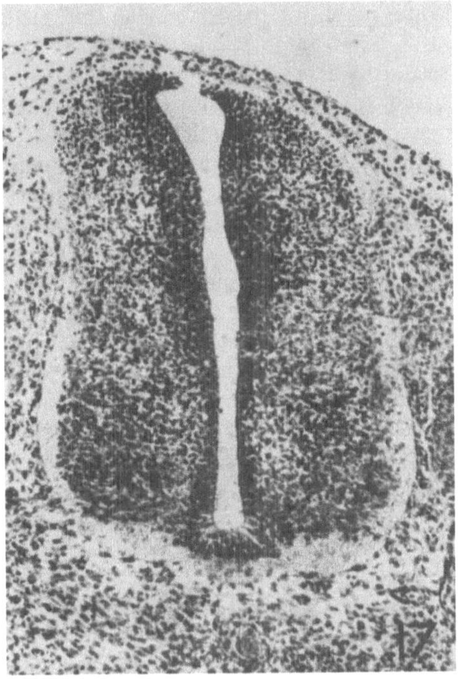

Figure 4.17 Rat fetus – day 14 of pregnancy. 6-Mercaptopurine, 60 mg/kg, on day 12. Spinal cord, marked necrosis in the mantle layer

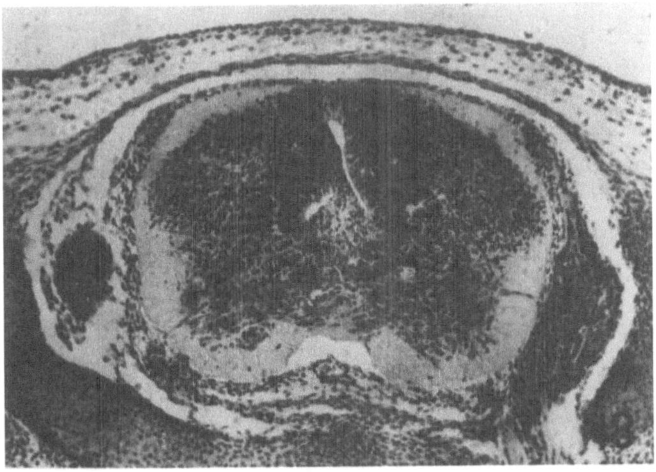

Figure 4.18 Rat fetus – day 17 of pregnancy. 6-Mercaptopurine, 60 mg/kg, on day 12. Spinal cord, normal architecture destroyed by regenerative changes

cell sparsity of some layers of the brain wall and clefts in the spinal cord could be observed.

A common feature of all the above cited observations is that reparative phenomena occurred following previous extensive cell death. It may be assumed, that in such conditions repair has to involve, besides the 'cleaning' process (by autolysis or by the activity of macrophages) also proliferation, an active production of new cell lines. A recent series of very thoroughly and systematically conducted morphological studies[77–79] demonstrated reparative cell proliferation in the spinal cord and in the brain of rat and chick embryos and fetuses, after the administration of FUdR. The dynamic investigation of successive events – among others by radioactive labelling – showed, that in spite of the extensive repair, a permanent shortage of differentiated neurons resulted. Moreover, the morphologically detected retardation of stratification and of cortical development has been verified by postnatal behaviour studies, which revealed various functional disburbances. To our knowledge, it is so far the most sophisticated experimental work with respect to repair in teratogenesis.

Finally, a recent observation made in our laboratory, in a study on eye anomalies induced by 6-aminonicotinamide[5]: if administered on day 15 of pregnancy, 6-aminonicotinamide induces, besides other changes, a marked vacuolar degeneration of the pigment epithelium (Figure 4.19). This regressive change, described and analysed in other tissues[80, 81], is due to the formation of growing 'perinuclear cysterns' (in fact between the layers of the nuclear membrane). It is assumed, that the process is triggered by the accumulation of osmotically active 6-phosphogluconate[81].

According to our findings, this pathological change of the pigment epithelium is maximal about two days after the treatment (on day 17 of pregnancy). Surprisingly, further control on days 18, 19 and 20 of pregnancy revealed a stepwise regression of degenerative phenomena attaining a real

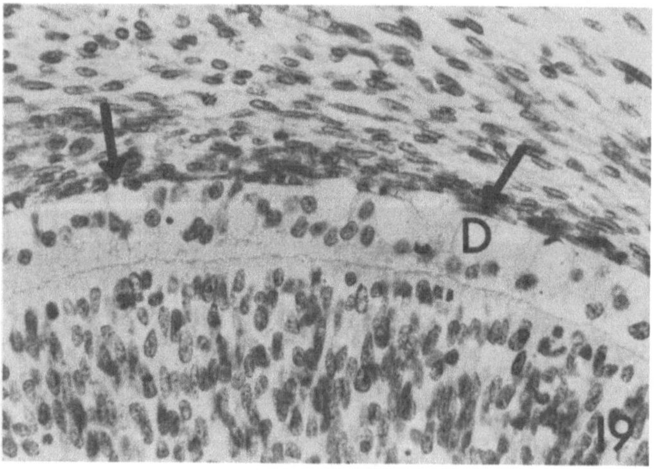

Figure 4.19 Rat fetus – day 17 of pregnancy. 6-Aminonicotinamide, 8 mg/kg, on day 15. Marked vacuolation of the pigment epithelium (see arrows)

restitutio ad integrum (at least at light microscopical level; further electron-microscopic investigations are just beginning) (Figures 4.20 and 4.21). In the 'recovered' pigment epithelium no increase of mitotic activity and no other signs which would have suggested the replacement of destroyed cells were detected.

In our opinion, this true reparative phenomenon observed in a fetal tissue, differentiated during the second half of pregnancy, is of considerable importance. It supports the idea that repair may be accomplished (perhaps first of all during later stages of development) also by compensation of intracellular metabolic disturbances, and by the stepwise normalization of intracellular 'husbandry'.

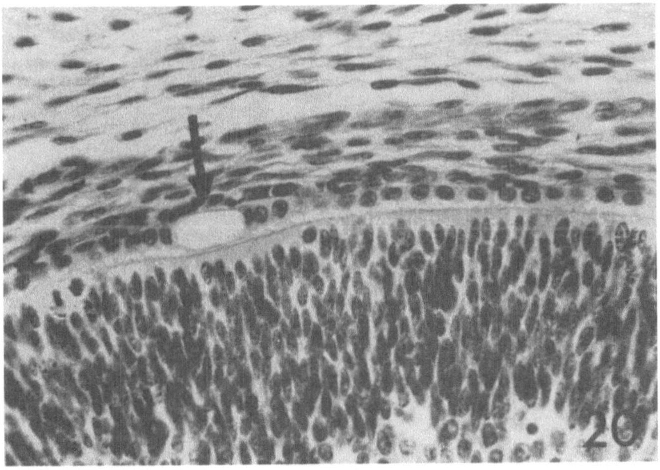

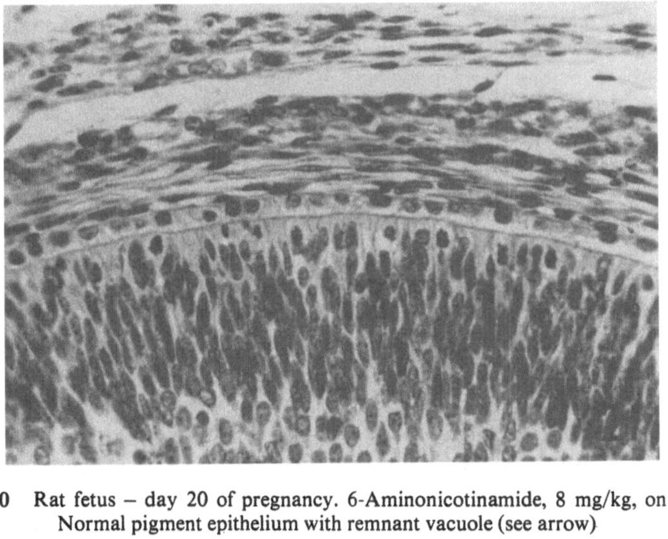

Figure 4.20 Rat fetus – day 20 of pregnancy. 6-Aminonicotinamide, 8 mg/kg, on day 15. Normal pigment epithelium with remnant vacuole (see arrow)

Figure 4.21 Rat fetus – day 20 of pregnancy. 6-Aminonicotinamide, 8 mg/kg, on day 15. Normal pigment epithelium

It is obvious, that as far as reparation in developmental pathology is concerned, many studies still remain to be carried out. Moreover, this direction of investigation may easily become one of the major topics of teratological research. Dreams are sometimes helpful in stimulating real activity, and it is not quite a dream that in the future one of the plausible pathways in the prevention of maldevelopment will be the strengthening of the reparative potencies during prenatal development.

SOME REMARKS CONCERNING THE EXTRAPOLATION OF EXPERIMENTAL RESULTS TO HUMANS

Since the teratological screening of drugs, of various chemicals and of other environmental factors having a supposed or potential noxious effect on human reproduction, has attracted the attention of an increasing number of specialists, the main and still unresolved problem has been the application of animal observations to man. Indeed, for many years no environmental agent with a proven noxious effect in animals could be incriminated with certainty in human prenatal pathology. On the other hand, the testing on animals of the few agents having a more or less clear-cut teratogenic effect on human offspring gave but a partial result. It is sufficient to mention here the well-known case of thalidomide testing in various animal strains.

Thus, a strange, contradictory and somehow disconcerting situation arose. On the one hand, teratological screening became an obligatory component of general, biological testing of new drugs and chemicals with the number of 'industrial' teratologists and of the substances tested increasing rapidly. Meanwhile, as a logical result of scepticism, of lowering confidence in the efficiency of current screening methodology, in the possibilities of extrapolation to man, new methods, principally new approaches, possessing a more precise predictive value, a more general validity including also human reproduction are searched for. (It is not the purpose of these short notes to discuss the already considerable number of experimental models, of *in vivo* and *in vitro* tests worked out during the last few years.) Of course, even those who emphasize the inadequacy of up-to-date teratological screening for the prevention of possible human teratogenesis are in agreement as to the other valuable data furnished by the extensive work in this field[82].

The front of (justified) scepticism has been firstly weakened, if not broken, by the appearance of a new teratological entity, the 'fetal alcohol syndrome'. We do not want to repeat well known data. Nevertheless, a short retrospection seems to be necessary.

In 1968 the first results from our laboratory concerning the effect of ethanol upon the developing chick embryo were reported[83,84].* In 1971 our investigations in this direction were extended to rats[7]. No one of these studies was undertaken with a direct purpose of 'industrial' screening, but as a complex methodology was applied (including also biometrical aspects) the results obtained were fairly significant. Both the experiments on chick and rat embryos revealed the obvious, an early noxious effect of ethanol intoxication

* The teratological study of 'daily' noxae (ethanol, caffeine, nicotine) was initially suggested in our laboratory by Prof. B. Menkes

upon development. In the conclusions of our first study we wrote: 'our results point out the possibility of alcoholic injury of the early chick embryo and of the early embryonic development generally (that means of human embryos too)'. In 1971 we were even more categorical: 'The similarity of results in two experimental models represents, in our opinion, a serious danger signal of prenatal risk of ethanol intoxication during early pregnancy in human'. At the 2nd Conference of the European Teratology Society (Prague, 1972) our main results concerning the prenatal effect of ethanol were presented[85] and evoked some scepticism as to the possibilities of extrapolation to man. In 1973, the first clinical cases of the 'fetal alcohol syndrome' were reported[4,86]. From the since constantly increasing number of reported investigations, several other experimental models[13,87–90], valuable casuistics, and an extensive epidemiological and clinico-anatomical study carried out in our country (to our knowledge one of the first and most serious attempts in this field)[91] are to be mentioned. This latter investigation, carried out in one of our well-known vineyard areas, used a complex, biostatistical, clinico-anatomical and epidemiological methodology and revealed strong correlations between alcohol consumption and several disturbances of pre- and postnatal development.

Thus, in a few years, ethanol as human (and experimental) teratogen entered the 'first-plan' of teratology, being mentioned among the most frequent and certain causal factors of dysmorphogenesis in man[92–95].

We considered it useful to draw attention to the 'alcohol problem' because of its general significance. If, as it happened in this case, experimental results may constitute a real danger signal for human teratology, screening on animals, even at the up to date methodological level, is justified. Scepticism, so useful in research, if pushed to extremes can be transformed into an impediment.

One more problem is worth pointing out. Our ethanol experiments were carried out on two taxonomically distant species. As is known, the current normals for teratological screening involve relatively related animals (lagomorphs and rodents). Perhaps one of the possible pathways to increase the efficiency of screening would be a change in this direction, by including into the test models also other, taxonomically more distant animals (e.g. amphibians or birds). In spite of several essential differences as to morphogenesis, metabolic pathways, etc., many basic developmental events, and cellular and extracellular functions are common or at least similar in all vertebrate embryos. Also, other trends for overcoming the handicaps of usual screening procedures, e.g. the application of cell and organ cultures, must also take into account important differences as compared with *in utero* conditions. It might be possible that detecting in 'distant' embryos some more general and common biological effects of the environmental agents investigated, the overwhelming complexity of factors influencing extrapolation could be somewhat reduced. This is but a tentatively made suggestion, a proposal in order to stimulate further thoughts and work.

ACKNOWLEDGEMENTS

We are indebted to Mrs M. Ciuguzan, Mrs M. Cosma, Mrs T. Logofăt and to Mrs E. Opris for their technical assistance during the experimental work reported, to Mrs M. Gherman for the photography and to Mrs S. Antal for the preparation of the manuscript.

The supply of 6-aminonicotinamide by Prof. H. Nishimura (Japan) and of Delagile by 'Usines Réunies de Produits Pharmaceutiques et Diététiques – Budapest (Hungary)' is kindly acknowledged.

References

1. Sandor, S. (1958). On the role of the reactivity of the infected organism in the development of tuberculosis. *Acta Morph. Acad. Sci. Hung.*, **9**, 89
2. Sandor, S. and Amels, D. (1971). A contribution to the methodology of teratological screening: Multiphasic screening. *Rev. Roum. Embryol. Cytol. Série Embryol.*, **8**, 67
3. Schultz, P. W., Reger, J. F. and Schultz, R. L. (1966). Effects of Triton Wh-1339 on the rat yolk sac placenta. *Am. J. Anat.*, **119**, 199
4. Jensh, R. P., Brent, R. L. and Barr, M. Jr. (1970). The litter effect as a variable in teratologic studies of the albino rat. *Am. J. Anat.*, **128**, 185
5. Sandor, S., Amels, D. and Checiu, M. (1978). 6-Aminonicotinamide induced eye defects in rats. *Rev. Roum. Morphol. Embryol. Physiol., Morphol. Embryol.* (In press)
6. Elias, S. and Sandor, S. (1965). A biostatistical study of the early morphogenesis of the chick embryo and of its appendages. *Rev. Roum. Embryol. Cytol. Série Embryol.*, **2**, 115
7. Sandor, S. and Amels, D. (1971). The action of ethanol on the prenatal development of albino rats. *Rev. Roum. Embryol. Cytol. Série Embryol.*, **8**, 105
8. Geelen, J. A. G. (1973). Vitamin A-induced anomalies in young rat embryos. *Acta Morphol. Neerl. Scand.*, **11**, 233
9. Marin-Padilla, M. (1966). Mesodermal alterations induced by hypervitaminosis A. *J. Embryol. Exp. Morphol.*, **15**, 261
10 Morriss, G. M. and Steele, C. E. (1974). The effect of excess vitamin A on the development of rat embryos in culture. *J. Embryol. Exp. Morphol.*, **32**, 505
11. Amels, D., Checiu, M. and Sandor, S. (1977). Contribution to the study of bisazo dyes-induced eye anomalies in rats. *Rev. Roum. Morphol. Embryol. Physiol., Morphol. Embryol.*, **23**, 93
12. Sandor, S. and Amels, D. (1975). The 'early malformative syndrome'. *Rev. Roum. Morphol. Embryol. Physiol., Morphol. Embryol.*, **21**, 21
13. Tittmar, H. (1974). Alcohol intoxication: a method for obtaining increased ethanol intake in gravid rats. *I.R.C.S.*, **2**, 1079
14. Turbow, M. M. (1966). Trypan blue induced teratogenesis of rat embryos cultivated in vitro. *J. Embryol. Exp. Morphol.*, **15**, 387
15. Theiler, K. (1967). Metameriestörungen und ihre Konsequenzen im Säugetierexperiment. *Z. Anat. Entwickl. Gesch.*, **126**, 31
16. Alexandrov, V. A. (1966). Mechanism of lethal effect of Myelosan upon rat embryos during early postimplantation stages (in Russian). *Ark. Anat. Ghist. Embryol.*, **51**, 13
17. Persianinov, L. S., Kiriuschenkov, A. P. and Skosyreva, A. M. (1974). Cultivation of rat embryos in vitro by New's method in pharmacoembryology (in Russian). *Ark. Anat. Ghist. Embryol.*, **66**, 93
18. Udalova, L. D. (1976). The study of postimplantation development of rat embryos after the influence of harmful agents on male gametes (in Russian). *Ark. Anat. Ghist. Embryol.*, **70**, 46
19. Baranov, V. S. (1966). Mechanism of the pathogenic effect of aminopterin upon embryogenesis in albino rat (in Russian). *Ark. Anat. Ghist. Embryol.*, **51**, 17
20. Merker, H. J. and Köhler, E. (1973). Electron-microscopical studies of the normal rat decidua under teratogenic conditions (actinomycin, methotrexate and protein free diet). Abstract. *Teratology*, **8**, 229

21. Golinsky, G. F. and Baranov, V. S. (1976). The mechanism of the pathogenic action of chloridine (daraprim) on albino rat embryos (in Russian). *Byull. Eksp. Biol. Med.*, **81**, 747
22. Amels, D. and Sandor, S. (1972). Multiphase screening of some known teratogenic drugs. *Rev. Roum. Embryol. Cytol. Série Embryol.*, **9**, 123
23. Wilson, J. G., Beaudoin, A. R. and Free, H. I. (1959). Studies on the mechanism of teratogenic action of trypan blue. *Anat. Rec.*, **133**, 155
24. Beck, F. and Lloyd, J. B. (1966). The teratogenic effects of azo dyes. In: D. H. M. Woollam (ed.). *Advances in Teratology*, **Vol. I**, pp. 131–193 (London: Logos Press)
25. Beck, F. (1970). Embryonic utilization of histiotroph. In: R. Bass, F. Beck, H. J. Merker, D. Neubert, B. Randhahn (eds.). *Metabolic Pathways in Mammalian Embryos during Organogenesis and its Modification by Drugs*, pp. 557–574. (Berlin: Freie Universität)
26. Lloyd, J. B. (1970). Histiotrophic nutrition – a target for 'macromolecular' teratogens in rats. In: R. Bass, F. Beck, H. J. Merker, D. Neubert, B. Randhahn (eds.). *Metabolic Pathways in Mammalian Embryos during Organogenesis and its Modification by Drugs*, pp. 575–592. (Berlin: Freie Universität)
27. Lloyd, J. B., Williams, F. E. and Beck, F. (1974). Placental function: a target for teratogens. *Biochem. Soc. Trans. 547th Meeting*, **2**, 702
28. Brent, R. L. (1966). The production of congenital malformations using tissue antisera IV. Evaluation of the mechanism of teratogenesis by varying the route and time of administration of anti-rat-kidney antiserum. *Am. J. Anat.*, **119**, 555
29. De Feo, V. Y. (1967). Decidualization. In: R. W. Wynn (ed.). *Cellular Biology of the Uterus*, pp. 191–290. (Amsterdam: North Holland Publ. Comp.)
30. Olenov, Ju. M. and Pushnitzina, A. D. (1960). On X-ray sensitivity of the central nervous system of mammalian embryos (in Russian). *Dokl. Akad. Nauk SSSR*, **84**, 405
31. Vickers, T. H. (1967). The thalidomide embryopathy in hybrid rabbits. *Br. J. Exp. Pathol.*, **48**, 107
32. Scott, W. J., Ritter, E. J. and Wilson, J. G. (1971). DNA synthesis inhibition and cell death associated with hydroxyurea teratogenesis in rat embryos. *Dev. Biol.*, **26**, 306
33. Menkes, B. and Sandor, S. (1956). Researches on the reactivity of the embryonic complexus to the tuberculons infection (in Rumanian). *Acad. R.P.R. Baza Timişoara, Stud. cerc. şt. Seria St. Med.*, **3**, 31
34. Menkes, B., Cotăescu, E., Oţetea, G. and Deleanu, M. (1956). Researches on the development of some malignant tumors, grafted onto the body or onto the chorionallantoic membrane of the chick embryo and on the reaction of the embryonic epithelia and mesenchyme to these grafts (in Rumanian). *Bul. şt. Secţia şt. Med.*, **8**, 307
35. Menkes, B. (1958). *Researches in Experimental Embryology*, Vol. I. (in Rumanian). (Bucureşti: Editura Academiei R.P.R.)
36. Menkes, B. and Sandor, S. (1965). Tumoral heterografts in the developing chick embryo. *Rev. Roum. Embryol. Cytol. Série Embryol.*, **2**, 57
37. Dyban, A. P. (1972). The mechanism of teratogenic action of drugs. Presented at the *2nd Conference European Teratology Society*, May 23–26, Prague
38. Schnall, B. S. and Smith, D. W. (1974). Nonrandom laterality of malformations in paired structures. *J. Pediatrics*, **85**, 509
39. Layton, W. M. Jr. and Haallesy, D. W. (1965). Deformity of forelimb in rats. Association with high doses of acetazolamide. *Science*, **149**, 306
40. Layton, W. M. (1974). The teratogenic effect of acetazolamide on mice with situs inversus viscerum. Abstract. *Teratology*, **9**, A-26
41. Land, P. W. and Polley, E. H. (1975). Aberrant retinofugal projections in rats with congenital eye defects. *Teratology*, **11**, 27A
42. Majima, A. (1970). Eye abnormalities in mouse embryos and fetuses caused by X-radiation of mothers. Processes of production in case of exposure with 200 r on the 8th day of pregnancy. *Nagoya J. Med. Sci.*, **24**, 171
43. Menkes, B., Sandor, S. and Ilieş, A. (1970). Cell death in teratogenesis. In: D. H. M. Woollam (ed.). *Advances in Teratology*, **Vol. IV**, pp. 169–215. (London: Logos Press)
44. Wendler, D. (1972). Der embryo-fetale Zelltod während der Normogenese und im Experiment. In: G. Uschmann (ed.) *Acta Historica Leopoldina*, **8**, pp. 7–295. (Leipzig: Johann Ambrosius Barth)

45. Saxén, L. and Rapola, J. (1969). *Congenital Defects*, pp. 129–31. (New York: Holt, Rinehart and Winston, Inc.)
46. Berndorfer, M. A. (1962). La régénération intra-uterine du bec-de-lièvre et de la division palatine. *Rev. de Stomatologie*, **63**, 356
47. Berndorfer, A. (1962). Intrauterine Regeneration der Missbildungen im klinischen Bilde. *Langenbecks Arch. klin. Chir.*, **299**, 729
48. Gottschewski, G. H. M. and Zimmermann, W. (1973). *Die Embryonal entwicklung des Hauskaninchens. Normogenese und Teratogenese.* (Hannover: Verl. M. & H. Schaper)
49. Nishimura, H., Takano, K., Tanimura, T. and Yasuda, M. (1968). Normal and abnormal development of human embryos: first report of the analysis of 1213 intact embryos. *Teratology*, **1**, 281
50. Nishimura, H. (1969). Incidence of malformations in abortions. In: F. C. Fraser and V. A. McKusick (eds.). *Congenital Malformations*, pp. 275–283. (Amsterdam-New York: Excerpta Medica Foundation)
51. Nishimura, H. (1973). Prenatal versus postnatal malformations based on the Japanese experience on induced abortions in the human being. In: R. J. Blandau (ed.). *Aging Gametes*, pp. 349–368. (Basel: S. Karger AG)
52. Nishimura, H. and Okamoto, N. (1976). *Sequential Atlas of Human Congenital Malformations.* (Tokyo: Igaku Shoin Ltd.)
53. Bellairs, R. (1971). *Developmental Processes in Higher Vertebrates.* (London: Logos Press Ltd.)
54. Waddington, C. H. (1952). *The Epigenetics of Birds.* (Cambridge: University Press)
55. Menkes, B., Miclea, C., Elias, S. and Deleanu, M. (1961). Researches on the formation of axial organs. I. Studies on the differentiation of the somites (in Rumanian). *Acad. R.P.R. Baza Timişoara, Stud. cerc. şt. Seria St. Med.*, **8**, 7
56. Menkes, B. and Miclea, C. (1962). Researches on the formation of axial organs. III. Possible recovery and reorganization of a mechanically dissociated presomitic axial mesoderm. Supranumerary somite formation (in Rumanian). *Acad. R.P.R. Baza Timişoara. Stud. cerc. şt. Med.*, **9**, 203
57. Menkes, B., Sandor, S., Elias, S. and Deleanu, M. (1969). Contributions to the problem of somitogenesis. *Rev. Roum. Embryol. Cytol. Série Embryol.*, **6**, 149
58. Sandor, S. (1972). Researches on the formation of axial organs in the chick embryo. VIII. Some aspects of regulation potencies during somitogenesis. *Rev. Roum. Embryol. Ctyol. Série Embryol.*, **9**, 113
59. Deuchar, E. M. (1975). Reconstitutive ability of axial tissue in early rat embryos after operations and culture in vitro. *J. Embryol. Exp. Morphol.*, **33**, 217
60. Smith, L. J. (1964). The effects of transsection and extirpation on axis formation and elongation in the young mouse embryo. *J. Embryol. Exp. Morphol.*, **12**, 787
61. Weiss, P. and Matoltsy, G. (1957). Absence of wound healing in young chick embryos. *Nature (Lond.)*, **180**, 854
62. Weiss, P. and Matoltsy, A. G. (1959). Wound healing in chick embryos *in vivo* and *in vitro*. *Dev. Biol.*, **1**, 302
63. Grenberg, T. F. (1960). Regeneration of epidermis in albino rat embryos (in Russian). *Ark. Anat. Ghist. Embryol.*, **38**, 45
64. Walshtrem, E. A. (1960). Pathogenesis of irradiation hazards and reparative processes in rat embryos after X-raying of rat females on the 10th day of pregnancy (in Russian). *Ark. Anat. Ghist. Embryol.*, **38**, 72
65. Jelínek, R. (1960). Development of the experimental exencephalia in the chick. *Československa Morfol.*, **8**, 368
66. Jelínek, R. and Doskočil, M. (1962). Mechanism of reparative changes following the embryonic cord injury. *Československa Morfol.*, **10**, 402
67. Jelínek, R. and Klika, E. (1963). *Experimental morphogenesis of some malformations of the C.N.S.* (in Czechoslovakian). (Praha: Statni Zdravotnicka Nakladatelstvi)
68. Källen, B. (1960). Embryological aspects of tumour genesis in the brain. In: D. B. Tower and J. P. Shadé (eds.). *Structure and Function of the Cerebral Cortex*, pp. 159–164. (Amsterdam: Elsevier Publishing Company)
69. Källen, B. (1960). Experimental neoplastic formation in embryonic brains. *J. Embryol. Exp. Morphol.*, **8**, 20

70. Menkes, B. and Alexandru, C. (1969). The effect of internal irradiation by injection of Au^{198}, on the neuroepithelia of the ventricular cavities of the chick embryo. *Rev. Roum. Embryol. Cytol. Série Embryol.*, **6**, 59
71. Menkes, B. and Alexandru, C. (1970). Investigations concerning the importance of the proliferating layer of the neural tube in the tectogenesis of the central nervous system. *Rev. Roum. Embryol. Cytol. Série Embryol.*, **7**, 81
72. Menkes, B. and Alexandru, C. (1973). Weitere Untersuchungen über experimentelle Gehirnblasenschrumpfung beim Hühnerembryo. *Rev. Roum. Embryol. Cytol. Série Embryol.*, **10**, 73
73. Menkes, B. and Sandor, S. (1977). Somitogenesis: Regulation, potencies, sequence determination and primordial interactions. In: D. A. Ede, J. R. Hinchliffe and M. Bells (eds.). *Vertebrate Limb and Somite Morphogenesis*, pp. 405–419. (Cambridge: University Press)
74. Miclea, C. and Arcan, A. (1969). Investigations regarding the effects of some cytostatics injected into the central channel of the chick embryo neural tube. *Rev. Roum. Embryol. Cytol. Série Embryol.*, **6**, 139
75. Sandor, S. and Fazakas-Todea, I. (1976). A new experimental model of overgrowth and consecutive exencephaly. *Rev. Roum. Morphol. Embryol. Physiol., Morphol. Embryol.*, **22**, 249
76. Adhami, H. and Noack, W. (1975). Histological effects of 6-mercaptopurine on the fetal rat central nervous system: a light-microscopic study. *Teratology*, **11**, 297
77. Crowley, K. K., Geelen, I. A. G. and Langman, J. (1978). Repair mechanism in the embryonic spinal cord after a chemical insult. *Teratology*, **17**, 1
78. Langman, J., Rodier, P., Webster, W., Crowley, K., Cardell, E. L. and Pool, R. (1975). The influence of teratogens on cellular and tissue behavior during the second half of pregnancy and their effect on postnatal behavior. In: D. Neubert and H. J. Merker (eds.). *New Approaches to the Evaluation of Abnormal Embryonic Development*, pp. 439–468. (Stuttgart: Thieme Verlag)
79. Langman, J. and Cardell, E. L. (1977). Cell degeneration and recovery of the fetal mammalian brain after a chemical insult. *Teratology*, **16**, 15
80. Merker, H. J. (1970). Morphological studies on the effect of 6-aminonicotinamide in rat embryos. In: R. Bass, F. Beck, H. J. Merker, D. Neubert, B. Randhahn (eds.). *Metabolic Pathways in Mammalian Embryos during Organogenesis and its Modification by Drugs*, pp. 395-397. (Berlin: Freie Universität)
81. Roetz, R. (1975). Histochemical and electronmicroscopical studies on the mechanism of the teratogenic action of 6-Aminonicotinamide (6-AN). In: D. Neubert and H. J. Merker (eds.). *New Approaches to the Evaluation of Abnormal Embryonic Development*, pp. 659–677. (Stuttgart: Georg Thieme Publishers)
82. Warkany, J. (1974). Problems in applying teratologic observations in animals to man. *Pediatrics*, **53**, 820
83. Sandor, S. and Elias, S. (1968). The influence of ethyl-alcohol on the development of the chick embryo. *Rev. Roum. Embryol. Cytol. Série Embryol.*, **5**, 51
84. Sandor, S. (1968). The influence of ethyl-alcohol on the developing chick embryo. II. *Rev. Roum. Embryol. Cytol. Série Embryol.*, **5**, 167
85. Sandor, S. and Amels, D. (1973). Multiphase analysis of the prenatal noxious action of some chemical compounds. In: E. Klika (ed.). *Acta Univ. Carolinae Medica Monographia* LVI–LVII, pp. 117–119. (Praha: Universita Karlova)
86. Jones, K. L., Smith, D. W., Ulleland, C. N. and Streissguth, A. P. (1973). Pattern of malformation in offspring of chronic alcoholic mothers. *Lancet*, i, 1267
87. Chernoff, G. F. (1977). The fetal alcohol syndrome in mice: an animal model. *Teratology*, **15**, 223
88. Kronick, J. B. (1976). Teratogenic effects of ethyl-alcohol, administered to pregnant mice. *Am. J. Obstet. Gynecol.*, **124**, 676
89. Schwetz, B. A., Leong, B. K. J. and Staples, R. E. (1965). Teratology studies on inhaled carbon monoxide and imbibed ethanol in laboratory animals. *Teratology*, **11**, 33A
90. Tittmar, H. G. (1977). Some effects of ethanol, presented during the pre-natal period, on the development of rats. *Br. J. Alcoholism*, **12**, 71
91. Dinculescu, V. (1977). *The Influence of Alcohol Consumption on the Conceptus.* Dissertation for Dr. Sci. (in Rumanian). (Timişoara: Library of Medical School)

92. Jones, K. L. and Smith, D. W. (1975). The fetal alcohol syndrome. *Teratology*, 12 1
93. Majewski, F., Bierich, J. R. Löser, H., Michaelis, R., Leiber, B. and Bettecken, F. (1976). Zur Klinik und Pathogenese der Alkohol-Embryopathie. *Münchener Med. Wochenschrift*, **118**, 1635
94. Majewski, F. (1977). Uber einige durch teratogene Noxen induzierte Fehlbildungen. *Mschr. Kinderheilk.*, 125, 609
95. Majewski, F., Bierich, J. R. and Michaelis, (1977). Diagnose: Alkoholembryopathie. *Deutsch. Ärztebl.-Ärztl. Mitteil.*, **74**, 1133

5
Cyclophosphamide treatment prior to implantation: the effects on embryonic development

H. SPIELMANN, H.-G. EIBS
AND URSULA JACOB-MÜLLER

INTRODUCTION

Research in teratology usually focuses on the period of organogenesis, because this is the most sensitive period for the induction of malformations in mammals[1,2]. Attempts to study drug actions on earlier periods of pregnancy have not been very successful, because the preimplantation embryo is remarkably resistant to teratogens. In the earliest study of this problem the treatment of cleaving rabbit eggs with purine analogues *in vivo*[3] has no effect on development until at or after implantation. Investigations on the effects of X-irradiation[4] during the first days of pregnancy in the rat increased neither gross congenital malformations nor fetal growth retardation at the end of pregnancy. There is, furthermore, no indication for an increased rate of abnormalities among the offspring from embryos cultured *in vitro* in the presence of various teratogens during the preimplantation period and subsequently transferred to foster mothers[5]. Additionally the fetuses from transplanted preimplantation mouse embryos on which different kinds of microsurgery had been performed[6] or which had been frozen to −269 °C for up to one year[7] never showed an increased malformation rate at term. The effect of teratogens on embryos during the preimplantation period, therefore, has been explained by Austin[8] as follows: 'The effect of teratogens on the cleavage embryo depends on the number of cells killed or inhibited: above a certain portion, the embryo dies; below that figure, the remaining cells multiply to replace those lost and subsequent development is essentially normal.'

Gottschewski[9,10] and co-workers[11], however, used a different approach to study this teratological problem. After treatment during the preimplantation period they inspected rabbit embryos during the period of organogenesis and not at term as usual. When cyclophosphamide (CPA) (10 or 40 mg/kg intravenously) was given to pregnant rabbits before implantation, these investigators found a high percentage of deformed fetuses on days 11, 17 and 30 post coitum (p.c.). It was, therefore, concluded that the teratogen penetrates into the blastocyst even before implantation and is changing normal development. Gottschewski[9,10] suggested to call an abnormality induced by this type of treatment 'blastopathia'.

Brock and Von Kreybig[12] repeated Gottschewski's studies in the rat. When CPA treatment was performed on day 3 p.c. and the embryos were examined during organogenesis (day 12), the number of implantation sites was not reduced compared to controls, indicating that the blastocysts did not degenerate before implantation. However, the number of resorbed and malformed embryos was significantly increased after 40 mg/kg CPA. Additionally the malformed embryos were considerably smaller than controls. On day 15 p.c. these investigators could only identify either resorbed or completely normal embryos. In other studies on the effects of CPA treatment during the preimplantation period in the mouse[13] and rabbit[14] increased resorption rates but no malformed embryos were found at term.

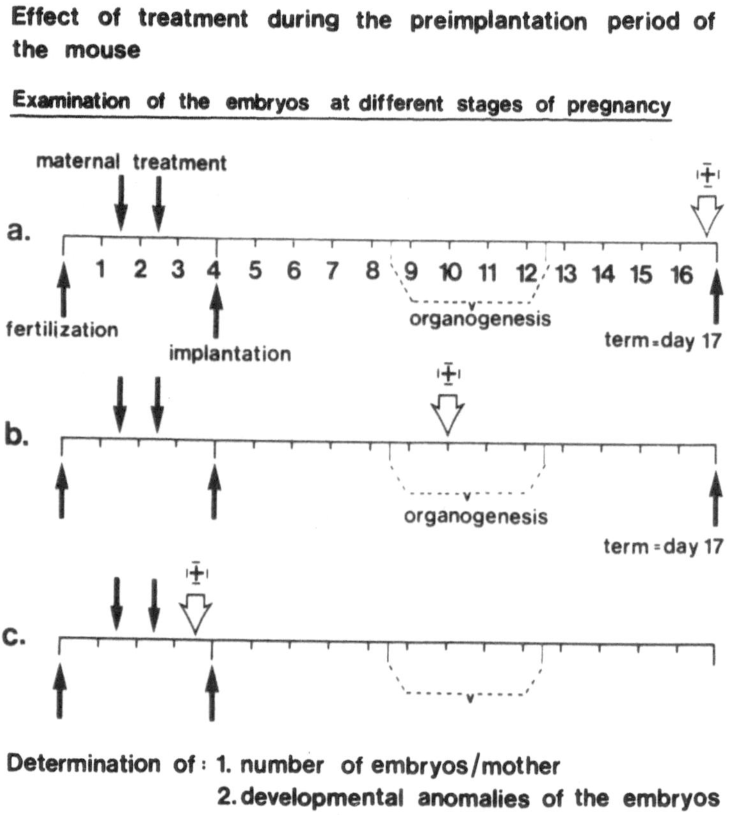

Figure 5.1 Diagram describing our screening procedure for developmental abnormalities at different stages of pregnancy after treatment during the preimplantation period in the mouse. Examination of the embryos after sacrifice of the mother can be performed at term (a), during organogenesis (b), or before implantation (c)

Methods for carrying out more detailed studies on the action of teratogens on embryos before and around implantation have been improved considerably during the past decade[15]. We have, therefore, repeated and extended the earlier investigations on the effects of CPA treatment during the preimplantation period in the rat[16] and mouse[17], to get further information on the mechanism of action of teratogens given before and around implantation. The basic teratological approach in our studies is outlined in Figure 5.1. After early CPA treatment the embryos were not only inspected at term (Figure 5.1a) but also during organogenesis (Figure 5.1b) and particularly before implantation (Figure 5.1c). We also used the embryo transplantation technique and the post-implantation culture of mouse and rat blastocysts to see whether CPA treatment of pregnant animals during the preimplantation period predominantly affects subsequent development of the embryo or the mother. Our investigations prove that the reaction of teratogens at this early stage of pregnancy is more complex than is generally assumed.

MATERIALS AND METHODS

Mating and treatment

Wistar rats of the strain SW-72 weighing 200 g (Winkelmann, Kirchborchen, Germany) and mice of the strain NMRI weighing 30 g (Zentralinstitut für Tierzucht, Hannover, Germany) were kept under a normal day/night cycle and placed with males overnight. The presence of sperm in vaginal smears the next morning indicated day 0 of pregnancy. Implantation occurs in our rats between day 4 and day 5 and in our mice between day 3 and day 4. Blastocysts were usually obtained at 2 p.m. on day 4 in the rats and on day 3 in the mice, which is 24 h after the time of treatment in both species. Cyclophosphamide (CPA), which was a gift from Prof. N. Brock (Asta-Werke AG, Chemical Factory Brackwede, Germany), was dissolved in distilled water and injected subcutaneously.

Handling of the embryos and morphological studies

The embryos were flushed from the uteri with Medium PB-1[18] supplemented with 0.4% BSA, freshly prepared from three-times quartz-distilled water. In the transplantation and *in vitro* culture experiments and also in the course of determination of the cell number the embryos were handled in the same medium. The cell number of the preimplantation embryos was determined by the method of Tarkowski[19], and the mitotic index (i.e. number of cells in mitosis as percentage of total cells) was calculated by the same method from the same preparations. To follow the decidual reaction, we injected 0.5 ml (in the rats) or 0.3 ml (in the mice) of a 1% Pontamine sky-blue solution into a tail vein 15 min before autopsy as described by Finn and Martin[20].

In the morphological studies the treated mothers were killed at different times during pregnancy, the uteri were excised and the embryos were removed from the implantation sites on days 11, 14, 17 and 20 in the rat and on days 10, 13 and 17 in the mouse. The embryos were checked for developmental anomalies under a stereo microscope (Carl Zeiss, Oberkochen, Germany) and their somite numbers were determined during organogenesis.

Embryo transplantation

In the embryo transfer experiments rat blastocysts were surgically transplanted in groups of five to one uterine horn of pseudopregnant females on day 4 of pseudopregnancy. The pseudopregnant recipients were anaesthetized with Evipan (hexobarbital) which was a gift from the Bayer AG (Pharmaceutical Company, Germany). Pseudopregnancy was induced by mating normal females with vasectomized males. The foster mothers were killed on day 20 of pregnancy and the success rate of the transplantations was determined by the number of resorbed and live embryos. The embryos were weighed, inspected for growth retardation and malformations and stained for skeletal abnormalities with Alizarin Red S according to Lorke[21].

Post-implantation development of mouse and rat blastocysts *in vitro*

Mouse and rat blastocysts were cultured in groups of ten in plastic culture dishes (NUNC, Nulcon, Denmark) without oil at 37 °C in a humidified 5% carbon dioxide in air atmosphere in one of the following media: Whitten's medium for ovum culture (WMOC)[22] supplemented with 0.3% BSA (WMOC–BSA); WMOC supplemented with 10% fetal calf serum (FCS, Gibco, USA) (WMOC—FCS); Eagle's minimum essential medium (MEM, Flow Laboratories, England) supplemented with 10% FCS (MEM–FCS); medium NCTC-109 (Difco, USA) supplemented with 10% FCS according to Sherman[23] (NCTC-109–FCS). When blastocysts were cultured for 120 h in the optimum medium NCTC-109–FCS, development proceeded via the following steps of differentiation[17,24]: hatching from the zona pellucida after 24–48 h, attachment to the surface of the culture dish after 36–60 h, and outgrowth of three characteristic cell types – a trophoblast layer with giant cells and an ICM consisting of the two germ layers ectoderm and endoderm, after 96–120 h. Success rates of development were calculated as percentage of embryos cultured under a given condition.

Photomicrographs of the embryos were taken on a Biovert photomicroscope (Reichert AG, Austria) at a magnification of × 130 on Ilford Pan F film (Ilford Co., England).

RESULTS

Lethality rate at term after CPA treatment 36 h before implantation

Groups of 10 pregnant animals received a single subcutaneous injection of 20, 40, 60 or 80 mg/kg CPA on day 2 (mouse) or on day 3 (rat). The maternal LD_{50} in our strains is 200 mg/kg in the mouse and 180 mg/kg in the rat. At term we found a resorption rate of 100% (Figure 5.1a; Table 5.1) after a single injection of 60 mg/kg CPA in the rat and in the mouse as previously described[16,17]. The resorption rate for untreated control animals in the two species was 10%. The number of implantation sites per animal was not influenced by the CPA treatment during the preimplantation period, indicating that the resorbed embryos have not necessarily died before implantation. The wet weight of living fetuses at term in the 20 and 40 mg/kg groups was not

Table 5.1 Effect of a single cyclophosphamide injection 24 h before implantation on the resorption rates at term in the rat and mouse

Cyclophosphamide dose (mg/kg)	*Term resorption rates*	
	Rat	*Mouse*
0	10%	8%
20	48%	35%
40	82%	78%
60	100%	100%
80	100%	100%

Maternal LD_{50} in our strains: rat 180 mg/kg, mouse 200 mg/kg. For each dosage level groups of 10 pregnant females were injected subcutaneously on day 3 in the rat and on day 2 in the mouse. Resorption rates were calculated as percentage of the total number of implantations found in pregnant females at term (day 17 in the mouse and day 20 in the rat; day 0 = first day after an overnight mating period)

reduced and no malformations could be detected after staining with Alizarin Red S for skeletal anomalies. When evaluating our studies on early CPA treatment at term, we found that the results obtained support the general assumption that embryos either die early or survive to term unharmed after treatment during the preimplantation period.

Changes in the number of dead and living embryos per animal throughout pregnancy after CPA treatment 36 h before implantation

Pregnant animals were treated with a single injection of 60 mg/kg CPA on day 3 in the rat and on day 2 in the mouse and the number of embryos and implantations was recorded at different times after treatment. This particular dose was chosen as it was the lowest dose causing a resorption rate of 100% at term. Around implantation, i.e. 24–48 h after treatment, the total number of embryos was calculated from the number of implantation sites visible after Pontamine sky blue injection and from the number of embryos (blastocysts) that could be flushed from the same uterus. During the following days of pregnancy the number of implantation sites could easily be determined. During organogenesis all implantation sites were dissected under a stereo microscope and inspected for living or dead embryos. Ten treated animals and ten controls were investigated on days 4, 5, 8, 11, 14 and 20 in the rat and on days 3, 4, 7, 10, 13 and 17 in the mouse. The results obtained in the rat experiments are shown in Figure 5.2. There are no significant effects of CPA treatment on the total number of implantations per female. In the treated and untreated groups there is a reduction in the number of embryos at the time of implantation. The most striking reduction in the number of living embryos in the treated group occurred during organogenesis, as indicated by the decrease from 70% to 10% between days 11 and 14. In an identical manner, CPA treatment on day 2 in the mouse significantly decreased the number of living embryos during organogenesis. To further investigate the teratogenic mechanisms inducing these late effects of CPA treatment prior to implantation, we

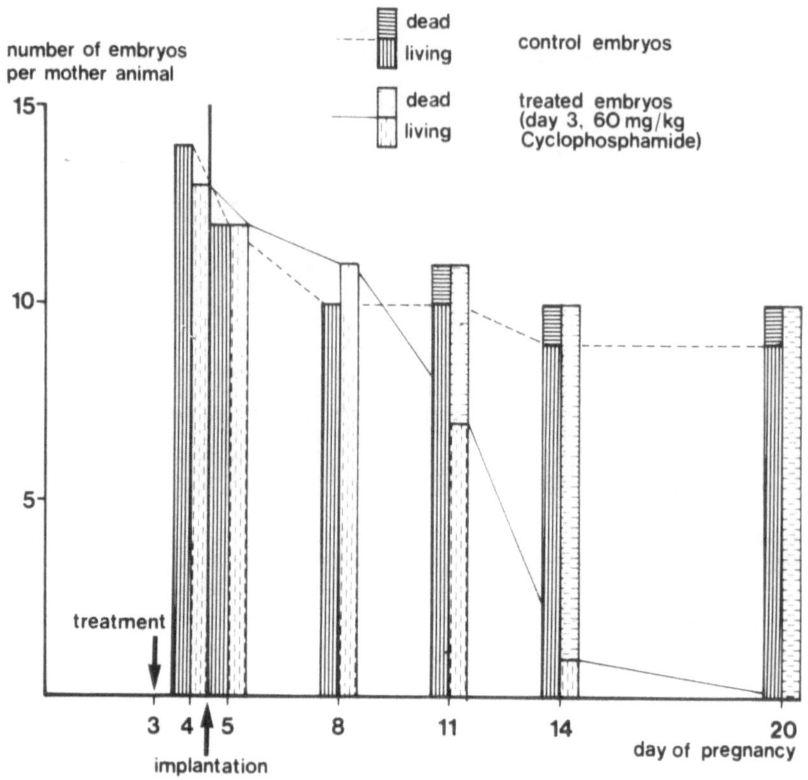

Figure 5.2 Influence of cyclophosphamide treatment 24–36 h before implantation on the number of embryos per female at different days of gestation in the rat. Columns represent means for 10 animals per day and per treatment group. At the time of implantation, determinations were performed on days 4, 4.5 and 5

decided to focus on two periods which are particularly important for normal development: the stage of organogenesis (Figure 5.1b) and the period of implantation (Figure 5.1c).

Development of embryos during organogenesis after maternal CPA treatment 36 h before implantation

We first tried to confirm observations of previous investigators[11, 12] on malformations of rodent embryos during organogenesis induced by CPA treatment during the preimplantation period. Among 50 litters of mouse[17] and rat[16] embryos carefully observed under the stereo microscope we never found malformations of brain or heart as described by Brock and v. Kreybig in the rat[12]. The embryos were retarded in development rather than grossly malformed[5] (Figure 5.3).

To further support this assumption we determined the somite numbers in treated and control embryos during organogenesis. The somite number per embryo was significantly lower in treated rat (Figure 5.4) and mouse[17]

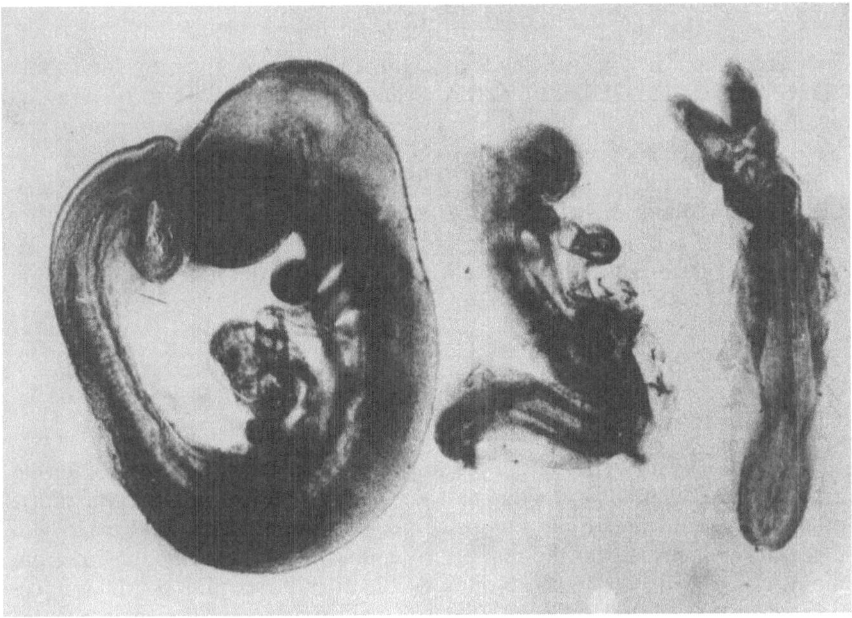

Figure 5.3 Rat embryos during organogenesis (day 11) after CPA treatment during the preimplantation period. The embryo on the left side is normal control, the mother of the two severely retarded embryos on the right side received 60 mg/kg CPA on day 3

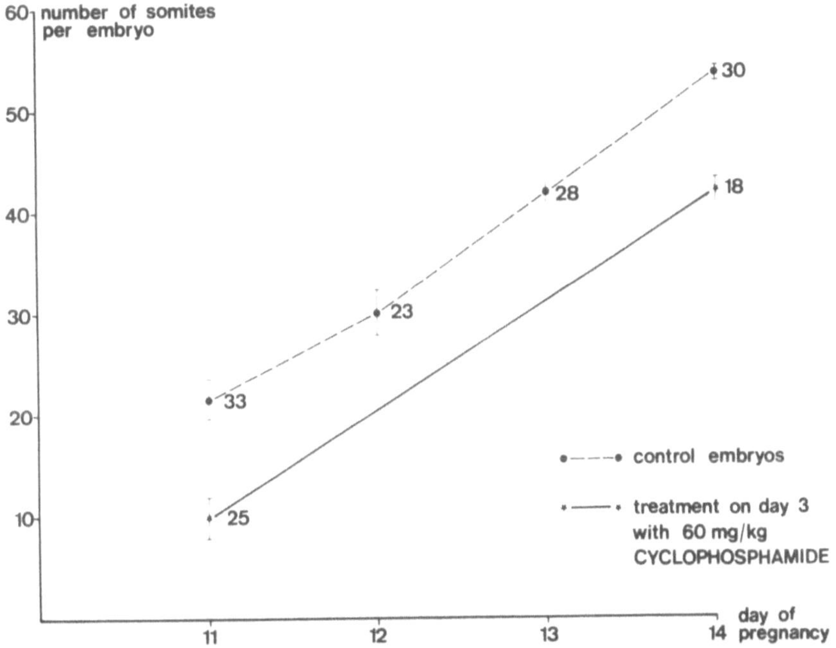

Figure 5.4 Influence of CPA treatment on day 3 of pregnancy on the somite number of rat embryos during organogenesis. Values are given ± standard deviation, the numbers indicate the amount of embryos used in each determination ●—●, control embryos; *—* embryos from mothers treated on day 3 with 60 mg/kg CPA

embryos. According to the developmental stage that is given by the somite number per embryo (Figure 5.4), the treated embryos seemed to be retarded by about 24 h in both species. Dry weight determinations of treated and untreated embryos gave the same results. Histological inspection of control and treated uteri during organogenesis[16] revealed a disturbed development of the placenta. Since light microscopical examination did not allow further elucidation of the action of the teratogen especially on the embryo, we attempted to get additional information by studying development of treated embryos and uteri at the time of implantation (Figure 5.1c).

Development of embryos at the time of implantation after maternal CPA treatment (60 mg/kg) 36 h before implantation

The developmental retardation of about 24 h in treated embryos during organogenesis could be explained by a delayed implantation. It was, therefore, of importance to determine the exact time of implantation for treated and untreated animals. From the number of embryos that could be flushed from the uteri and from the implantations visible after Pontamine sky blue injection, there was no indication of a delayed implantation in the treated groups, neither in the mouse[17] nor in the rat[16]. From these results we conclude that the retardation of treated embryos during organogenesis is not caused by a delayed implantation.

To find out if the abnormal development of rat and mouse embryos during organogenesis after maternal treatment with 60 mg/kg CPA on day 3 (rat) or day 2 (mouse) is caused by an action of the drug predominantly on the mother (decidual reaction) or on the embryo or on both, we studied the development

Table 5.2 Cell number of rat and mouse embryos at implantation 24 h after cyclophosphamide treatment (s.c.) of the mother

Developmental stage	*Time of treatment*	*Dose* (mg/kg)	*Cell number per embryo*	*Number of determinations*
Rat morula	0	0	14.6 ± 4.2	47
Rat blastula	0	0	27.7 ± 4.5	48
,, ,,	day 3	20	19.9 ± 3.9*	43
,, ,,	day 3	40	17.3 ± 4.3*	48
,, ,,	day 3	60	14.2 ± 5.8*	36
Mouse morula	0	0	25.9 ± 4.8	18
Mouse blastula	0	0	42.3 ± 11.3	25
,, ,,	day 2	20	34.5 ± 13.0	17
,, ,,	day 2	40	21.2 ± 3.7*	35
,, ,,	day 2	60	18.4 ± 4.6†	43
,, ,,	day 2	80	15.9 ± 2.8†	63

Determinations were performed at 2 p.m. on day 4 in the rat and on day 3 in the mouse, implantation occurs between day 4 and day 5 in the rat and between day 3 and day 4 in the mouse
* Significantly lower than cell number of control blastocysts at $p < 0.01$ (Wilcoxon-test)
† Significantly lower than cell number of control blastocysts and morulae at $p < 0.01$ (Wilcoxcon-test)

of treated embryos before implantation. Groups of 10 pregnant females were treated with single injections of CPA on day 3 in the rat and on day 2 in the mouse, and the uteri were flushed 24 h later. The number of blastocysts per female and the blastocyst/morula relation were not reduced in any of the treated groups. However, the cell numbers of the treated blastocysts were significantly lower than those of control blastocysts (Table 5.2), although the mitotic index[16] was not affected by CPA treatment. At the highest CPA concentrations, the cell number of the blastocysts is identical with that of normal morulae of the two species. Blastulation, which is the first step of morphological differentiation in embryonic development, is at the same time not affected by CPA treatment. The considerable reduction in the cell number of the treated blastocysts indicates that CPA or one of its metabolites interfered with embryonic development during the 24 h or preimplantation development. Further information on the target tissues of CPA treatment during the preimplantation period was obtained from transplantation experiments.

Transplantation experiments in the rat after maternal treatment 36 h before implantation

After treatment during the preimplantation period, transplantation of preimplantation embryos allows to distinguish whether an agent predominantly affects the mother or directly affects the embryo[5,25–27]. To determine if CPA treatment during the preimplantation period has an effect on embryonic development after implantation, we transferred blastocysts on day 4 (24 h after maternal treatment with 60 mg/kg CPA) to pseudopregnant recipients on day 4 of pseudopregnancy[27]. To check for effects of CPA treatment on the mother, we furthermore transplanted untreated blastocysts on day 4 to pseudopregnant recipients on the same day of pregnancy which had been treated with 60 mg/kg of CPA on day 3. Untreated blastocysts transplanted to untreated pseudopregnant recipients on day 4 of pregnancy served as controls.

The results of the transfer experiments are given in Table 5.3. This table shows that under our conditions of embryo transfer in untreated control

Table 5.3 Influence of maternal cyclophosphamide treatment (day 3; 60 mg/kg) on development of rat blastocysts after transplantation to pseudopregnant recipients on day 4*

	Untreated control	*Blastocyst treated recipient untreated*	*Blastocyst untreated recipient treated*
A. Number of embryos transferred	120	145	115
B. Implantations (percentage of A)	102 (85%)	130 (90%)	75 (65%)
C. Living embryos (percentage of A)	56 (47%)	17 (12%)	14 (12%)
D. Resorptions (percentage of A)	46 (38%)	113 (78%)	61 (53%)

* Blastocysts were transferred 24 h after treatment of the mother or the recipient to one uterine horn of a recipient which was killed on day 20 for determination of success rates

experiments 47% of the blastocysts developed into viable fetuses at term. This success rate is sufficient to allow conclusions on the results in the treated groups. Since the percentage of living fetuses at term was identical in both treated groups (12%), the teratogen seems to act on the preimplantation embryo to the same extent as on the mother. No indications of malformations could be found among the living term fetuses from the control transplantations nor from the transplantations with pretreated embryos or recipients. We, therefore, have to conclude that CPA when given to rats on day 3 of gestation interferes with the development of the embryo before implantation and also with the decidual reaction of the uterus.

During organogenesis, histological examinations of the uteri from transplantation experiments also indicated effects of the early CPA treatment on both the embryo and the uterus[16].

Further investigations on the differentiation of CPA treated preimplantation embryos of the two species were performed in recently established *in vitro* culture systems that allow embryonic development beyond the time of implantation[28–30].

Characteristics of the *in vitro* culture of mouse and rat blastocysts beyond implantation

The routine culture of blastocysts *in vitro* during the time of implantation has so far only been reported for the mouse[28–30]. We, therefore, first tested the differentiation of normal mouse blastocysts of our strain NMRI during a 96 h culture period in several standard culture media. In our medium for the routine culture of preimplantation mouse embryos, WMOC–BSA, only hatching and attachment of the blastocysts to the surface of the culture dish occurred. A low degree of trophoblast outgrowth and inner cell mass (ICM) development was achieved in MEM–BSA medium. When BSA was replaced in MEM by FCS, differentiation and growth of the trophoblast were improved. However, in NCTC-109–FCS medium growth and differentiation of mouse blastocysts were significantly better than in any of the media tested.

Development of mouse blastocysts in NCTC-109–FCS medium, which is used for routine culture experiments in our laboratory, is characterized by the following time-dependent steps of differentiation[24] (Table 5.4): after 24 h 9% of the blastocysts have hatched; after 48 h 84% have attached to the surface of the culture dish, and 32% show trophoblast outgrowth; after 72 h 82% have an ICM, and 52% have an ICM consisting of two germ layers (endoderm and ectoderm, i.e. extensive ICM); after 96 h the ICM of 87% of the blastocysts consists of two germ layers (Figure 5.5a).

In contrast to preimplantation mouse embryos, which can successfully be cultured *in vitro* from fertilization up to the late blastula stage in many laboratories, preimplantation rat embryos can only be cultured *in vitro* for one or two cleavage divisions[27,31]. It is, therefore, not surprising that the successful differentiation of rat blastocysts under *in vitro* culture conditions has so far not been reported. Similar to our experiments with mouse blastocysts we first tested the development of rat embryos in the following media: WMOC–BSA, MEM–BSA, MEM–FCS, and NCTC-109–FCS. Again growth and differen-

Table 5.4 Development of mouse and rat blastocysts *in vitro* after 24–96 h in medium NCTC-109–FCS

Time in medium (h)	Hatching (%)	Development and differentiation: Trophoblast outgrowth (%)	ICM (%)	Extensive ICM (2 germ layers) (%)
MOUSE				
24	9 ± 2	0	0	0
48	89 ± 8	32 ± 20	0	0
72	96 ± 4	93 ± 5	82 ± 11	52 ± 23
96	96 ± 4	93 ± 5	90 ± 10	87 ± 10
RAT				
24	34 ± 33	0	0	0
48	73 ± 33	48 ± 35	27 ± 16	0
72	83 ± 25	65 ± 28	30 ± 22	1 ± 4
96	88 ± 19	81 ± 21	37 ± 13	7 ± 6

Success rates of development are given as percentages of all embryos cultured in 30 determinations (mouse) and 24 determinations (rat) with 10 embryos/group

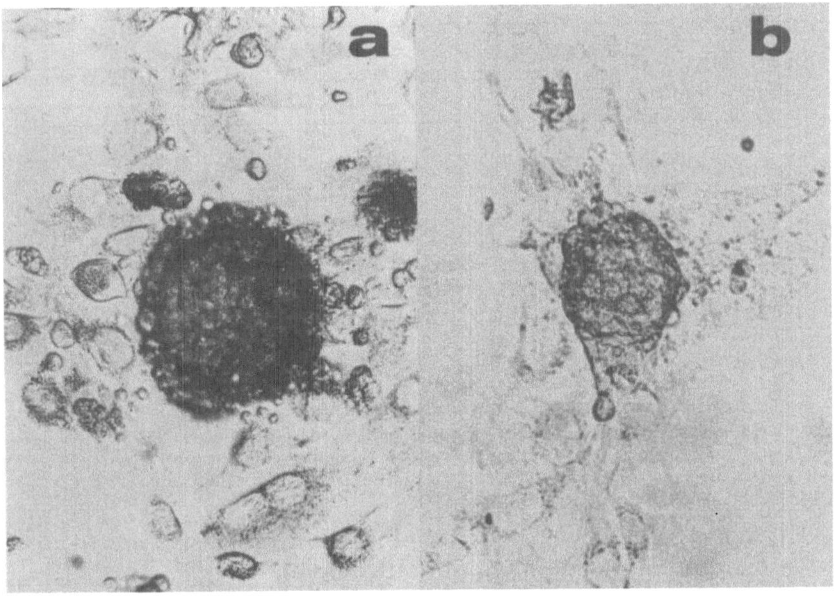

Figure 5.5 *In vitro* development of mouse and rat blastocysts in medium NCTC-109–FCS (culture period 96 h): a, one well-developed mouse blastocyst consisting of an ICM (centre) with two germ layers (endoderm and ectoderm) and a trophoblast layer (with giant cell nuclei) filling the rest of the picture; b, one well-developed rat blastocyst, the morphological structures are identical to (a). For abbreviations, see Materials and Methods. Magnification × 91, reduced for reproduction

tiation of rat blastocysts was significantly better in NCTC-109–FCS than in any of the media tested, but there are some important differences compared to the development of mouse blastocysts. As we had expected, the success rate of embryos reaching the different steps of differentiation *in vitro* were significantly lower in the rat than in the mouse. ICM development was particularly poor in the rat (Table 5.4), since only 37% of the embryos developed an ICM and in less than 10% the ICM consisted of two germ layers (Table 5.4; Figure 5.5b).

In vitro development of hatched rat blastocysts was generally much better than that of normal rat blastocysts still surrounded by their zona pellucida. Since pronase treatment as suggested by Pienkowski *et al.*[29] did not improve the results, we tested the developmental potential of rat blastocysts flushed from the uteri at different times on day 4. At 2 p.m. all blastocysts were still surrounded by the zona and development was rather poor; at 10 p.m. all blastocysts had already hatched from the zona and their development was significantly better; at 6 p.m. about 50% of the blastocysts had hatched and the development of hatched and unhatched blastocysts was as good as in the 10 p.m. group. All rat blastocysts used in the culture experiments (Tables 5.4 and 5.5) were, therefore, obtained at 6 p.m.

Table 5.5 **Effect of cyclophosphamide (CPA) treatment *in vivo* on differentiation of mouse and rat blastocysts *in vitro***

CPA dose (mg/kg)	*No. of blastocysts* (100%)	*Hatching* (%)	*Extensive trophoblast growth* (%)	*ICM* (%)	*ICM 2 germ layers* (%)
MOUSE					
0	300	96	93	90	87
20	119	89*	82*	73†	61‡
40	73	81	71*	48†	25†
60	184	62†	50†	31*	11†
80	62	30*	22‡	8†	0†
RAT					
0	260	88	81	37	7
30	139	81	79	5‡	0+
60	131	40†	32†	0†	0

Culture period 96 h. Significance levels were determined by the χ^2 test separately for each step of differentiation by comparing the growth rate (as percentage of blastocysts cultured) at every CPA dose with the growth rate at the next lower CPA dose. * $p<0.05$; † $p<0.01$; ‡ $p<0.001$

Post-implantation development of mouse and rat blastocysts *in vitro* after maternal CPA treatment prior to implantation

Mice were treated with CPA on day 2 of pregnancy and rats on day 3. Twenty-four hours later the blastocysts were transferred to the culture medium NCTC-109–FCS. In this culture system embryos from CPA-treated mothers showed a dose-dependent inhibition of all steps of differentiation (Table 5.5). In the mouse the percentage of embryos that underwent ICM differentiation into two germ layers was the most sensitive parameter[17]. It was already significantly decreased at the lowest CPA concentration (20 mg/kg), which had no significant effect on the cell number of the blastocysts at the beginning of the *in vitro* culture period on day 3 (Table 5.2).

In the rat where sufficient development of an ICM consisting of two germ layers could not be achieved (Table 5.4), development of the ICM was the most sensitive parameter after CPA treatment *in vivo* (Table 5.5). Again development of treated blastocysts was significantly inhibited at CPA concentrations that hardly affected the cell number of treated blastocysts at the beginning of the *in vitro* culture (Table 5.2). However, in both species trophoblast development was not significantly inhibited at the lower CPA concentrations, indicating that trophoblast giant cells are less sensitive to teratogens than ICM cells from which the embryo finally develops. Previous studies in other laboratories on the development of mouse blastocysts *in vitro* in the presence of substances that are interfering with nucleic acid metabolism have also revealed a higher sensitivity of ICM cells than of trophoblast cells[32–35].

DISCUSSION

Early teratogenicity of cyclophosphamide (CPA)

Our investigations on CPA treatment during early pregnancy[16,17] clearly demonstrate that information on the teratogenic, toxic and lethal effects of different agents on preimplantation embryos cannot be fully obtained when evaluation of the resulting developmental abnormalities is performed only at term, since the damaged embryos rarely survive the total prenatal period up to birth. From studies on abnormal development during organogenesis according to Gottschewski[10] we gained a considerable amount of information on the effect of CPA treatment of pregnant mice and rats about 36 h prior to implantation. We confirmed the findings of Brock and Von Kreybig[12] that after CPA treatment of pregnant rats during the preimplantation period the embryos die in the course of organogenesis. Closer examination of embryonic development during this developmental period revealed, in contrast to the earlier findings, that treated embryos were not malformed before being resorbed but were retarded in development by about 24 h, as indicated by the somite number and the wet weight of the embryos. The 'blastopathy'[10] that was induced by our CPA treatment is, therefore, characterized by a retarded embryolethal effect and not by any malformations. After our type of treatment, even though it is performed before implantation, the embryos neither die before or during implantation nor survive up to term, but survive up to organogenesis. The common view on drug action before implantation has, therefore to be modified.

Today one can go a step further than the earlier investigators since methods have been developed to study embryonic development before and around implantation[15,27]. The new methods enabled us to determine the exact time of implantation. Early CPA treatment in our studies had no delaying effect on implantation. In addition, we were able to find a definite effect of the drug on embryonic development before implantation. There are several reports on morphological abnormalities in preimplantation embryos after treatment of the mother, e.g. when using X-rays[36] and a lead-supplemented diet[37] in the mouse or a zinc-deficient diet[38] in the rat. However, in our CPA studies the morphology of preimplantation mouse and rat embryos was not affected as far as blastulation is concerned but the cell number was reduced at the same time. This effect of CPA on *in vivo* development of preimplantation embryos between the eight-cell stage (time of treatment) and the blastocyst stage indicates that blastulation, the first step of morphological differentiation[39], does not depend on the presence of a particular number of cells and that embryonic cells, having lost the capacity to divide at a normal rate, can still form the blastocyst cavity. The effects of CPA on subsequent development of mouse and rat embryos *in vitro* might be explained by a retarded clearance of the alkylating agent or one of its metabolites from the blastocyst cavity[5,16,17].

The embryo transplantations, and also the histological examination of the uteri of pretreated recipients, demonstrated an effect of the alkylating drug on the decidual reaction of the uterus. The survival rate of pretreated blastocysts after transplantation was also reduced at term. This result provides additional evidence for a direct effect of the drug or one of its metabolites on the embryo. On day 11 the histological examination of the uteri of untreated hosts to which treated embryos had been transplanted points in the same direction, since the placentae showed signs of retarded or abnormal development but the decidua was not affected[16]. So far, our investigations give no indication whether the retarded development of the embryos is due to an action of the drug predominantly on the embryo or to secondary effects caused by the inhibited development of the decidua[16].

Investigations on the teratogenic effect of CPA during organogenesis indicated that the unmetabolized compound was teratogenic[40,41]. This was somewhat surprising, since the alkylating activity of the drug is due to its metabolites[42,43]. More recent studies on the teratogenic mechanisms of CPA indicate that its metabolites are acting on embryos during organogenesis in the mouse by alkylation of embryonic macromolecules[44,45].

Chemically induced chromosome aberrations in the mouse produce embryonic mortality in the pre- as well as in the post-implantation period[46]. Malformations could not be found in such embryos during organogenesis, though chromosome anomalies were easily detectable. Since CPA is a very potent mutagenic agent[47], the abnormal development of our treated embryos could be caused by somatic mutations and subsequent chromosomal imbalance.

Recent autoradiographic studies in our laboratory provide evidence for a premature induction of the decidual reaction after CPA in the rat following a single injection of the drug on day 3 of pregnancy[48]. The induction of superovulation by 50 mg/kg of CPA in the rat as reported by Russell *et al.*[49] supports the assumption that CPA is interfering with the hormonal control of the normal development of the decidua. In a similar fashion the effects of lead treatment at dose levels inhibiting embryonic development before[37] and during implantation[50] seem to be due to a maternal hormonal imbalance[51] induced by the treatment.

The embryo transfer technique and embryo culture beyond implantation *in vitro* in experimental teratology

For the evaluation of the effects of teratogens on pre-implantation embryos the *in vitro* culture of *in vivo* treated blastocysts beyond implantation as described in this and previous reports[15–17,52] has several advantages when compared to the embryo transfer technique. Inconsistent success rates obtained even in experienced laboratories are the general disadvantage of transplantation experiments. Most investigators, therefore, include in their calculations only those experiments in which the recipient animals carry implantations at term[15]. All animals which received blastocysts but which are not found pregnant at term are generally excluded although they often represent up to 50% of the transfers. There is, furthermore, no general agreement on whether it is correct to calculate the success rates at term only from the living fetuses or also from the resorbed embryos. At the end of pregnancy, treated and subsequently transplanted embryos were not found to be malformed[15,16] but were found to be either alive and normal or dead[26,52,53]. The success rate of transfer experiments can, therefore, be determined only with these two parameters. It consequently takes a long time and many pregnant animals to test the effect of a single dose of a teratogen in transplantation experiments. It is, therefore, financially almost impossible to use this method for establishing dose–response relations for several doses of a teratogen.

The *in vitro* system requires fewer embryos, and it is faster and more precise because maternal factors and individual variations are not involved. It also gives clearcut dose–response relations, which are difficult to obtain when treated embryos are transferred. We have demonstrated the advantages of the *in vitro* approach in detail for embryos treated *in vivo* (with CPA[15,17]) or *in vitro*[54] prior to being cultured during the implantation period. The same *in vitro* system has proved sensitive enough to allow investigations of early genetically determined abnormalities[55,56] and effects of *in vitro* exposure to ultraviolet irradiation[57], to X-irradiation[58], to [^{3}H]thymidine[32,59], and to metabolic inhibitors[33–35] in differentiating mouse blastocysts. The system has also been used as part of *in vitro* approaches to improve screening tests for the detection of dominant lethal mutations in the mouse[60,61]. In an investigation comparable to our studies on CPA, Wide[62] recently used the *in vitro* system to elucidate effects of inorganic lead on mouse blastocysts *in vitro* after treatment *in vivo* and *in vitro*. So far, in the rat similar investigations have not been reported. The present report, however, demonstrates that the *in vitro* culture beyond implantation holds promise for teratological investigations on preimplantation rat embryos.

The *in vitro* approach only analyses the direct effects of the teratogen on the embryo up to the early egg cylinder stage[63] and does only indirectly allow an assessment of a disturbed maternal physiology. The detection of maternal effects of the early treatment still requires transplantation experiments. Transplantation of embryos is also necessary in experiments to test whether or not treatment of preimplantation embryos during *in vitro* culture can induce malformations at term.

ACKNOWLEDGEMENT

Our investigations were supported by grants of the Deutsche Forschungsgemeinschaft awarded to the Sonderforschungsbereich 29 'Embryonale Entwicklung und Differenzierung'.

References

1. Neubert, D., Merker, H.-J. and Kwasigroch, T. E. (1977). *Methods in Prenatal Toxicology. Evaluation of Embryotoxic Effects in Experimental Animals.* (Stuttgart: Thieme Verlag)
2. Wilson, J. G. (1973). *Environment and Birth Defects.* (New York and London: Academic Press)
3. Adams, C. E., Hay, M. F. and Lutwak-Mann, C. (1961). The action of various agents upon the rabbit embryo. *J. Embryol. Exp. Morphol.*, **9**, 468
4. Brent, R. L. and Bolden, B. T. (1968). Indirect effect of X-irradiation on the first day of gestation. *Radiat. Res.*, **36**, 563
5. Spielmann, H. (1976). Embryo transfer technique and action of drugs on the preimplantation embryo. In: A. Gropp and K. Benirschke (eds.). *Current Topics in Pathology*, Vol. **62**, pp. 87–103 (Berlin, Heidelberg: Springer-Verlag)
6. Gardner, R. L. (1971). Manipulations on the blastocyst. In: G. Raspé (ed.). *Advances in the Biosciences*, **6**, pp. 279–296 (Oxford, Braunschweig: Pergamon Press, Vieweg Verlag)
7. Whittingham, D. G., Leibo, S. P. and Mazur, P. (1972). Survival of mouse embryos frozen to −196° and −269 °C. *Science*, **178**, 411
8. Austin, C. R. (1973). Embryo transfer and sensitivity to teratogenesis. *Nature* (*Lond.*), **244**, 333
9. Gottschewski, G. H. M. (1963). Das Entstehen bestimmter äusserer Missbildungen als Folge exogener Reize beim Säugetier. *Med. Welt*, **50**, 2545
10. Gottschewski, G. H. M. (1964). Mammalian blastopathies due to drugs. *Nature* (*Lond.*), **201**, 1232
11. Gottschewski, G. H. M. and Zimmermann W. (1963). Auslösung von Blastopathien beim Säugetier durch Cyclophosphamid und Thalidomid. *Naturwissenschaften*, **50**, 525
12. Brock, N. and von Kreybig, Th. (1964). Experimentaller Beitrag zur Prüfung teratogener Wirkungen von Arzneimitteln an der Laboratoriumsratte. *Naunyn-Schmiedeberg's Arch. Exp. Path. Pharmak.*, **249**. 117
13. Gebhardt, D. O. E. (1970). The embryolethal and teratogenic effects of cyclophosphamide on mouse embryos. *Teratology*, **3**, 273
14. Fritz, H. and Hess, R. (1971). Effects of cyclophosphamide on embryonic development in the rabbit. *Agents and Actions*, **2**. 83
15. Spielmann, H. and Eibs, H. G. (1978). Recent progress in teratology: a survey of methods for the study of drug actions on the preimplantation embryo. *Drug Res.*, **28**, 1733
16. Spielmann, H., Eibs, H. G. and Merker, H.-J. (1977). Effects of cyclophosphamide treatment on the development of rat embryos after implantation. *J. Embryol. Exp. Morphol.*, **41**, 65
17. Eibs, H. G. and Spielmann, H. (1977). Inhibition of post-implantation development of mouse blastocysts *in vitro* after cyclophosphamide treatment *in vivo*. *Nature* (*Lond.*), **270**, 54
18. Whittingham, D. G. and Wales, R. G. (1969). Storage of two-cell mouse embryos *in vitro*. *Aust. J. Biol. Sci.*, **22**, 1065
19. Tarkowski, A. K. (1966). An air drying method for chromosome preparations from mouse eggs. *Cytogenetics*, **8**, 394
20. Finn, C. A. and Martin, L. (1972). Temporary interruption of the morphogenesis of deciduomata in the mouse uterus by actinomycin D. *J. Reprod. Fertil.*, **31**, 353
21. Lorke, D. (1965). Embryotoxische Wirkungen an der Ratte. *Naunyn-Schmiedeberg's Arch. Exp. Path. Pharmak.*, **250**, 360
22. Whitten, W. K. (1971). Nutrient requirements for the culture of preimplantation embryos *in vitro*. In: G. Raspé (ed.). *Advances in the Biosciences*, **6**, pp. 129–141. (Oxford, Braunschweig: Pergamon Press, Vieweg Verlag)

23. Sherman, M. I. (1975). Long term culture of cells derived from mouse blastocysts. *Differentiation*, **3**, 51
24. Eibs, H. G., Spielmann, H., Häegele, M. and Klose, J. (1979). Effects of steroid sex hormones on the development of mouse embryos *in vitro* and *in vivo*. In: T. V. N. Persaud (ed.). *Advances in the Study of Birth Defects*, vol. 7, pp. 113–137 (Lancaster: MTP Press Limited)
25. Finn, C. A. and Bredl, J. C. S. (1973). Studies on the development of the implantation reaction in the mouse uterus: influence of actinomycin D. *J. Reprod. Fertil.*, **34**, 247
26. Bell, P. S. and Glass, R. H. (1975). Development of the mouse blastocyst after actinomycin D treatment. *Fertil. Steril.*, **26**, 449
27. Eibs, H. G. and Spielmann, H. (1977). Preimplantation embryos, Part II: culture and transplantation. In: D. Neubert, H.-J. Merker and T. E. Kwasigroch (eds.). *Methods in Prenatal Toxicology. Evaluation of Embryotoxic Effects in Experimental Animals*, pp. 221–230. (Stuttgart: Thieme Verlag)
28. Spindle, A. I. and Pedersen, R. A. (1973). Hatching, attachment, and outgrowth of mouse blastocysts *in vitro*: Fixed nitrogen requirements. *J. Exp. Zool.*, **186**, 305
29. Pienkowski, M., Solter, D. and Koprowski, H. (1974). Early mouse embryos: Growth and differentiation *in vitro*. *Exp. Cell Res.*, **85**, 424
30. Sherman, M. I. (1974). *In vivo* and *in vitro* differentiation during early mammalian embryogenesis. *Front. Rad. Therapy*, **9**. 28
31. Spielmann, H. (1975). Different patterns of energy metabolism in the rat and mouse zygote. *J. Reprod. Fertil.*, **42**, 391
32. Ansell, J. D. and Snow, M. H. L. (1975). The development of trophoblast *in vitro* from blastocysts containing varying amounts of inner cell mass. *J. Embryol. Exp. Morphol.*, **33**, 177
33. Sherman, M. I. and Atienza, S. B. (1975). Effects of bromodeoxyuridine, cytosine arabinoside and colcemid upon *in vitro* development of mouse blastocysts. *J. Embryol. Exp. Morphol.*, **34**, 467
34. Rowinski, J., Solter, D. and Koprowski, H. (1975). Mouse embryo development *in vitro*: Effects of inhibitors of RNA and protein synthesis on blastocysts and post-blastocyst embryos. *J. Exp. Zool.*, **192**, 133
35. Glass, R. H., Spindle, A. I. and Pedersen, R. A. (1976). Differential inhibition of trophoblast outgrowth and inner cell mass growth by actinomycin D in cultured mouse embryos. *J. Reprod. Fertil.*, **48**, 443
36. Russell, L. B. and Montgomery, C. S. (1966). Radiation-sensitivity differences within cell-division cycles during mouse cleavage. *Int. J. Radiat. Biol.*, **10**, 151
37. Jaquet, P., Leonard, A. and Gerber, G. B. (1976). Action of lead on early divisions of the mouse embryo. *Toxicology*, **6**, 129
38. Hurley, L. S. and Shrader, R. E. (1975). Abnormal development of preimplantation rat eggs after three days of maternal dietary zinc deficiency. *Nature (Lond.)*, **254**, 427
39. Epstein, C. J. (1975). Gene expression and macromolecular synthesis during preimplantation embryonic development. *Biol. Reprod.*, **12**, 82
40. Gibson, J. E. and Becker, B. A. (1968). Effect of phenobarbital and SKF 525-A on the teratogenicity of cyclophosphamide in mice. *Teratology*, **1**, 393
41. Gibson, J. E. and Becker, B. A. (1971). Teratogenicity of structural truncates of cyclophosphamide in mice. *Teratology*, **4**, 141
42. Brock, N. (1967). Pharmacologic characterization of cyclophosphamide (NCS-26271) and cyclophosphamide metabolites. *Cancer Chemother. Rep.*, **51**, 315
43. Sladek, N. E. (1973). Evidence for an aldehyde possessing alkylating activity as the primary metabolite of cyclophosphamide. *Cancer Res.*, **33**, 651
44. Murthy, V. V., Becker, B. A. and Steele, W. J. (1973). Effects of dosage, phenobarbital, and 2-diethylamino-2, 2-diphenylvalerate on the binding of cyclophosphamide and/or its metabolites to DNA, RNA, and protein of the embryo and liver in pregnant mice. *Cancer Res.*, **33**, 664
45. Short, R. D. and Gibson, J. E. (1974). ^{14}C-cyclophosphamide alkylation of mouse embryo macromolecules. *Proc. Soc. Exp. Biol. Med.*, **145**, 620
46. Basler, A., Buselmaier, B. and Röhrborn, G. (1976). Elimination of spontaneous and

chemically induced chromosome aberrations in mice during early embryogenesis. *Hum. Genet.*, **33**, 121

47. Röhrborn, G. and Buckel, U. (1976). Investigations on the frequency of chromosome aberrations in bone marrow cells of chinese hamsters after simultaneous application of caffeine and cyclophosphamide. *Hum. Genet.*, **33**. 113
48. Herken, R., Eibs, H. G. and Spielmann, H. (1978). Premature induction of the decidual reaction after cyclophosphamide treatment in the rat. (In preparation)
49. Russell, W. R., Walpole, A. L. and Labhsetwar, A. P. (1973). Cyclophosphamide: induction of superovulation in rats. *Nature* (*Lond.*), **241**, 129
50. Wide, M. and Nilsson, O. (1977). Differential susceptibility of the embryo to inorganic lead during preimplantation in the mouse. *Teratology*, **16**, 273
51. Jacquet, P., Gerber, G. B., Leonard, A. and Maes, J. (1977). Plasma hormone levels in normal and lead-treated pregnant mice. *Experientia*, **33**, 1375
52. Spielmann, H., Eibs, H. G., Jacob, U., Nagel, D. and Gregg, C. T. (1978). Teratological studies on effects of carbon-13 incorporation into preimplantation mouse embryos on development after implantation *in vivo* and *in vitro*. In: T. A. Baillie (ed.). *Stable Isotopes. Applications in Pharmacology, Toxicology and Clinical Research*, pp. 217–225. (London: Macmillan Press)
53. Fisher, D. L. and Smithberg, M. (1972). Early and late effects of *in vitro* exposure of preimplantation mouse embryos to trypan blue. *Teratology*, **6**, 159
54. Spielmann, H., Eibs, H. G., Nagel, D. and Gregg, C. T. (1976). The effect of carbon-13 incorporation into preimplantation mouse embryos on development before and after implantation. *Life Sci.*, **19**, 633
55. Pedersen, R. A. (1974). Development of lethal yellow (A^y/A^y) mouse embryos *in vitro*. *J. Exp. Zool.*, **188**, 307
56. Wudl, L. H. and Sherman, M. I. (1976). *In vitro* studies of mouse embryos bearing mutations at the t locus: t^w5 and t^{12}. *Cell*, **9**, 523
57. Pedersen, R. A. and Cleaver, J. E. (1975). Repair of UV-damage to DNA of implantation-stage mouse embryos *in vitro*. *Exp. Cell Res.*, **95**, 247
58. Goldstein, L. S., Spindle, A. I. and Pedersen, R. A. (1975). X-ray sensitivity of the preimplantation mouse embryo *in vitro*. *Radiat. Res.*, **62**, 276
59. Snow, M. H. L., Aitken, J. and Ansell, J. D. (1976). Role of the inner cell mass in controlling implantation in the mouse. *J. Reprod. Fertil.*, **48**, 403
60. Goldstein, L. S. and Spindle, A. I. (1976). Detection of X-ray induced dominant lethal mutations in mice: an *in vitro* approach. *Mutat. Res.*, **41**, 289
61. Bürki, K. and Sheridan, W. (1978). Expression of TEM-induced damage to postmeiotic stages of spermatogenesis of the mouse during early embryogenesis. I. Investigations with *in vitro* culture. *Mutat. Res.*, **49**, 259
62. Wide, M. (1978). Effect of inorganic lead on the mouse blastocyst *in vitro*. *Teratology*, **17**, 165
63. Wiley, L. M. and Pedersen, R. A. (1977). Morphology of mouse egg cylinder development *in vitro*: a light and electron microscopic study. *J. Exp. Zool.*, **200**, 389

6
Effects of steroid sex hormones on the development of early mouse embryos *in vitro* and *in vivo*

H.-G. EIBS, H. SPIELMANN, MARGRET HÄGELE AND J. KLOSE

INTRODUCTION

Clinical studies

The teratogenicity of steroid sex hormones is a matter of controversy. Several recent clinical and experimental studies suggest that the hormones can induce congenital abnormalities in sex organs of male[1] and female[2–5] offspring. Other studies question their involvement in the production of neural tube defects[6–8], the transposition of great vessels[9–13], VACTERL and DiGeorge syndromes[10,14,15] and limb malformations[16–18]. In addition, epidemiological studies fail to agree on the subject[19–27], but this may be due to lack of uniformity in route, frequency and dose of administration of the hormones, as well as incomplete information about the stage of pregnancy during treatment.

Animal studies *in vivo*

Various agents with oestrogenic or progestonic activity have been reported to be teratogenic or embryotoxic when administered in relatively high doses to pregnant experimental animals during organogenesis. In none of these studies were the steroid hormones administered before day 7 (mice) or day 8 (rats), as it is generally assumed that if the drugs are given before this time no malformations, but only embryolethality, will result[28]. Treatment with these drugs resulted in increased resorption rates[29–33], disturbed sexual differentiation[31,34], cleft palate[29,33,35], defective abdominal wall[35], open eyelid[32], minor skeletal abnormalities[32], and exencephaly[32].

In vitro studies

It has been reported that the exposure of mammalian preimplantation embryos to oestrogens or progestins *in vitro* inhibits normal embryonic development (mouse[36,37]; rabbit[38–40]). However, the drug concentrations used in these experiments are not comparable to dosages that would be obtained after *in vivo* treatment.

We are particularly interested in early pregnancy, since women often receive sex hormones during the first weeks of pregnancy. In the present study we have examined the effects of maternal treatment especially with the steroid sex hormone cyproterone acetate during early pregnancy of mice. This agent is an antiandrogen with strong progestonic properties and it is reported to have weak glucocorticoid potency[41]. The combination of cyproterone acetate and ethinylestradiol has recently been introduced as an oral contraceptive in Europe. We have particularly focused on the following problems:

(a) Influence of maternal treatment on embryonic *in vitro* development around the time of implantation.
(b) Occurrence of teratogenic effects at the end of pregnancy after treatment before the time of implantation.
(c) To test various hypotheses on the teratogenic effects of steroid sex hormones after treatment during pregnancy, e.g. their effects on hormonal balance or Wilkins' hypothesis[42] suggesting that the occurrence of abnormalities is due to an error in maternal metabolism. The current hypotheses of steroid sex hormone action and their relation to teratogenesis will be outlined in more detail.

MATERIALS AND METHODS

Drug administration

Cyproterone acetate (CA; Schering, Berlin, Germany), medroxyprogesterone-acetate (MPA; Upjohn, Kalamazoo, USA) and oestrone (Sigma, München, Germany) were dissolved in benzyl benzoate (10% w/v) and diluted with castor oil to obtain a volume of 0.1 ml/33 g mouse.

Pregnant mice received one single subcutaneous injection of a steroid hormone suspension on one day between days 1 and 11. Control animals received either a solvent control injection (benzyl benzoate/castor oil) or remained untreated.

For the exposure of preimplantation mouse embryos to steroid hormones *in vitro* the hormones were dissolved in ethanol and added to the culture medium.

Bromodeoxyuridine (BrdU; Sigma, München, Germany), methylnitrosourea (MNU; Ferak, Berlin, Germany) and cyclophosphamide (Asta, Bielefeld, Germany) were dissolved in distilled water and injected intraperitoneally.

Animals

Nullipara NMRI mice (Hannover) weighing 30–34 g were placed with males overnight. The day when a vaginal plug was present was designated day 0 of pregnancy.

In vitro culture

Preimplantation mouse embryos were obtained by flushing oviducts on day 2 of pregnancy (four- and eight-cell embryos) or uteri on day 3 (morulae and blastocysts) with culture medium. Embryos from day 2 of pregnancy – mostly four-cell-embryos – were placed in Falcon plastic Petri dishes containing Whitten's medium[43] with 3% bovine serum albumin (W-BSA) and incubated at 37 °C in a humidified 5% CO_2 atmosphere[44]. Normally 86–100% of the four-cell and eight-cell embryos develop into blastocysts during a culture period of 48 h. Morulae and blastocysts which had developed either *in vivo* or *in vitro* by the above procedure were incubated in culture medium NCTC-109 with 10% fetal calf serum (NCTC-109 FCS) for 120 h[45]. During this culture period the following events occur[44] (Figure 6.1): hatching out of the zona pellucida; attachment on the surface of

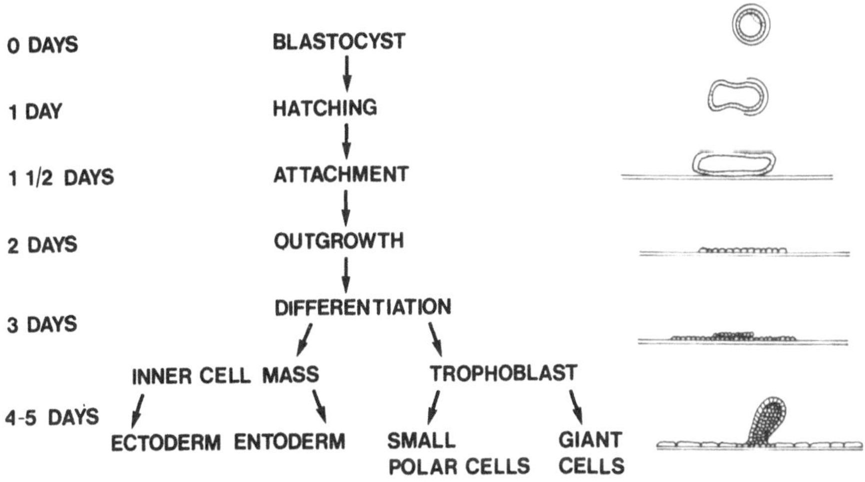

Figure 6.1 Diagram describing differentiation of mouse blastocysts *in vitro* in medium NCTC-109 supplemented with 10% fetal calf serum during a 120 h culture period. The first column shows the time in culture, the second one the steps of development and differentiation of the blastocysts and the third one gives light microscopic examples of the embryos

the Petri dish; outgrowth over the surface of the Petri dish; differentiation into inner cell mass (ICM) and trophoblast layer. The ICM cells subsequently differentiate into ectoderm- and entoderm-layers and the trophoblast into a polar layer, containing small trophoblast cells and trophoblast giant cells corresponding to the mural trophoblast[46]. The limit of the culture period is 120 h since after this time embryonic cells begin to disaggregate and to degenerate.

Incubation and preparation of embryos for determination of qualitative protein synthesis

Embryos were grown in W-BSA. L-[^{35}S]methionine (New England Nuclears; specific activity about 600 Ci/mmol) was added to medium to obtain a final concentration of 300 μCi/ml. Blastocyst- and morula-stage embryos from untreated mothers and from mothers treated with 30 mg/kg CA 24 h earlier were incubated in the radioactive medium for 2 h. Between 60 and 100 embryos were labelled in one experiment. After exposure to L-[^{35}S]methionine the embryos were washed 10 times in cold medium. Ten microlitres of medium containing the embryos were added to 30 μl of lysis buffer (9.5 M urea, 1.6% ampholine pH 5–7 (LKB, Stockholm, Sweden), 0.4% ampholine pH 3–10 (Serva, Heidelberg, Germany), 5% mercaptoethanol (Serva, Heidelberg, Germany)). For determination of quantitative incorporation, samples were prepared as described by Van Blerkom and Brockway[47]. Labelled proteins were separated by two-dimensional polyacrylamide gel electrophoresis[48,49]. After isoelectric focusing on a cylindrical gel the proteins were separated by discontinuated polyacrylamide slab gels. Subsequently, the slab gels were prepared for autoradiography according to Laskey and Mills[50]. Each spot detected on the autoradiographs represents one polypeptide into which methionine had been incorporated. The incorporation of radiolabelled [^{35}S]methionine into the embryonic protein was measured by counting the trichloroacetic acid-precipitable fraction of the embryos.

Examination of the fetuses

For teratological investigations pregnant mice were killed by cervical dislocation on day 17 of pregnancy. Implantation sites were counted in each uterine horn. Each fetus recorded as dead or alive was examined for external malformations and weighed individually. The fetuses were fixed in Bouin's solution and examined by Wilson's razor blade technique[51] after a second careful examination of the limbs and tails.

Statistics

In general, each treated group was taken as the experimental unit and data were expressed as average weight per fetus or as average number of living fetuses or embryos per dam. Experimental groups were compared by the Wilcoxon test. However, some figures contain standard deviations. Success rates of embryonic development during *in vitro* culture and rates of malformations at term are expressed as percentage of total groups for both treated and control embryos. These groups were then compared by the χ^2-test.

RESULTS

Examination of embryos at the end of the preimplantation period

Effect of early maternal treatment on development of preimplantation embryos in vivo

Treatment with CA (5 to 900 mg/kg), MPA (20 mg/kg) or oestrone (0.03 μg/mouse) on days 1 or 2 of pregnancy did not affect either the number of embryos or the number of blastocysts per mother up to 100 mg/kg CA. However, simultaneous treatment with 0.03 μg oestrone per mouse and 20 mg/kg CA reduced the average number of blastocysts per dam (Figure 6.2). Table 6.1

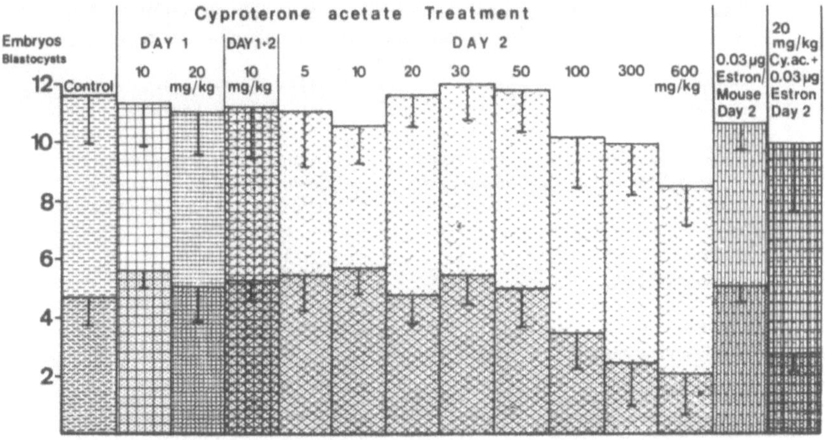

Figure 6.2 Total number of living embryos (morulae and blastocysts; long columns) and blastocysts (short columns) per pregnant mouse on day 3 of pregnancy after treatment with steriod sex hormones. Columns represent means ± standard deviation. The number of embryos or blastocysts per mother was only reduced at the highest concentrations of cyproterone acetate and when cyproterone acetate (20 mg/kg) and oestrone (0.03 μg/mouse) were given simultaneously

Table 6.1 Cell number and mitotic index of mouse blastocysts and morulae 24 h after cyproterone acetate treatment on day 2 of pregnancy

Dose (mg/kg)	*Number of embryos*	*Developmental stage*	*Cell number per embryo*	*Mitotic cells per embryo*
0	26	Blastocyst	31.1 ± 2.0	1.8
0	11	Morula	23.8 ± 2.2	1.0
30	25	Blastocyst	31.4 ± 2.6	1.6
30	23	Morula	22.5 ± 2.5	0.9

shows that after a dose of 30 mg/kg CA neither the cell number of morulae or blastocysts nor the number of mitoses per embryo was reduced as compared to control embryos. These data indicate that the proliferation capacity of the embryonic cells is not influenced by progestin treatment.

Changes in the pattern of protein synthesis

Maternal treatment with CA did not influence the rate of protein synthesis in morulae and blastocysts (CPM/embryo). However, the qualitative pattern of protein synthesis determined by two-dimensional polyacrylamide gel electrophoresis was changed in embryos after CA treatment (30 mg/kg). In different runs the occurrence or intensity (but never the position) of some proteins with an isoelectric point near pH 5.5 was changed (Table 6.2). The total number of detectable protein spots was more than 500. These changes in protein synthesis, induced by maternal treatment, did not correspond to changes in protein synthesis due to the development from morulae to blastocysts, since the protein spots that were changed by the development from morulae

Table 6.2 Effect of maternal cyproterone acetate treatment (30 mg/kg) on the pattern of specific protein synthesis in mouse morulae and early blastocysts

		Effect of treatment	
Type of variation in specific protein spots	*Untreated blastocysts vs. morulae*	*Treated vs. control morulae*	*Treated vs. control blastocysts*
Appearance	1	1	1
Disappearance	1	3	—
Increase	4	2	2
Decrease	—	2	4
Positional change	—	—	—
Total changed	6	8	7

to blastocysts have isoelectric points higher than pH 6 and have preponderably lower molecular weights ($< 10^5$).

In vitro *culture experiments after treatment* in vivo

Mouse blastocysts can be cultured routinely for up to 120 h. This is also generally true for morulae. Therefore the culture system can aid in the study of pharmacological and toxicological effects of drugs on the embryo after maternal treatment during the preimplantation period, since it permits examination of embryos during periods of growth and differentiation (Figure 6.1) which cannot be observed by *in vivo* methods[52, 53].

After maternal treatment with 5–900 mg/kg CA no significant effect on the *in vitro* development of mouse morulae and blastocysts could be seen up to 72 h (Tables 6.3 and 6.4). The trophoblast layer was expanded and the cells could be divided into developing mural trophoblast giant cells and small polar trophoblast cells. In the centre, the optically dense ICM layer could be distinguished from the trophoblast cells. However, fewer embryos from CA-treated mothers developed an ICM with two germ layers (ectoderm and entoderm) by 96–120 h in culture (Tables 6.3 and 6.4). To test whether the reduced ICM development was due to a relative oestrogen deficiency, oestrone (0.03 μg/mouse) was administered along with CA (20 mg/kg) on day 2 of pregnancy. At the end of the culture period embryonic development was reduced to about the same level as was seen after treatment with CA only. Single treatment with 0.03 μg oestrone/mouse or MPA (10 or 30 mg/kg) produced the same effect on embryonic development (Tables 6.3 and 6.4).

In vitro *exposure to steroid hormones*

To exclude the possibility that the reduced embryonic development in the *in vitro* system was due to an effect on the embryos prior to the time of culture via an alteration in maternal metabolism, previously untreated embryos (mostly four-cell stage) were removed on day 2 of pregnancy and exposed to the gestagens CA or MPA during an *in vitro* period of 48 h. While high concentrations (30 mg/l medium) of the two hormones were directly embryotoxic, lower concentrations (e.g. 3 mg/l medium) did not affect the develop-

Table 6.3 Effect of maternal steroid sex hormone treatment on the *in vitro* development of mouse blastocysts in medium NCTC-109

Substance	*Dose*	*Day of pregnancy*	*Number of blastocysts*	*Hatched* %	*Outgrowth* %	*Expanded trophoblast layer* %	*Expanded inner cell mass* %	*ICM with two germ layers* %
Control			146	98.8	98.8	97.5	82.3	75.9
Cyproterone acetate	5 mg/kg	2	42	100.0	100.0	100.0	71.4	42.9†
Cyproterone acetate	10 mg/kg	2	99	86.2	86.2†	79.3†	42.3†	33.2†
Cyproterone acetate	20 mg/kg	2	118	95.8	95.8	91.9	63.6†	43.2†
Cyproterone acetate	30 mg/kg	2	39	100.0	94.9	94.9	66.7	41.1†
Cyproterone acetate	100 mg/kg	2	44	90.9	88.6*	84.1*	50.0†	40.9†
Cyproterone acetate	300 mg/kg	2	34	88.2	88.2*	85.3*	50.0†	38.2†
Cyproterone acetate	10 mg/kg	1	80	95.0	95.0	88.8*	63.8*	43.8†
Cyproterone acetate	10 mg/kg	1 + 2	54	100.0	100.0	92.6	59.3†	25.9†
Cyproterone acetate + oestrone	20 mg/kg 0.03 μg/mouse	2	55	98.2	98.2	92.7	63.6*	41.8†
Oestrone	0.03 μg/mouse	2	64	98.4	98.4	90.6	65.6*	37.5†
Medroxyprogesterone acetate	10 mg/kg	2	39	100.0	100.0	100.0	84.6	53.9*
Medroxyprogesterone acetate	30 mg/kg	2	54	96.3	96.3	96.3	66.7	40.7†

* $p < 0.01$
† $p < 0.001$

Table 6.4 Effect of maternal steroid sex hormone treatment on the *in vitro* development of mouse morulae in medium NCTC-109

Substance	*Dose*	*Day of pregnancy*	*Number of morulae*	*Hatched* %	*Outgrowth* %	*Expanded trophoblast layer* %	*Expanded inner cell mass* %	*ICM with two germ layers* %
Control			122	90.2	92.6	88.5	50.8	36.9
Cyproterone acetate	5 mg/kg	2	64	87.5	87.5	75.0	37.5	25.0
Cyproterone acetate	20 mg/kg	2	168	82.7	82.1*	81.0	46.4	23.8
Cyproterone acetate	30 mg/kg	2	65	86.2	83.1	76.9	35.4	16.9*
Cyproterone acetate	300 mg/kg	2	91	57.1†	65.9†	57.1†	28.6†	19.8*
Cyproterone acetate	10 mg/kg	1	39	46.2†	46.2†	35.9†	12.8†	5.1†
Cyproterone acetate	10 mg/kg	1 + 2	28	78.6	78.6	78.6	28.6	21.4
Cyproterone acetate + oestrone	20 mg/kg 0.03 μg/mouse	2	112	97.3	97.3	92.9	45.5	20.5*
Oestrone	0.03 μg/mouse	2	67	85.1	79.1*	76.1	46.3	23.9
Medroxyprogesterone acetate	10 mg/kg	2	58	74.1*	77.6*	75.9	31.0	13.8*
Medroxyprogesterone acetate	30 mg/kg	2	76	81.6	81.6	76.3	34.2	14.5*

* $p < 0.01$
† $p < 0.001$

Table 6.5 Effect of ***in vitro*** incubation of mouse four-cell embryos in the presence of steroid sex hormones on ***in vitro*** development around implantation

Substance	*Concentration* mg/l	*Number of four-cell embryos* A	*Developed into blastocysts* B	(% from A)	*Hatched and attached* % from B	*Outgrowth* % from B	*Expanded trophoblast layer* % from B	*Expanded inner cell mass* % from B	*ICM with two germ layers* % from B
Control		81	72	88.9	95.8	98.6	98.6	73.6	45.8
Cyproterone acetate	3.0	84	78	92.9	96.2	96.2	96.2	52.6*	29.5
Medroxyprogesterone acetate	3.0	84	80	95.2	95.0	95.0	95.0	36.3†	20.0†
Cyproterone acetate	30.0	49	8	16.3†					
Medroxyprogesterone acetate	30.0	48	21	43.8†					

* $p < 0.01$
† $p < 0.001$

ment into blastocysts (Table 6.5). After 48 h the embryos were transferred into the hormone-free blastocyst culture medium NCTC-109 and cultured for an additional 120 h. At the end of this period ICM development and differentiation were reduced when compared to control embryos that had, at the four-cell stage, been exposed only to the alcohol vehicle.

Teratological investigations near term

Teratology after treatment on day 2 of pregnancy

Treatment with a single injection of CA (5–900 mg/kg) on day 2 of pregnancy neither increased the resorption rate at any dose nor did it affect the average fetal weight when examined on day 17. Externally visible malformations of extremities and other bone malformations occurred sporadically, without regard to the dose of drug used. The rate of exencephaly rose from 0.4% in control fetuses to 1.5% after treatment with CA on day 2 (Table 6.6). No clear dose–response could be observed for this type of malformation. The razor blade sectioning method according to Wilson revealed an increase in the rate of occurrence of various internal malformations (Table 6.6; Figure 6.3). The highest frequency could be observed in the incidence of cleft palate (Figure 6.4a). In control fetuses cleft palates are rare in our NMRI strain (0.4%). After CA treatment this malformation could be observed in 0–42% (dose-dependently). Malformations of the urinary tract also occurred with a high frequency. The number of fetuses with anomalies of the urinary tract was 127. Seventy per cent had an enlargement of the pelvic cavity (Figure 6.4b), 21% had an enlarged bladder cavity (Figure 6.4c), 1.6% showed ectopia of the kidney, and 33% had distinctly different sizes of the kidney. Some malformations in the vascular system were also detected (10 cases in the treated groups). Dislocation of the heart associated with dislocation of the great vessels was especially prominent (90%) (Figure 6.4d). A ventricular septal defect was detected only once. Transposition of great vessels was not observed. Retarded fusion and development of parietal and occipital bones were not taken into consideration, as these irregularities also occurred at an inconsistent rate in control fetuses.

Teratology after treatment on one single day between days 1 and 11

No single period of peak sensitivity to a dose of CA (30 mg/kg) was detected for all organ systems affected (Table 6.7; Figure 6.5). The rate of cleft palate increased slowly and irregularly to a peak on day 11. This corresponds to what has been previously reported[28]. We, furthermore, found a tentative maximum in kidney and urinary bladder malformations after treatment on days 5–6 and an occurrence of extensively enlarged tracheal and bronchial lumina (Figure 6.6a) after treatment on days 8–11 (Table 6.7; Figure 6.5). After treatment on days 8–10 a diaphragmatic hernia was visible in 10 cases (Figure 6.6b).

Co-teratological investigations

To test whether the teratogenic effect of CA could be enhanced by co-treatment with other known teratogens, various drugs (BrdU, CPA, MNU)

Table 6.6 Frequency of malformations in mouse fetuses on day 17 after maternal treatment with cyproterone acetate on day 2 of pregnancy

Treatment	*Dose* mg/kg	*Number of dams*‡	*Number of fetuses*	♀	♂	*Cleft palate* %	*Urinary tract malformation* %	*Exencephaly* %	*Heart malformations* %	*Other* %	*Malformed fetuses* %
Untreated control		42	489	257	232	0.2	0.6	0.6	0.0	0.4	1.8
Solvent control		34	581	287	294	0.5	4.0	0.5	0.0	0.5	5.0
All control		76	1070	544	526	0.4	2.4	0.5	0.0	0.5	3.6
Cyproterone acetate	5	14	106	53	53	2.8	1.9	0.9	0.0	0.0	5.7
Cyproterone acetate	10	8	108	57	51	0.0	0.9	3.6	0.0	0.0	4.6
Cyproterone acetate	20	20	260	130	130	0.0	1.9	1.2	0.4	1.2	4.6
Cyproterone acetate	30	15	137	62	75	5.1	5.8	0.7	0.7	0.7	12.4
Cyproterone acetate	50	8	87	43	44	0.0	10.3	1.2	0.0	0.0	11.5
Cyproterone acetate	75	7	61	32	29	0.0	16.4	0.0	0.0	3.3	19.7
Cyproterone acetate	100	28	302	140	162	8.3	10.9	2.3	0.7	1.3	23.5
Cyproterone acetate	300	8	85	44	41	21.2	20.0	0.0	3.5	3.5	48.2
Cyproterone acetate	600	8	91	39	52	33.0	24.2	1.1	3.3	2.2	63.7
Cyproterone acetate	900	4	36	15	21	41.7	27.8	2.8	0.0	5.6	63.9
All CA treated	5–900	120	1273	615	658	7.7†	10.0†	1.5	0.8*	1.3	20.3†

* $p < 0.01$
† $p < 0.001$
‡ At least 10 animals were used for each group on day 2 of pregnancy

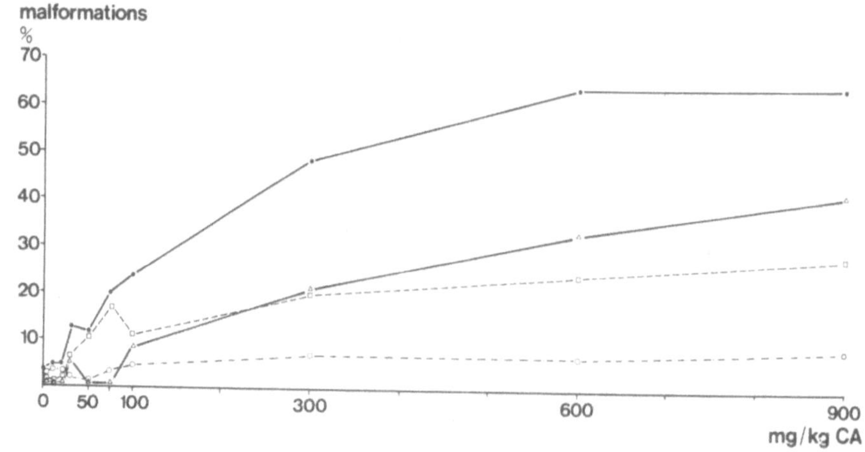

Figure 6.3 Malformations of mouse fetuses after cyproterone acetate (CA) treatment on day 2 of pregnancy (10–900 mg/kg); the rates of malformations were detected in razor blade slices (Wilson's technique[51]). The fetuses were examined near term (day 17); ●——● malformed fetuses; △——△ cleft palate; □---□ urinary tract malformations; ○-·-·-○ other malformations

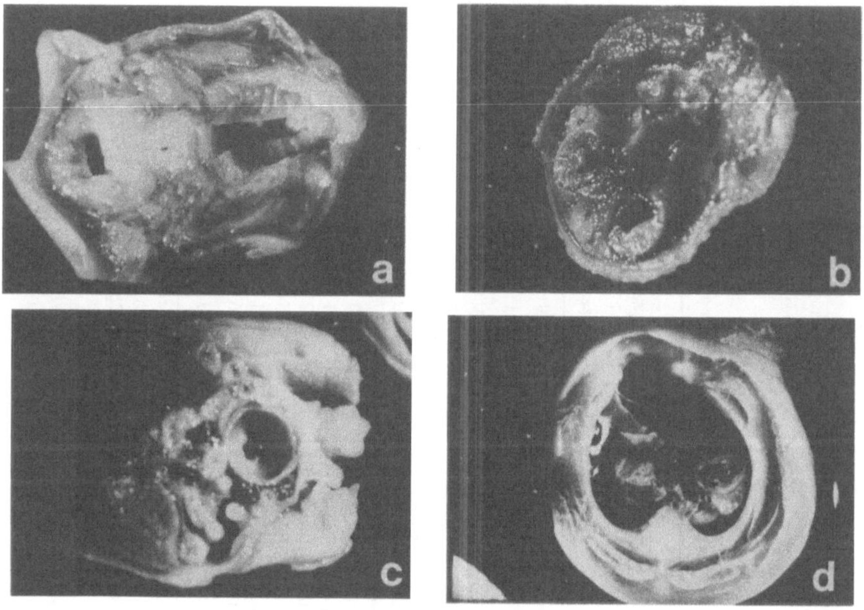

Figure 6.4 Razor blade slices (Wilson's technique[51]) of mouse fetuses near term (day 17) showing examples of internal abnormalities: a, Cleft palate; b, Enlargement of the pelvis of ureter. In the right kidney the pelvis is enlarged and the parenchyma is reduced: c, Enlargement of the urinary bladder; d, Heart distorted to the left thorax. (Caudocephalad view)

Table 6.7 Frequency in malformations in mouse fetuses on day 17 after a single maternal injection with 30 mg/kg cyproterone acetate between days 1 and 11 of pregnancy

Treatment	Day	Number of dams*	Number of fetuses	♀	♂	Cleft palate %	Urinary tract malformation %	Respiratory tract malformation %	Exencephaly %	Heart malformation %	Other %	Malformed fetuses %
Control		76	1070	544	526	0.4	2.4	0.2	0.5	0.0	0.3	3.6
CA 30 mg/kg	1	9	81	32	49	1.2	0.0	0.0	1.2	0.0	1.2	3.6
CA 30 mg/kg	2	15	137	62	75	5.1	5.8	0.0	0.7	0.7	0.7	12.4
CA 30 mg/kg	3	5	69	35	34	1.5	10.1	0.0	0.0	0.0	1.5	13.1
CA 30 mg/kg	4	4	50	25	25	0.0	16.0	0.0	2.0	0.0	0.0	18.0
CA 30 mg/kg	5	4	26	14	12	3.9	26.9	0.0	0.0	3.9	0.0	34.6
CA 30 mg/kg	6	5	63	30	33	9.5	28.6	1.6	0.0	1.6	0.0	38.1
CA 30 mg/kg	7	5	61	29	32	0.0	14.8	1.6	0.0	0.0	0.0	16.4
CA 30 mg/kg	8	6	53	25	28	11.3	7.6	17.0	0.0	0.0	3.8	35.9
CA 30 mg/kg	9	6	75	36	39	2.7	5.3	17.3	0.0	1.3	1.3	26.7
CA 30 mg/kg	10	5	65	28	37	27.7	1.5	9.2	0.0	0.0	7.7	43.1
CA 30 mg/kg	11	6	76	34	42	35.5	13.2	4.0	0.0	0.0	0.0	50.0
CA 100 mg/kg	8	4	53	27	26	11.3	13.2	24.5	0.0	0.0	0.0	45.3
CA 100 mg/kg	9	4	47	23	24	8.5	6.4	2.1	2.1	0.0	2.1	21.3

* At least 7 animals were used for each group, but not all mice were found to be pregnant on the day of sacrifice

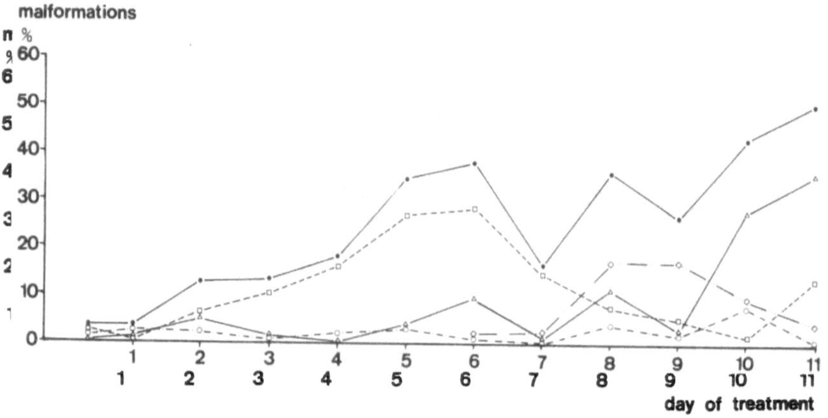

Figure 6.5 Malformations of mouse fetuses after a single injection of 30 mg/kg cyproterone acetate (CA) between days 1 and 11 of pregnancy; rates of malformations – detected in razor blade slices (Wilson's technique[51]). The fetuses were examined near term (day 17); ●——● malformed fetuses; △——△ cleft palate; □– – –□ urinary tract malformations; ◇–· · ·–◇ respiratory tract malformations; O–·–·–O other malformations

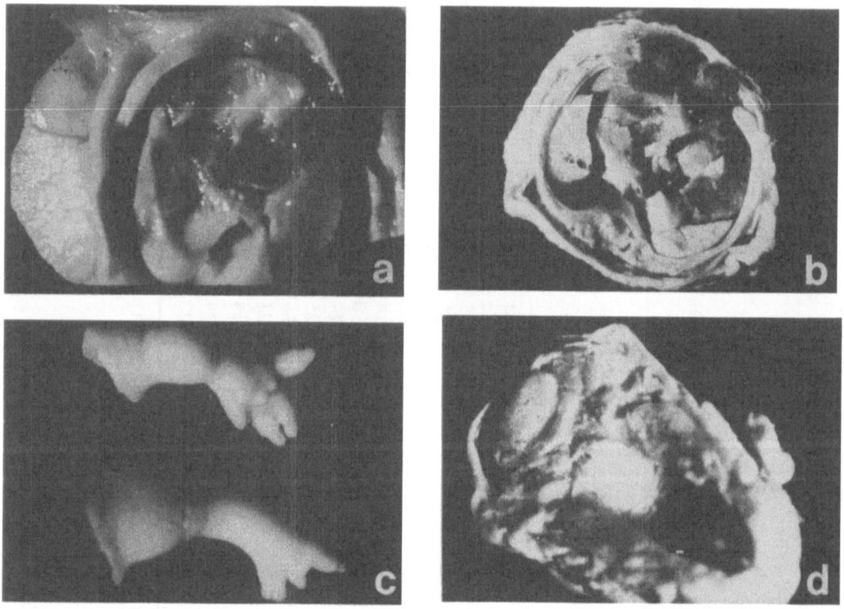

Figure 6.6 Razor blade slices (Wilson's technique[51]) of mouse fetuses near term (day 17) showing examples of internal abnormalities: a, Expanded right bronchus; b, Diaphragmatic hernia. Liver parenchyma is protruded into the frontal thorax: c, Syndactylia; d, Palatal slit

were administered subsequent to CA. After the combined treatment with CA on day 2 and CPA or MNU (5 and 10 mg/kg on days 6–12) we could not detect a significant difference in the rate of externally visible malformations when fetuses were examined on day 17. After a single injection of 500 mg/kg BrdU between days 6 and 10, large numbers of fetuses with exencephaly, enlargement of the oesophagus or stomach and limb deformities (Figure 6.6c) were found (Table 6.8). However, the rate of these malformations was not influenced by the previous treatment with 30 mg/kg CA on day 2 of pregnancy (Table 6.8).

In the same manner, the rates of BrdU-induced malformations were not altered when 30 mg/kg CA was given once between days 2 and 7 and BrdU was injected on day 7 (Table 6.9). The frequency of malformations was independent of BrdU-treatment and was identical to that found after CA treatment alone (Table 6.7; Figure 6.5).

The rate of palatal slits (not clefts; Figure 6.6d) in exencephalic fetuses was significantly increased after the application of both substances. Exencephaly associated with cleft face only occurred after a combined CA–BrdU treatment.

DISCUSSION

Investigation of steroid sex hormones for teratogenic potential during the first days of gestation is of considerable practical importance because women often receive these substances early in pregnancy. If pregnancy occurs during oral contraceptive therapy the embryo is exposed to comparatively low hormone doses. However, relatively high doses are administered in other situations, e.g. irregularities of sexual cycle, hormonal pregnancy tests, depot contraception or dermatological indications. Furthermore, experimental evidence obtained with steroid sex hormone treatment before complete formation of the placenta can better be extrapolated to the human situation, since in primates, during later stages of pregnancy, the placenta takes over hormonal pregnancy control by synthesizing chorionic gonadotropins and different steroids with sexual hormonal properties[54].

The hormones in this study were administered subcutaneously. This route of administration was recommended by Andrew and Staples[33]. Since CA is a long acting hormone with maximum dose reaching the embryonic compartment (in man) 24 h after application[55] we administered a single injection. Several treatment times were used to determine specific periods of teratogenic susceptibility.

Effect of steroid sex hormone treatment on preimplantation embryos

The number of embryos per mother and the development into blastocysts were not altered after maternal CA treatment. The cell number and mitotic index of embryos exposed to the hormone was not altered. These results indicate that proliferation was not affected and the drug was not directly cytotoxic.

Hatching from the zona pellucida and attachment onto the surface of the Petri dish (homologous to the embryonic activity involved in implantation)

Table 6.8 Frequency of malformations in mouse fetuses on day 17 after maternal treatment with cyproterone acetate (CA; 30 mg/kg) on day 2 and the subsequent application of bromodeoxyuridine (BrdU; 500 mg/kg) between days 6 and 10

Treatment (Day of BrdU application)	*Number of fetuses*	♀	♂	*Cleft palate* %	*Urinary tract* %	*Respiratory tract* %	*Exencephaly* %	*Exencephaly with* *Cleft face* %	*Exencephaly with* *Open eye* %	*Exencephaly with* *Palatal slit* %	*Digestive tract* %	*Other malformations‡* %	*Total malformations* %
BrdU (6)	23	12	11	0.0	0.0	47.8	4.4	0.0	4.4	0.0	8.7	4.4	52.2
BrdU (6) + CA	34	15	19	2.9	5.9	38.2	5.9	0.0	5.9	2.9	8.8	5.8	58.8
BrdU (7)	117	56	61	0.0	6.0	5.1	18.0	0.0	10.3	3.6	1.7	1.7	30.8
BrdU (7) + CA	52	26	26	1.9	9.6	5.8	29.0	3.9	9.6	0.0	0.0	1.9	42.3
BrdU (8)	55	26	29	0.0	5.5	0.0	92.7	0.0	65.5	29.1	0.0	3.6	96.4
BrdU (8) + CA	57	29	28	1.8	14.0	0.0	82.5	5.3	59.7	66.7	0.0	3.5	84.2
BrdU (9)	29	18	11	3.5	3.5	3.5	0.0	0.0	0.0	0.0	24.1	10.4	44.8
BrdU (9) + CA	69	34	35	7.3	1.5	10.1	1.5	0.0	0.0	0.0	14.5	5.9	31.9
BrdU (10)	62	42	20	61.3	4.8	59.7	0.0	0.0	0.0	0.0	8.1	0.0	82.3
BrdU (10) + CA	34	12	22	38.2	0.0	41.2	0.0	0.0	0.0	0.0	26.5	2.9	67.7
BrdU (all)	286	154	132	13.6	4.9	19.2	29.0	0.0	17.1	2.5	5.6	2.9	57.7
BrdU + CA (all)	246	116	130	8.5	6.5	17.5	24.6	2.0*	16.7	17.1†	8.9	4.0	54.9

* $p < 0.05$
† $p < 0.001$
‡ Diaphragmatic hernia, limb and heart malformations

Table 6.9 Frequency of malformations in mouse fetuses on day 17 after maternal treatment with bromodeoxyuridine (BrdU; 500 mg/kg) on day 17 and the previous single application of cyproterone acetate (CA; 30 mg/kg) between days 2 and 7

Treatment	*Day of CA application*	*Number of fetuses*	♀	♂	*Cleft palate* %	*Urinary tract* %	*Respiratory tract* %	*Exencephaly* %	*Exencephaly with* *Cleft face* %	*Exencephaly with* *Open eye* %	*Exencephaly with* *Palatal slit* %	*Digestive tract* %	*Other malformations‡* %	*Total malformations* %
BrdU + CA	2	52	26	26	1.9	9.6	5.8	29.0	3.9	9.6	0.0	0.0	1.9	42.3
BrdU + CA	3	58	34	24	1.7	1.7	0.0	17.2	0.0	13.8	8.6	3.5	1.7	22.4
BrdU + CA	4	46	27	19	6.5	8.7	0.0	15.2	0.0	4.4	8.7	4.4	6.6	37.0
BrdU + CA	5	38	20	18	2.6	21.1	0.0	23.7	0.0	21.1	0.0	5.3	0.0	42.1
BrdU + CA	6	69	32	37	11.6	7.3	2.9	29.0	1.5	23.2	23.2	2.9	0.0	50.7
BrdU + CA	7	50	24	26	4.0	0.0	16.0	10.0	0.0	6.0	6.0	0.0	0.0	44.0
BrdU + CA	all	313	163	150	5.1	7.4	3.2	21.1	1.0	13.4	11.5	2.6	1.6	39.0
BrdU	—	117	56	61	0.0*	6.0	5.1	18.0	0.0	10.3	2.6†	1.7	1.7	30.8

* $p < 0.05$
† $p < 0.01$
‡ Limb and heart malformations

were not inhibited after treatment. However, by the end of the culture period ICM cells from CA-treated embryos had not developed to the same extent as in control embryos. The same effect could be seen after MPA treatment, after the simultaneous application of CA and oestrone or after oestrone treatment only. This suggests that the inhibition of embryonic development *in vitro* was neither due to a specific CA effect nor to a relative oestrogen deficiency. To exclude the possibility that alterations of embryonic *in vitro* development were exclusively mediated by changed maternal metabolism or by metabolites of the hormones administered, previously untreated embryos were exposed to hormones *in vitro*. The success rate determined by the number of embryos which were viable after 120 h of culture was lower for blastocysts cultured *in vitro* from the four-cell stage than for *in vivo* developed blastocysts[56]. However, the profile of embryonic development was similar to that after maternal treatment. High concentrations (30 mg/l) of the gestagens CA and MPA prevented cleavage of the embryonic cells and most of the embryos degenerated at the four-cell stage. Similar antimitotic effects of steroid sex hormones (progesterone[36–39]; oestrogen[37,40]) have been reported. In our studies, high concentrations of any sex steroids added to blastocyst cultures inhibited hatching of the blastocysts, as well as the attachment of embryos to the surface of the Petri dish, suggesting that at very high concentrations implantation processes were affected.

When cultured in the presence of high concentrations of progestins (30 mg/l medium) crystals of the hormones became visible microscopically in the embryo. Probably sex steroids are also accumulated in the embryo, e.g. in the blastocoele (for rabbits[57]), after maternal progestin treatment without any effect on embryonic development before steroid metabolism occurs in the mouse embryo, i.e. not before day 7 of pregnancy in mice[58,59]. To test this hypothesis we looked for possible changes in biochemical or genetic properties. Protein synthesis was examined via autoradiography using two-dimensional polyacrylamide gel electrophoresis[48,49]. Preimplantation development was accompanied by a change in the pattern of protein synthesis as recently reported by other investigators[60,61]. Our results also suggest that progestin treatment induced changes in the rates of synthesis of specific embryonic proteins. The proteins that were altered in their rate of synthesis have yet to be identified. The absence of positional changes of newly synthesized proteins may indicate that no mutated genes were detectable during our period of investigation.

Induction of malformations by early sex steroid treatment

The dramatic effect of sex steroid treatment on *in vitro* development of mouse blastocysts without a consistent dose–response relationship raises the question of the teratological relevance of these findings. After treatment during the preimplantation period cleft palate and malformations of the urinary tract revealed a clear dose–response. Also, heart defects and exencephaly were frequently produced. According to Andrew and Staples[33] the appearance of cleft palate might be due to a glucocorticoid action of CA. Pinsky and DiGeorge[62] reported that the frequency of cleft palate in the mouse after maternal treatment with steroid hormones is a teratogenic index of glucocorticoid potency. The application of CA (30 mg/kg) on a single day of pregnancy between days 1 and 11 showed a peak for the days of pregnancy which are reported to be most sensitive to the occurrence of cleft palate (day 11). However, recent epidemiological studies indicated a relation between an increased frequency of cleft palates and steroid sex hormone treatment during pregnancy in humans[22,26]. The question whether in our experiments cleft palate induced after early CA treatment were due to a retarded clearance of the drug has to be solved by metabolic and pharmacokinetic investigations on the maternal and embryological compartments of pregnant mice in additional experiments.

Urinary tract malformations (enlargement of the total kidney volume or the pelvic or bladder cavity) after progestin treatment have been found in one other study using experimental animals[63], but in our studies the incidence of these malformations revealed an unexpected teratological profile with a peak on day 5 and day 6. Clinical and epidemiological reports of urinary tract malformations after steroid sex hormone treatment during pregnancy have recently been published[22,64]. Additional histological investigations are needed to determine if the predominant urinary tract malformation (enlargement of the pelvic cavity) is due to some externally invisible genital tract malformation.

Limb defects, malformations of the heart and of the central nervous system were searched for and identified in retrospective clinical studies[6–13,16–18]. In our investigation, torsion of the heart to the extreme left of the thorax with associated shortening of the great vessels predominated over all other heart deformities. Transposition of the great vessels, however, was not observed in our razor blade slices. CNS malformations observed in our studies appeared to be different from those in clinical reports; exencephaly was the only malformation found and it was never associated with spina bifida, myelocele or meningocele.

Our *in vitro* studies with preimplantation embryos showed that CA treatment affects embryonic protein metabolism *in vitro*. However, a decision whether the teratogenic effects in fetuses near term are induced by a direct alteration of embryonic development or are mediated by influences on maternal metabolism can only be evaluated conclusively by studying the effects of transplantation of treated embryos to pseudopregnant untreated foster mothers[65]. Our transplantation experiments have not been completed.

Coteratogenic investigations

Sex hormones have been suspected to support carcinogenic activity of other chemicals, via an ability of the hormones to alter the rate of proliferation in target organs[66,67]. CA treatment during the preimplantation period does not alter the rate of mitosis in the early mouse embryo. Gaudin *et al.*[68,69] have suggested that sex steroids have cocarcinogenic activities since they inhibit DNA repair replication in mammalian cells. Possibly, by a similar mechanism, the hormones might exhibit coteratogenic activity. To examine this problem BrdU, a rather specific teratogen which affects genetic material, was chosen since its teratogenic properties have been extensively investigated[70–72].

The occurrence or increase in incidence of certain specific malformations (cleft face, palatal slits) was statistically significant. However, the rates of malformations specific to the two teratogens (e.g. malformed urinary tract for CA treatment days 5–6, exencephaly for BrdU-treatment days 7–8) were not increased by the combined administration of the two substances. On the other hand, cleft face and palatal slits were always associated with exencephaly. Therefore, conclusive evidence for a definitive coteratogenic action, due to an interaction on genetic material, is lacking in the present study.

ACTIONS OF STEROID SEX HORMONES AT THE MOLECULAR LEVEL: CURRENT HYPOTHESES AND RELATIONS TO TERATOGENESIS

Most drugs which induce congenital anomalies have been reported to exert their effects via one of the following mechanisms of action[28]:

(a) Cytostatic agents are cytotoxic when given at the proper dose and produce necrosis, especially in rapidly proliferating cells.

(b) Reversible inhibition of proliferation occurs, for example, after hydroxyurea application.

(c) Alteration of genetic material, by means of a somatic mutation, can occur after treatment with alkylating agents or irradiation in low doses.

(e) Inhibitors of RNA-polymerase are able to inhibit relatively selective gene expression.

(f) Different teratogens, e.g. thalidomide, are suspected to inhibit enzymes which are crucial for certain phases of cytodifferentiation.

(g) Changed membrane properties are suggested to be at least partly responsible for the teratogenic activity of vitamin A, retinoic acid and perhaps glucocorticoids in rodents and rabbits.

The properties of steroid sex hormones, contrary to other definite or suspected teratogens, are rather complex:

(a) The cellular action is derived from specific binding of hormone receptor complexes to DNA and nuclear non-histone proteins, and unspecific binding (masking, reserve) to DNA and histone proteins[73,74]. The specific bindings are assumed to cause control – largely via stimulation of RNA-polymerases I and III – of gene induction, gene expression, modification of gene products on different levels, and proliferation of specific cell types (although activity of RNA-polymerases is not increased after sex steroid application[75]). These effects could alter actual gene expression or stimulate metabolism either in the maternal genital tract or in the embryo which would be harmful to normal embryonic development.

(b) Irreversible binding of steroid hormones to protein and DNA has been reported by Bolt and Kappus[76] especially for synthetic oestrogens and gestagens. Their results might give an explanation for different teratological properties of different sex steroids. However, in view of these results, it remains unclear, why under our conditions, positional changes of protein spots in two-dimensional electrophoresis were not detectable.

(c) Steroid sex hormones have been suspected to be cocarcinogenic, although an effect via specific inhibition of DNA repair capacity[68,69] was shown to be unlikely[77]. A coteratogenic effect might be responsible for undesired sex steroid effects. However, our results of the combined treatment with CA and substances affecting genetic material (BrdU, MNU, CPA) suggest that no interactive effect on DNA alteration is present. The above mentioned multiple malformations cannot be associated with additional effects on DNA damage.

(d) A possible effect on membrane properties mediated for example by glucocorticoid effects have been mentioned before[33,78]. However, the supposed link between cleft palate and sex steroid application during pregnancy in humans[22,26] and the recent finding of palatal glucocorticoid receptors in human fetuses[79] require a re-evaluation of this hypothesis.

(e) The possibility of a hormonal imbalance after treatment cannot be excluded by our experiments. This effect would agree with former studies on the maintenance of pregnancy in rodents after castration and incomplete maintenance of hormone levels[80–82]. However, this concept seems to be unlikely since the profiles of specific teratological events in our experiments are similar to those found after treatment with other teratogens. Furthermore, the types of malformations and the high number of resorptions in hormone deficient pregnancies are not identical with those after hormone treatment without castration.

The evaluation of teratogenic mechanisms of steroid sex hormones is not conclusive at the present time. Many biochemical and teratogenic investigations have to be performed to permit an ultimate statement. It is always difficult to extrapolate from animal experiments and apply the results to the human situation[32,33,35]. However, results such as those obtained in this study should contribute to the understanding of the mode of action of steroid hormone treatment during pregnancy. The widespread use of steroid sex hormones for prolonged or short periods during or prior to a possible pregnancy necessitates that appropriate precautions are taken. Although many reports during the last decade could not conclusively document a relationship between the use of steroid sex hormones and teratogenicity in humans, our results support the demand for a better controlled use of these drugs.

ACKNOWLEDGEMENTS

The authors thank Dr T. E. Kwasigroch and Dr R. G. Skalko for helpful discussion. We also wish to express our appreciation to Mrs U. Jacob-Müller for expert technical assistance and to Mrs E. Gottschalk for her work in revision of this manuscript. This work was supported by grants of Deutsche Forschungsgemeinschaft awarded to the Sonderforschungsbereich 29 'Embryonale Entwicklung und Differenzierung'.

References

1. Heinonen, O. P., Sloane, D. and Shapiro, S. (1977). *Birth Defects and Drugs in Pregnancy.* (Littleton, Mass: Publishing Sciences Group)
2. Bongiovanni, A. M., DiGeorge, A. M. and Grumbach, M. M. (1959). Masculinization of the female infant associated with estrogenic therapy alone during gestation. Four cases. *J. Clin. Endocrinol.*, **19**, 1004
3. Wilkins, L. (1960). Masculinization of female fetus due to use of orally given progestins. *J. Am. Med. Assoc.*, **172**, 1028
4. Hagler, S., Schultz, A., Hankin, H. and Kunstadter, R. H. (1963). Fetal effects of steroid therapy during pregnancy. *Am. J. Dis. Children*, **106**. 586
5. Voorhess, M. L. (1967). Masculinization of the female fetus associated with norethindrone–mestranol therapy during pregnancy. *J. Pediatr.*, **71**. 128
6. Gal, I., Kirman, B. and Stern, J. (1967). Hormonal pregnancy tests and congenital malformation. *Nature (Lond.)*, **216**, 83
7. Gal, I. (1972). Risks and benefits of the use of hormonal pregnancy test tablets. *Nature (Lond.)*, **240**, 241
8. Laurence, M., Miller, M., Vowles, M., Evans, K. and Carter, C. (1971). Hormonal pregnancy tests and neural tube malformations. *Nature (Lond.)*, **233**. 495
9. Mitchell, S. C., Sellmann, A. H., Westphal, M. C. and Park, J. (1971). Etiologic correlates in a study of congenital heart disease in 56109 births. *Am. J. Cardiol.*, **28**. 653
10. Nora, J. J. and Nora, A. H. (1973). Birth defects and oral contraceptives. *Lancet*, i, 941
11. Levy, E. P., Cohen, A. and Fraser, F. C. (1973). Hormone treatment during pregnancy and congenital heart defects. *Lancet*, i, 611
12. Mulvihill, J. J., Mulvihill, C. G. and Neill, C. A. (1974). Congenital heart defects and prenatal sex hormones. *Lancet*, i. 1168
13. Yasuda, M. and Miller, J. R. (1975). Prenatal exposure to oral contraceptives and transposition of the great vessels in man. *Teratology*, **12**. 239
14. Kaufmann, R. L. (1973). Birth defects and oral contraceptives. *Lancet*, i, 1396
15. Balci, S., Say, B., Pirnar, T. and Hicsonmez, A. (1973). Birth defects and oral contraceptives. *Lancet*, ii, 1098
16. Janerich, D. T., Piper, J. M. and Glebatis, D. M. (1973). Hormones and limb-reduction deformities. *Lancet*, ii. 96
17. Janerich, D. T., Piper, J. M. and Glebatis, D. M. (1974). Oral contraceptives and congenital limb-reduction defects. *N. Engl. J. Med.*, **291**, 697
18. Jaffe, P., Liberman, M. M., McFadyen, I. and Valman, H. B. (1975). Incidence of congenital limb-reduction deformities. *Lancet*, i, 526
19. Peterson, W. F. (1969). Pregnancy following oral contraceptive therapy. *Obstet. Gynecol.*, **34**, 363
20. Robinson, S. C. (1971). Pregnancy outcome following oral contraceptives. *Am. J. Obstet. Gynecol.*, **109**, 354
21. Janerich, D. T. (1975). The pill and subsequent pregnancies. *Lancet*, i, 681
22. Harlap, S., Prywes, R. and Davies, A. M. (1975). Birth defects and estrogens and progesterones in pregnancy. *Lancet*, i. 682
23. Royal College of General Practitioners Oral Contraceptive Study. (1976). The outcome of pregnancy in former oral contraceptive users. *Br. J. Obstet. Gynecol.*, **83**. 608
24. Nora, J. J. and Nora, A. H. (1976). Prospective study of infants born to mothers receiving exogenous hormones in pregnancy. *Teratology*, **13**. A-32
25. Goujard, J. and Rumeau-Rouquette, C. (1977). First trimester exposure to progestagen/estrogen and congenital malformations. *Lancet*, i, 482
26. Greenberg, G., Inman, W. H. W., Weatherall, J. A. C., Adelstein, A. M. and Haskey, J. C. (1977). Maternal drug histories and congenital abnormalities. *Br. Med. J.*, **2**, 853
27. Heinonen, O. P., Slone, D., Monson, R. R., Hook, E. B. and Shapiro, S. (1977). Cardiovascular birth defects and antenatal exposure to female sex hormones. *N. Engl. J. Med.*, **296**, 67
28. Wilson, J. G. (1973). *Environment and Birth Defects.* (New York and London: Academic Press)
29. Nishihara, G. (1958). Influence of female sex hormones in experimental teratogenesis. *Proc. Soc. Exp. Biol. Med.*, **97**. 809

30. Wei Cheng, D. (1959). Effect of progesterone and estrone on the incidence of congenital malformations due to maternal vitamin E deficiency. *Endocrinology*, **64**, 270
31. Roy, S. K. and Kar, A. B. (1967). Foetal effect of norethynodrel in rats. *Ind. J. Exp. Biol.*, **5**, 14
32. Gidley, J. T., Christensen, H. D., Hall, I. H., Palmer, K. H. and Wall, M. E. (1970). Teratogenic and other effects produced in mice by norethynodrel and its 3-hydroxymetabolites. *Teratology*, **3**, 339
33. Andrew, F. D. and Staples, R. E. (1977). Prenatal toxicity of medroxyprogesterone acetate in rabbits, rats, and mice. *Teratology*, **15**, 25
34. Revesz, C., Chappel, C. I. and Gaudry, R. (1960). Masculinization of female fetuses in the rat by progestational compounds. *Endocrinology*, **66**, 140
35. Takano, K., Yamamura, H., Suzuki, M. and Nishimura, H. (1966). Teratogenic effect of chlormadinone acetate in mice and rabbits. *Proc. Soc. Exp. Biol. Med.*, **121**, 455
36. Whitten, W. K. (1957). The effect of progesterone on the development of mouse eggs *in vitro*. *J. Endocrinol.*, **16**, 80
37. Kirkpatrick, J. F. (1971). Differential sensitivity of preimplantation mouse embryos *in vitro* to estradiol and progesterone. *J. Reprod. Fertil.*, **27**, 283
38. Daniel, J. C. (1964). Some effects of steroids on cleavage of rabbit eggs *in vitro*. *Endocrinology*, **75**, 706
39. Daniel, J. C. and Levy, J. D. (1964). Action of progesterone as a cleavage inhibitor of rabbit ova *in vitro*. *J. Reprod. Fertil.*, **7**, 323
40. McGaughey, R. W. and Daniel, J. C. (1966). Effect of estradiol-17β on fertilized rabbit eggs *in vitro*. *J. Reprod. Fertil.*, **11**, 325
41. Neumann, F. (1977). Pharmacology and potential use of cyproterone acetate. *Horm. Metab. Res.*, **9**, 1
42. Wilkins, L., Jones, H. W., Holman, G. H. and Stempfel, R. S. (1958). Masculinization of the female fetus associated with administration of oral and intramuscular progestins during gestation: non-adrenal female pseudohermaphrodism. *J. Clin. Endocrinol.*, **18**, 559
43. Whitten, W. H. (1971). Nutrient requirements for the culture of preimplantation embryos *in vitro*. In: G. Raspé (ed.). *Advances in the Biosciences*, **6**, pp. 129–139. (Oxford, Braunschweig: Pergamon Press, Vieweg Verlag)
44. Eibs, H.-G. and Spielmann, H. (1977). Preimplantation embryos. Part II: Culture and transplantation. In: D. Neubert, H.-J. Merker and T. E. Kwasigroch (eds.). *Methods in Prenatal Toxicology*, pp. 221–230. (Stuttgart: Thieme Verlag)
45. Sherman, M. I. (1975). Long term culture of cells derived from mouse blastocysts. *Differentiation*, **3**, 51
46. Wiley, L. M. and Pedersen, R. A. (1977). Morphology of mouse egg cylinder development *in vitro*: a light and electron microscopic study. *J. Exp. Zool.*, **200**, 389
47. Van Blerkom, J. and Brockway, G. O. (1975). Qualitative patterns of protein synthesis in the preimplantation mouse embryo. I. Normal pregnancy. *Dev. Biol.*, **44**. 148
48. O'Farrell, P. H. (1975). High resolution two-dimensional electrophoresis of proteins. *J. Biol. Chem.*, **250**, 4007
49. Klose, J. (1975). Protein mapping by combined isoelectric focusing and electrophoresis of mouse tissues. *Humangenetik*, **26**, 231
50. Laskey, R. A. and Mills, A. D. (1975). Quantitative film detection of ^3H and ^{14}C in polyacrylamide gels by fluorography. *Eur. J. Biochem.*, **56**, 335
51. Wilson, J. G. (1965). Embryological considerations in teratology. In: J. G. Wilson and J. Warkany (eds.). *Teratology: Principles and Techniques*, pp. 251–277. (Chicago: University of Chicago Press)
52. Eibs, H.-G. and Spielmann, H. (1977). Inhibition of post-implantation development of mouse blastocysts *in vitro* after cyclophosphamide treatment *in vivo*. *Nature* (*Lond.*), **270**, 54
53. Spielmann, H. and Eibs, H.-G. (1978). Recent progress in teratology: a survey of methods for the study of drug actions during the preimplantation period. *Arzneim. Forsch. Drug Res.*, **28**, 1733
54. Tuchmann-Duplessis, H. (1974). Teratogenic screening methods and their application to contraceptive products. *Acta Endocrinol.*, **185 (Suppl.)**, 203
55. Jentsch, D., Schulz, V. and Wendt, H. (1976). Pharmakokinetik von Cyproteronacetat bei

gesunden Probanden nach intramuskulärer und oraler Applikation. *Arzneim. Forsch.*, **26**, 914

56. Spielmann, H., Eibs, H.-G., Jacob, U. and Nagel, D. (1978). Teratological studies on effects of carbon-13 incorporation into preimplantation mouse embryos on development after implantation *in vivo* and *in vitro*. In: T. A. Baillie (ed.). *Stable Isotopes. Application in Pharmacology, Toxicology and Clinical Research*, pp. 217–225. (London: Macmillan Press)
57. Borland, R. M., Erickson, G. F. and Ducibella, T. (1977). Accumulation of steroids in rabbit preimplantation blastocysts. *J. Reprod. Fertil.*, **49**, 219
58. Sherman, M. I. and Atienza, S. B. (1977). Production and metabolism of progesterone and androstenedione by cultured mouse blastocysts. *Biol. Reprod.*, **16**, 190
59. Antila, E., Koskinen, J., Niemelä, P. and Saure, A. (1977). Steroid metabolism by mouse preimplantation embryos *in vitro*. *Experientia*, **33**, 1374
60. Van Blerkom, J., Barton, S. C. and Johnson, M. H. (1976). Molecular differentiation in the preimplantation mouse embryo. *Nature (Lond.)*, **259**, 319
61. Johnson, M. H., Handyside, A. H. and Braude, P. R. (1977). Control mechanisms in early mammalian development. In: M. H. Johnson (ed.). *Development in Mammals, Vol. 2*, pp. 67–97. (Amsterdam: North-Holland Publishing Company)
62. Pinsky, L. and DiGeorge, A. M. (1965). Cleft palate in the mouse: a teratogenic index of glucocorticoid potency. *Science*, **147**, 402
63. Andrew, F. D., Christensen, H. D., Williams, T. L., Thompson, M. G. and Wall, M. E. (1973). Comparative teratogenicity of contraceptive steroids in mice and rats. *Teratology*, **7**, A-11
64. Roberts, I. F. and West, R. J. (1977). Teratogenesis and maternal progesterone. *Lancet*, ii, 982
65. Spielmann, H. (1976). Embryo transfer technique and action of drugs on the preimplantation embryo. *Curr. Topics Pathol.*, **62**, 87
66. Forsberg, J.-G. (1975). Late effects in the vaginal and cervical epithelia after injections of diethylstilbestrol into neonatal mice. *Am. J. Obstet. Gynecol.*, **121**, 101
67. Robboy, S. J., Scully, R. E., Welch, W. R. and Herbst, A. L. (1977). Intrauterine diethylstilbestrol exposure and its consequences. *Arch. Pathol. Lab. Med.*, **101**, 1
68. Gaudin, D., Gregg, R. S. and Yielding, K. L. (1971). DNA repair inhibition: a possible mechanism of action of co-carcinogens. *Biochem. Biophys. Res. Commun.*, **45**, 630
69. Gaudin, D., Guthrie, L. and Yielding, K. L. (1974). DNA repair inhibition: a new mechanism of action of steroids with possible implications for tumor therapy. *Proc. Soc. Exp. Biol. Med.*, **146**, 401
70. Skalko, R. G., Packard, D. S., Schwendimann, R. N. and Raggio, J. F. (1971). The teratogenic response of mouse embryos to 5-bromodeoxyuridine. *Teratology*, **4**, 87
71. Packard, D. S., Menzies, R. A. and Skalko, R. G. (1973). Incorporation of thymidine and its analogue, bromodeoxyuridine, into embryos and maternal tissues of the mouse. *Differentiation*, **1**, 397
72. Packard, D. S., Skalko, R. G. and Menzies, R. A. (1974). Growth retardation and cell death in mouse embryos following exposure to the teratogen bromodeoxyuridine. *Exp. Mol. Pathol.*, **21**, 351
73. Gorski, J. and Gannon, F. (1976). Current models of steroid hormone action: a critique. *Annu. Rev. Physiol.*, **38**, 425
74. O'Malley, B. W., Schwartz, R. J. and Schrader, W. T. (1976). A review of regulation of gene expression by steroid hormone receptors. *J. Steroid Biochem.*, **7**, 1151
75. Warner, C. M. and Tollefson, C. M. (1977). The effect of estradiol on RNA synthesis in preimplantation mouse embryos cultured *in vitro*. *Biol. Reprod.*, **16**, 627
76. Bolt, H. M. and Kappus, H. (1974). Irreversible binding of ethynylestradiol metabolites to protein and nucleic acids as catalyzed by rat liver microsomes and mushroom tyrosinase. *J. Steroid Biochem.*, **5**, 179
77. Cleaver, J. E. and Painter, R. B. (1975). Absence of specificity in inhibition of DNA repair replication by DNA-binding agents, cocarcinogens, and steroids in human cells. *Cancer Res.*, **35**, 1773
78. Fraser, F. C. and Fainstat, T. (1951). Production of congenital defects in the offspring of pregnant mice treated with cortisone. *Pediatrics*, **8**, 527

79. Goldman, A. S., Shapiro, B. H. and Katsumata, M. (1978). Human foetal palatal corticoid receptors and teratogens for cleft palate. *Nature* (*Lond.*), **272**, 464
80. Jost, A. (1964). Embryopathies d'origine hormonale. *Bruxelles Med.*, **44**, 245
81. Poulson, E., Robson, J. M. and Sullivan, F. M. (1965). Embryopathic effects of progesterone deficiency. *J. Endocrinol.*, **31**, xxviii
82. Kroc, R. L., Steinetz, B. G. and Beach, V. L. (1958–59). The effects of estrogens, progestagens, and relaxin in pregnant and non-pregnant laboratory rodents. *Ann. N.Y. Acad. Sci.*, **75**, 942

7
Hormonal teratogenesis in mammary glands of the mouse

K. HOSHINO

INTRODUCTION

The embryogenesis of the mammary glands of the mouse has been studied and also reviewed[1-3]. On the other hand, hormonal teratogenesis of the mouse mammary gland has not been extensively studied, except for the excellent work reported by Raynaud[3].

In our laboratory, mammary teratogenesis caused by various hormones has been investigated in mice. The mammary glands of neonatal mice are too small and fragile to investigate morphologically using whole mount preparation. Therefore, in our series of experiments, prenatally treated young mice were raised until three weeks of age and then their mammary glands were examined. The postnatal growth of mammary glands in untreated immature mice was first studied and used as the control data for the experimental work.

POSTNATAL MAMMARY GROWTH IN UNTREATED MICE

Mammary glands were examined in their whole mount preparations. Normal male and female CC-F_2 young were used. The greatest dimension of each mammary gland was measured by a pair of calipers as shown in Figure 7.1, and its width was measured on the line crossing at right angle to the line along the greatest dimension.

All male mice did not have nipples (namely, athelia), whereas all female mice had nipples. Mammary duct systems were present in all the mammary glands of the female mice, whereas some mammary glands of male mice lacked mammary parenchyma (amastia) as well. The fifth (puboinguinal) pairs of mammary glands were often damaged during preparation of the whole mount preparation, including skinning procedures. Therefore, the data concerning the fifth pairs and other damaged mammary glands were excluded from analysis.

The incidence of amastia in 48 normal male mice is summarized in Table 7.1. Mammary glands of the second (axillary) and fourth (abdominal) pairs were significantly more frequently absent than those of the first (cervical) pairs.

Table 7.1 The incidence of absence of mammary glands in normal male CC-F_2 mice

Location of mammary glands	*Incidence of absent mammary glands*	
	No.	*%*
1st pair	6/48	12.5
2nd pair	15/48	31.2
3rd pair	13/48	27.1
4th pair	18/48	37.5

Statistical analyses: 1st < 2nd ($p < 0.05$); 1st < 4th ($p < 0.005$)

The greatest dimensions (Figure 7.1) and widths of the mammary glands were measured in 613 glands of male mice and in 827 glands of female mice at varying ages after birth ranging from 2 days to 87 days of age.

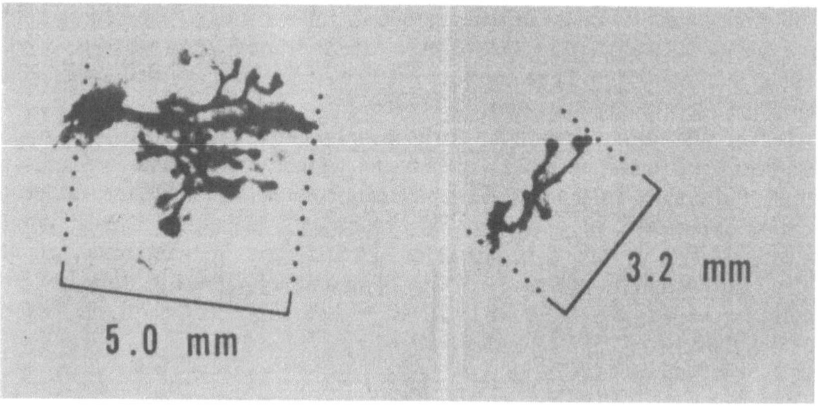

Figure 7.1 Measurements of the greatest dimension of growth extent of mammary glands in mice

In female mice, mammary glands increased in size steadily from birth, but levelled off from 18 to 21 days of age. From 21 days of age, the mammary glands grew rapidly and steadily until young adulthood (Table 7.2). Also in male mice, mammary growth levelled off from 18 days of age and resumed from the fifth week of age. After the eighth week of age, the mammary glands did not increase in size in males (Table 7.3). From this data, it was decided to take the measurements of the mammary glands of the experimental mice at 3 weeks of age.

Table 7.2 Growth extent of mammary glands of normal female mice at different ages

Group no.	Age		No. of m.gl.	Greatest dimension (mm)		Width (mm)	
	Week	Day		Mean ± S.E.	Range	Mean ± S.E.	Range
I	1	2–6	49	2.7 ± 0.02	1.6–3.8	1.4 ± 0.05	0.6–2.5
II	2	8–14	52	3.2 ± 0.07	1.8–5.0	1.6 ± 0.06	0.6–2.7
III	3	18	64	4.0 ± 0.11	1.8–6.5	2.3 ± 0.08	1.1–4.0
IV	3	21	212	3.9 ± 0.06	1.3–5.7	2.1 ± 0.04	0.5–4.0
V	4	23	41	4.6 ± 0.14	2.9–7.0	2.4 ± 0.12	1.1–4.0
VI	4	25	84	5.0 ± 0.10	2.0–6.8	2.9 ± 0.10	0.5–5.0
VII	4	28	46	6.0 ± 0.29	2.8–10.8	3.8 ± 0.26	1.8–8.8
VIII	5	29–35	89	9.0 ± 0.28	4.7–16.8	5.2 ± 0.21	2.0–12.5
IX	6	36–42	82	12.9 ± 0.41	4.6–20.0	7.9 ± 0.11	1.8–15.9
X	7	45–49	52	16.8 ± 0.58	7.8–25.0	10.3 ± 0.74	3.6–18.5
XI	8–9	53–61	56	17.3 ± 0.46	7.5–30.6	9.4 ± 0.47	3.8–18.3
		Total	827				
	Statistical analyses:			I < II	($p<0.001$)	I < II	($p<0.025$)
				II < III	($p<0.001$)	II < III	($p<0.001$)
				IV < V	($p<0.001$)	IV < V	($p<0.005$)
				V < VI	($p<0.05$)	V < VI	($p<0.005$)
				VI < VII	($p<0.001$)	VI < VII	($p<0.001$)
				VII < VIII	($p<0.001$)	VII < VIII	($p<0.001$)
				VIII < IX	($p<0.001$)	VIII < IX	($p<0.001$)
				IX < X	($p<0.001$)	IX < X	($p<0.001$)

Table 7.3 Growth extent of mammary glands of normal male mice at different ages

Group no.	*Age*		*No. of m.gl.*	*Greatest dimension* (mm)		*Width* (mm)	
	Week	*Day*		*Mean* ± *S.E.*	*Range*	*Mean* ± *S.E.*	*Range*
I	1	2–6	53	1.4 ± 0.06	0.5–2.8	0.8 ± 0.04	0.2–1.7
II	2	8–14	32	1.4 ± 0.04	0.6–3.3	0.7 ± 0.07	0.2–1.6
III	3	18	31	1.7 ± 0.08	0.9–2.8	0.9 ± 0.07	0.2–1.8
IV	3	21	109	1.7 ± 0.06	0.4–2.7	0.8 ± 0.04	0.1–2.4
V	4	25–28	38	1.7 ± 0.12	0.4–3.4	0.9 ± 0.10	0.2–2.7
VI	5	31–34	32	2.1 ± 0.21	0.8–3.9	0.9 ± 0.07	0.2–1.8
VII	6	38–42	55	3.4 ± 0.18	0.8–7.3	1.6 ± 0.13	0.3–6.0
VIII	7	44–47	31	3.6 ± 0.30	1.1–9.5	1.8 ± 0.22	0.3–4.0
IX	8	50–55	56	3.4 ± 0.29	0.6–8.6	1.7 ± 0.18	0.2–6.7
X	9	58–63	39	3.1 ± 0.29	0.9–7.0	1.6 ± 0.21	0.3–5.8
XI	10	66–70	37	2.9 ± 0.22	0.5–8.0	1.2 ± 0.16	0.3–5.8
XII	11	75	31	3.1 ± 0.24	1.1–6.9	1.5 ± 0.15	0.5–3.7
XIII	12	80	31	3.0 ± 0.23	1.1–5.2	1.4 ± 0.13	0.4–3.3
XIV	13	87	38	3.9 ± 0.31	1.4–10.0	2.0 ± 0.18	0.6–4.8
		Total	613	II < III	($p < 0.05$)	II < III	($p < 0.05$)
Statistical analyses:				VI < VII	($p < 0.001$)	VI < VII	($p < 0.001$
						XIII < XIV	($p < 0.01$)

MAMMARY TERATOGENESIS BY 17β-OESTRADIOL BENZOATE

There are two types of structural anomalies in mouse mammary glands, athelia and amastia. However, the mammary gland with athelia is of a typical male type. A typical female type of mammary gland consists of a nipple and its connected duct system.

Experimental set-up

As shown in Table 7.4, the influence of 17β-oestradiol benzoate (OB) upon the progress of pregnancy in CC-F_2 hybrid mice was studied. Among 34 surviving litters, the young in 33 litters were raised until 3 weeks of age and killed. Morphological configurations of their mammary glands were examined by whole mount preparations and summarized in Table 7.5.

Table 7.4 Influence of 17β-oestradiol benzoate upon the progress of pregnancy in CC-F_2 hybrid mice

Daily dose (μg)	*Day of pregnancy when infected subcutaneously*	*Number of cases*	*Abortion*	*Stillbirth*	*Surviving litter*
None		6	0	0	6 (100%)
Sesame oil 0.02 ml	9 10 11 12 13 14	5	0	0	5 (100%)
OB 0.05 μg	8 9	11	5 (45.5%)	0	6 (54.5%)
OB 0.05 μg	9 10	7	6 (85.7%)	0	1 (14.3%)
OB 1.0 μg	11 12	11	0	0	11 (100%)
OB 1.0 μg	13 14	8	1 (12.5%)	2 (25.0%)	5 (62.5%)

Table 7.5 Mammary teratogenesis by 17β-oestradiol benzoate in CC-F_2 hybrid mice

Prenatal hormonal treatment	In female mice					In male mice				
	No. of mammary glands	*No nipple no duct*	*+ Nipple + duct*	*No nipple + duct*	*+ Nipple no duct*	*No. of mammary glands*	*No nipple no duct*	*+ Nipple + duct*	*No nipple + duct*	*+ Nipple no duct*
No injection	144	0	144 (100%)	0	0	176	56 (32%) a	0	120 (68%) A	0
Sesame oil Days 9–14	120	0	120 (100%)	0	0	136	36 (27%) b	0	100 (74%) B	0
OB 0.5 μg Days 8–9	160	5 (3%) x	153 (96%) y	0	2 (1%)	112	52 (46%) c	1 (1%)	59 (53%) C	0
OB 1.0 μg Days 11–12	224	0	224 (100%)	0	0	184	75 (41%) d	0	109 (59%) D	0
OB 1.0 μg Days 13–14	120	0	120 (100%)	0	0	104	39 (38%)	0	65 (63%)	0

Statistical differences: x and y are significantly different from others in each category:
$a < c, b < c, a < d, b < d, A < C, B < C, A < D, B < D$

In the female, two daily injections of 0.5 μg OB on days 8 and 9 of gestation caused a significant decrease in the number of mammary glands of a typical female type. It was due to the adverse effect of this hormone which inhibited the growth and development of the nipples and also the duct system, resulting in athelia together with amastia (five cases) and two other cases of a very rare form of mammary glands. The latter consisted only of the nipple without the duct system. No adverse effect was exerted even when 1.0 μg of OB was administered daily either on days 11 and 12 or days 13 and 14.

In the male, the most detrimental effects were exerted by OB when administered on days 8 and 9. The number of cases of amastia together with athelia were significantly increased over the controls. When 1.0 μg of OB was injected daily on days 11 and 12, the same anomaly was increased significantly over the controls which were treated with sesame oil alone. Interestingly, a typical female type of mammary gland developed in one of the male mice so treated (Figure 7.2).

The persisting mammary glands in these experimental mice were measured at their greatest dimension (Table 7.6). The mammary gland in mice which were exposed to 0.5 μg OB on days 8 and 9 were greatest among those examined. Coincidentally, the body weights of these mice were also heavier than others. A similar tendency was observed in the male mice. In some of both sexes, nipple development was accelerated (Table 7.6).

These data indicate that exposure of the fetuses to OB on days 8 and 9 of gestation was most detrimental to mammary embryogenesis. However, once the mammary glands survived on prenatal exposure to OB, their postnatal growth was greater than either the controls without OB exposure or those treated at later periods of pregnancy.

Raynaud[3] reviewed in detail the teratogenic effects of oestrogens upon mammary glands in mice. The premature development of the nipple and a total or partial suppression of the development of the mammary anlagen were the main causes of mammary malformations in mice. Raynaud also described microscopic anomalies occurring in fetal mammary glands exposed to oestrogens. Administering much larger doses of oestradiol dipropionate, either

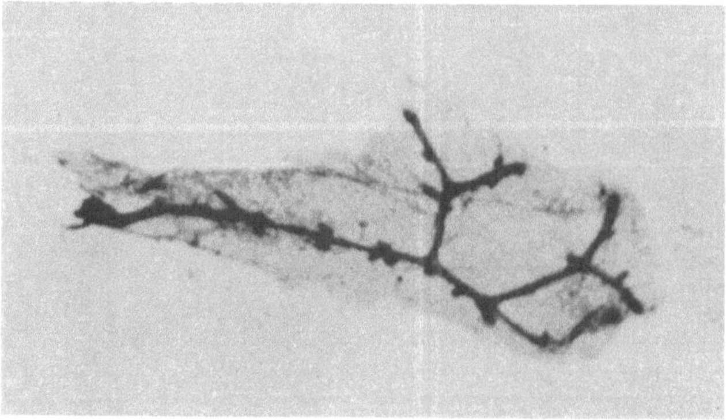

Figure 7.2 A female type mammary gland having a nipple found in a male mouse treated prenatally with 17β-oestradiol benzoate

Table 7.6 Greatest dimension of persisting mammary glands and body weights of 3-week-old mice prenatally exposed to 17β-oestradiol benzoate

Prenatal hormonal treatment	*In female mice*			*In male mice*		
	Greatest dimension of mammary gland	*No. of glands*	*Body weight* (range)	*Greatest dimension of mammary gland*	*No. of glands*	*Body weight* (range)
No injection	3.5 ± 0.08	144	7.6 g (6.0–11.0)	1.6 ± 0.06	120	7.7 g (6.5–9.5)
Sesame oil Days 9–14	3.5 ± 0.08	120	7.9 g (7.0–11.5)	1.5 ± 0.06	100	8.7 g (7.0–12.0)
OB 0.5 μg Days 8–9	3.7 ± 0.10	151	8.8 g (6.0–11.0)	2.0 ± 0.11	60	9.1 g (8.0–10.5)
OB 1.0 μg Days 11–12	3.7 ± 0.06	223	8.9 g (7.1–11.0)	1.6 ± 0.06	109	8.4 g (7.0–10.0)
OB 1.0 μg Days 13–14	3.6 ± 0.07	120	7.9 g (7.0–9.0)	1.5 ± 0.08	65	8.2 g (7.0–9.0)

directly into the fetuses (40 to 150 μg per fetus) or to pregnant mice (50 to 1000 μg per mother), a great number of mammary malformations were induced. Raynaud[3] stated that the mammary anlagen were very sensitive to oestrogen between days 12 and 14, but apparently he did not inject the hormone earlier than day 12. In our experiment, the mammary anlagen seemed to be more sensitive to oestrogen on days 8 and 9 than on later days of pregnancy.

MAMMARY TERATOGENESIS BY PROGESTERONE

Experimental set-up

Young of the CBA strain of mice which had been exposed to progesterone (10 mg daily from days 7 to 13 via subcutaneous injections to pregnant mothers) and survived were killed, skinned, and examined for teratogenesis in their mammary glands. Those of the untreated control CBA young were used as controls.

Observations and discussion

The frequency of the absence of mammary glands (athelia together with amastia) in the control and the experimental male young is summarized in Figure 7.3. In the controls, the frequency of total absence of mammary glands was highest in the second (axillary) pairs, followed by the third (thoracic) and the fourth (abdominal) pairs, and least in the first (cervical) pairs.

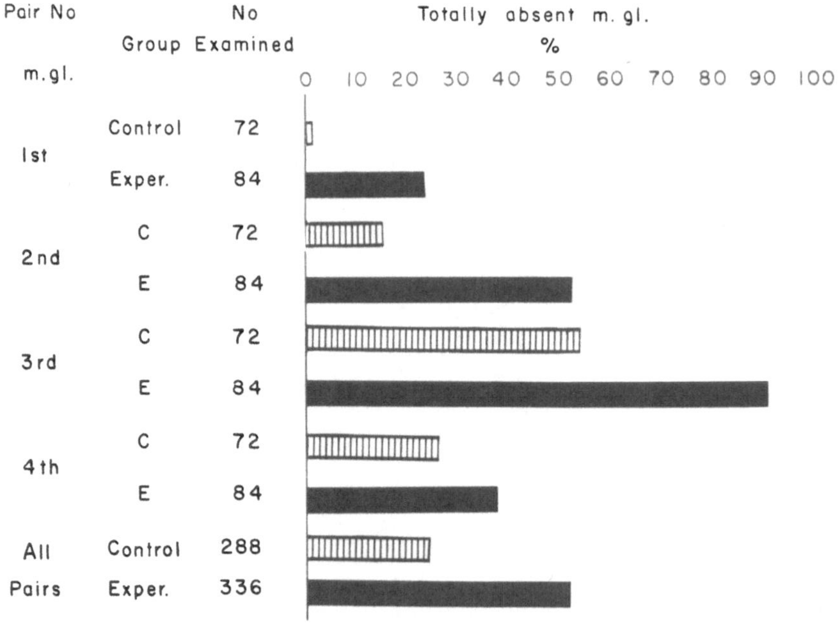

Figure 7.3 Occurrence of absence of mammary glands in male mice exposed prenatally to progesterone and in controls

This tendency remained after exposure of the mammary anlagen *in utero* to progesterone, although the fetal development of the second pairs was most severely arrested. Prenatally administered progesterone did not impair fetal development of the mammary glands in the female young.

In the persisting mammary glands, various types of structural configurations were detected. In addition to those found following prenatal exposure of the mammary anlagen to oestrogen, a new type of configuration was observed. This type of mammary gland consisted of the nipple and the mammary duct system, but they were not connected with each other (Figure 7.4). The frequency of the different types of structural configurations of the persisting mammary glands of male (Table 7.7) and female (Table 7.8) mice is summarized.

In male mice, the male type, consisting of the duct system without the nipple, was significantly decreased particularly in the third (thoracic) and the second (axillary) pairs. On the other hand, the female type, consisting of the nipple and its connected duct, was significantly increased. In addition, the rare type, namely the presence of the nipple alone, was remarkably increased particularly in the third and second pairs. Another type consisting of the nipple and its disconnected duct system occurred in the first and the fourth pairs. These results indicate stimulation of the fetal development of the nipple and the feminization of mammary embryogenesis in male mice by transplacentally administered progesterone.

In the female mice, all the mammary glands had their nipples, but their

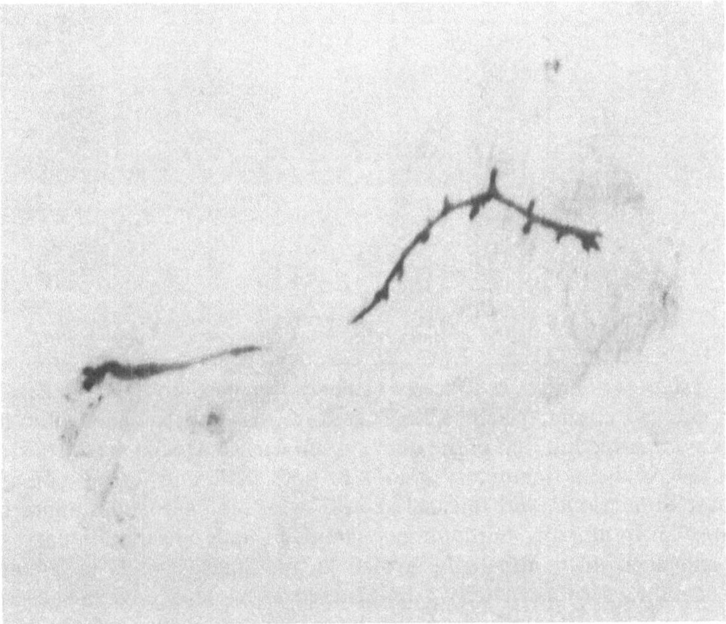

Figure 7.4 A malformed mammary gland, consisting of a nipple and a disconnected duct system in a male mouse prenatally treated with progesterone

Table 7.7 **Frequency of different types of configuration of persisting mammary glands in male CBA mice exposed prenatally to progesterone**

Pair no. m.gl.	*Group*	*Total no. of m.gl.*	I *no nipple + duct*	II *+ nipple no duct*	III *+ nipple disconnected*	IV *+ nipple + duct*
1st	Control	71	97.2%	0	0	2.8%
	Exper.	64	65.6%	6.3%	7.8%	20.3%
2nd	Control	61	98.4%	0	0	1.6%
	Exper.	40	57.5%	27.5%	0	15.0%
3rd	Control	33	100%	0	0	0
	Exper.	7	28.6%	57.1%	0	14.3%
4th	Control	53	100%	0	0	0
	Exper.	52	94.2%	0	1.9%	3.9%
All pairs	Control	218	98.6%	0	0	1.4%
	Exper.	163	71.1%	11.7%	3.7%	13.5%

Table 7.8 Frequency of different types of configuration of mammary glands in female CBA mice exposed prenatally to progesterone

Pair no. m.gl.	*Group*	*Total no. of m.gl.*	I *no nipple + duct*	II *+ nipple no duct*	III *+ nipple disconnected duct*	IV *+ nipple + duct*
1st	Control	52	0	0	0	100%
	Exper.	88	0	0	0	100%
2nd	Control	52	0	0	0	100%
	Exper.	88	0	0	1.1%	98.9%
3rd	Control	52	0	0	0	100%
	Exper.	88	0	5.7%	2.3%	92.0%
4th	Control	52	0	0	0	100%
	Exper.	88	0	0	0	100%
All pairs	Control	208	0	0	0	100%
	Exper.	352	0	1.4%	0.9%	97.7%

duct system was either completely (amastia) or partially (disconnected from their nipples) absent (Table 7.8). Therefore, the development of the nipples was not impaired but that of the duct system was somewhat arrested.

In the persisting mammary glands of both male and female young, their greatest dimensions were measured at 3 weeks of age and summarized in Figure 7.5. In all pairs of mammary glands in male mice, mammary growth was enhanced. In contrast, the growth of mammary glands in female mice was not altered by prenatally administered progesterone when these glands were examined postnatally. Because the mammary glands were not examined at birth, it is not known whether the fetal development of mammary glands of the male mice was impaired throughout their fetal periods or was stimulated

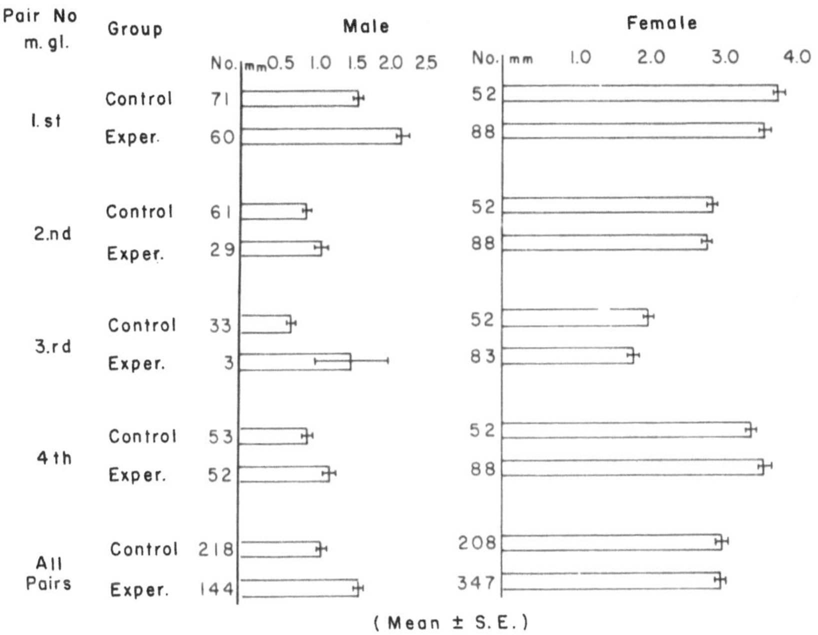

Figure 7.5 Greatest dimension of mammary glands of male and female CBA mice prenatally exposed to progesterone

during the late stages of gestation. Different from the mammary glands of female young, mammary growth of male mice was much accelerated. it is not known whether this acceleration was induced by the increased sensitivity of the mammary glands of male mice after intra-uterine exposure to progesterone or indirectly by the endocrine functions of the male mice altered *in utero* by prenatal progesterone treatment.

To the best of our knowledge, no reports comparable to the present experiments have been published.

MAMMARY TERATOGENESIS BY TESTOSTERONE PROPIONATE

Masculinization in mammary organogenesis was observed following prenatal treatments with androgen either directly into the fetuses or by injections to pregnant mice. Androgen inhibits the formation of the nipple and the development of some mammary anlagen[3]. Teratogenic effects of TP upon the mouse mammary anlagen were studied and reported previously from our laboratory[4]. A brief summary of this data is presented.

Experimental set-up

Single subcutaneous injections of 5 mg TP were given to CC-F_1 pregnant mice either 12, 15, 16 or 17 days after the detection of vaginal plugs. The mammary glands of the surviving young were examined at 3 weeks of age. In one experiment, duct-segments of vitally stained mammary glands, each 0.6 mm long, were isografted into the fourth mammary gland-free fat pads of the recipient female mice by a 'quantitative transplantation technique'[5]. The two donor CC-F_1 female mice had been prenatally treated at day 12 of pregnancy with 5 mg TP. Their mammary glands were masculinized and lacked nipples. As controls, the other two donor CC-F_1 females received no treatment. Mammary duct-segments obtained from the androgen prenatally treated donors were transplanted into the right side and those from the controls into the left side of the fourth mammary gland-free fat pads of the isologous CC-F_1 recipient mice. The recipients were mated with their siblings and killed shortly after parturition for morphological and histological examinations of the grafted mammary tissues.

Table 7.9 Frequency of nipples in seventy-seven 3-week-old female mice after single injections of 5 mg testosterone propionate were given to their mothers on different days of pregnancy

Day of pregnancy when androgen was given	*No. of persisting nipples / No. of mammary glands examined**		*No. of persisting nipples / No. of persisting mammary glands†*
12	2/72	2.8% (a)	3.8% (A)
15	167/176	94.9% (b)	97.1% (B)
16	161/200	80.5% (c)	86.1% (C)
17	168/168	100.0% (d)	100.0% (D)
Nontreated control	212/212	100.0%	100.0%

Statistical differences ($p<0.05$) between: (A) < (B), (a) < (b), (A) < (C), (a) < (c), (A) < (D), (a) < (d), (C) < (B), (c) < (b), (B) < (D), (b) < (d), (C) < (D), and (c) < (d)

* Numbers including both persisting and absent glands were examined

† Actual numbers of mammary glands are given in Table 7.10

Cited from Reference 4

Table 7.10 Frequency of the persisting mammary glands in 3-week-old male (64) and female (77) mice after single injections of 5 mg testosterone propionate were given to their mothers on different days of pregnancy

Day of pregnancy when androgen was given	*No. of persisting mammary glands / No. of mammary glands examined* — *Males*		*Females*	
12	55/80	68.8% (a)	53/72	73.6% (A)
15	107/128	83.6% (b)	172/176	97.7% (B)
16	106/160	66.3% (c)	187/200	93.5% (C)
17	102/144	70.8% (d)	168/168	100.0% (D)
Nontreated control	78/109	71.6% (e)	212/212	100.0% (E)

Statistical differences ($p<0.05$) ($\nless$ means the difference is not significant): (a) < (b), (c) < (b), (d) < (b), (e) < (b), (a) $\nless$ (A), (e) $\nless$ (A), (A) < (B), (A) < (C), (A) < (D), (C) < (B), (C) < (D)

Cited from Reference 4

Table 7.11 Frequency of the absence of mammary glands of the different pairs in 3-week-old male (64) and female (77) mice after single injections of 5 mg testosterone propionate were given to their mothers on different days of pregnancy

Day of pregnancy when androgen was given	*Males*					*Females*				
	No. of glands absent	*Percentage*				*No. of glands absent*	*Percentage*			
		1st pair	*2nd pair*	*3rd pair*	*4th pair*		*1st pair*	*2nd pair*	*3rd pair*	*4th pair*
12	25	16.0	40.0 (1)	12.0 (2)	32.0	19	26.3	21.1	15.8	36.8
15	21	9.5 (3)	42.9 (4)	28.6	19.0	4	25.0	0	50.0	25.0
16	54	14.8 (5)	40.8 (6)	14.8 (7)	29.6	13	7.7 (18)	53.9 (19)	7.7 (20)	30.7
17	42	11.9 (8)	40.5 (9)	11.9 (10)	35.7 (11)	0	—	—	—	—
Total	142	13.4 (12)	40.8 (13)	15.5 (14)	30.3 (15)	36	19.4	30.6	16.7	33.3
Nontreated control	31	12.9 (16)	35.5 (17)	19.4	32.2	0	—	—	—	—

Statistical differences ($p < 0.05$) between: (2) < (1), (3) < (4), (5) < (6), (7) < (6), (8) < (9), (10) < (9), (8) < (11), (10) < (11), (12) < (13), (14) < (13), (12) < (15), (14) < (15), (16) < (17), (18) < (19), and (20) < (19)

Cited from Reference 4

Observations and discussion

The formation of the nipples of male CC-F_1 mice was lacking in both control and treated groups. That of female young was also arrested by TP administered prenatally on day 12, but not on day 17 of gestation (Table 7.9).

In the male young, mammary anlagen persisted more often when exposed prenatally to TP on day 15 of pregnancy, whereas no enhancing effects of TP were observed when TP was given either on day 12, 16 or 17 (Table 7.10). In contrast, in the female young, fewer mammary anlagen persisted when TP was given on day 16 or earlier of gestation. When TP was given on day 12 of pregnancy, only 73.6% of the mammary glands in female young persisted, the number comparable to that in male mice of either the corresponding experimental group or the control group. TP given on day 15 was less detrimental to the development of the mammary duct system of female young than when injected on day 16 or day 12 of gestation (Table 7.10).

The differences in the occurrence of amastia among the four different pairs of mammary glands in male young were not altered by prenatal administration of TP. In control female young, no amastia was observed. However, after prenatal TP treatments, this male type of tendency was observed in the overall incidence of amastia in female young (Table 7.11).

Mammary growth of the female young prenatally exposed to TP on day 12 of pregnancy was arrested and remained smaller in size comparable to the

Table 7.12 Two-dimensional extent of mammary growth of 3-week-old male (64) and female (77) mice after single injections of 5 mg testosterone propionate were given to their mothers on different days of pregnancy

Day of pregnancy when androgen was given	*Male young*		*Female young*	
	Length	*Width*	*Length*	*Width*
	Mean ± *SE** (mm)		*Mean* ± *SE* (mm)	
12	1.7 ± 0.09 (1) (55)†	0.6 ± 0.05 (a) (55)	1.8 ± 0.08 (I) (53)	0.8 ± 0.05 (A) (53)
15	2.0 ± 0.07 (2) (107–1‡)	1.0 ± 0.04 (b) (107–1)	4.3 ± 0.03 (II) (172–3)	2.5 ± 0.07 (B) (172–3)
16	2.1 ± 0.07 (3) (106–1)	1.0 ± 0.04 (c) (105–1)	3.9 ± 0.08 (III) (187–4)	2.3 ± 0.06 (C) (187–4)
17	1.9 ± 0.07 (4) (102)	0.9 ± 0.04 (d) (102)	4.0 ± 0.07 (IV) (168–3)	2.4 ± 0.05 (D) (168–3)
Nontreated control	1.7 ± 0.06 (5) (109)	0.8 ± 0.04 (e) (109)	3.9 ± 0.06 (V) (212)	2.1 ± 0.04 (E) (212)

Statistical differences ($p < 0.05$) ($\nless$ means the difference is not significant): (1) < (2), (1) < (3), (1) $\nless$ (4), (1) $\nless$ (I), (5) $\nless$ (I), (5) < (2), (5) < (3), (5) $\nless$ (4), (I) < (II), (I) < (III), (I) < (IV), (I) < (V), (III) < (II), (IV) < (II), (V) < (II), (a) < (e), (a) < (b), (a) < (c), (a) < (d), (e) < (b), (e) < (c), (A) < (B), (A) < (C), (A) < (D), (A) < (E), (C) < (B), (E) < (B), (E) < (C), and (E) < (D)

* Standard error of mean

† Number of mammary glands measured

‡ Number to be subtracted indicates the number of mammary glands which have been damaged during skinning and omitted from record of measurements

Cited from Reference 4

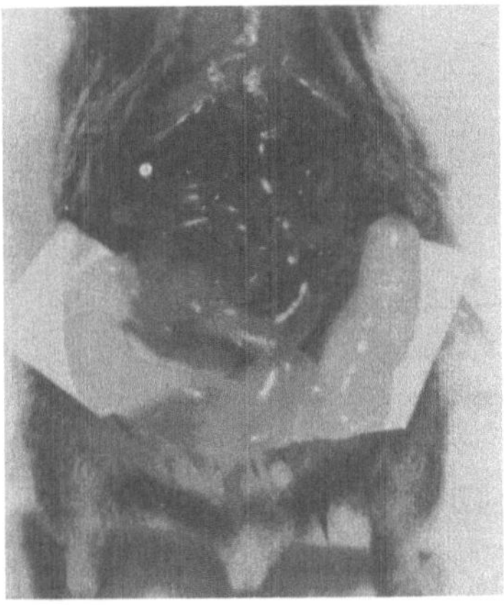

Figure 7.6 The uterine horns of a 3-week old female mouse exposed to 5 mg of testosterone propionate treatment on day 12 of fetal life via the mother are tremendously distended with serous fluids. No vaginal orifice was detected. × 1.875

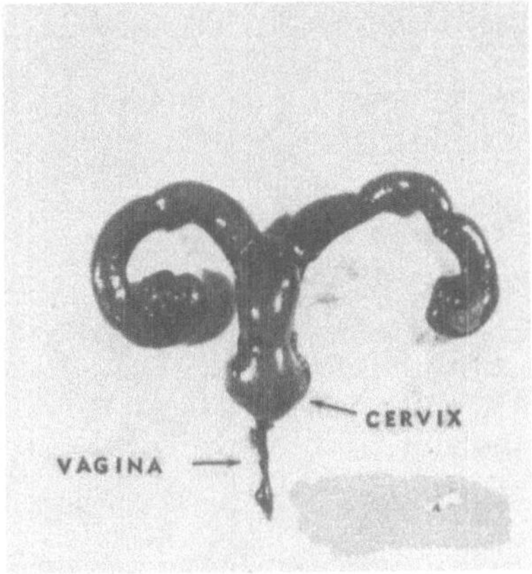

Figure 7.7 The distended uterine horns and atrophic lower vagina of the donor mouse described in the legend of Figure 7.8 are shown after trypan blue injection. Cited from Reference 4. × 1.5

size of mammary glands of male young in either the untreated control group or the experimental group prenatally exposed to TP on day 12 of pregnancy (Table 7.12). The growth of mammary glands persisting following prenatal exposure to TP on day 15 of pregnancy was significantly enhanced in female young (Table 7.12).

The organogenesis of the female genital tracts was arrested by TP given on day 12 of pregnancy. The uterine horns were all extremely distended and filled with serous fluid (Figure 7.6), and atresia vaginae were present (Figure 7.7).

Transplantability of mammary duct-segments was not impaired by prenatal exposure of mammary anlagen to TP on day 12 of pregnancy when the persisting mammary glands were isografted at 3 weeks of age (Table 7.13).

Morphological features of mammary grafts obtained from the donors prenatally exposed to TP were not distinguishable from those of the controls.

Table 7.13 Recovery of mammary glands, transplanted from 68-day-old donor mice that had been either exposed *in utero* through mothers given 5 mg testosterone propionate on day 12 of gestation or unexposed to androgen, from the female hosts shortly after their parturition

	No. of successfully grafted and milk-secreting mammary grafts /		*No. of mammary transplants*
	Series 1	*Series 2*	*Total*
Androgen-treated grafts	7/10 (a)	10/11(b)	17/21(c)
Untreated control grafts	4/10 (A)	11/11 (B)	15/21 (C)

There are no statistically significant differences between: (a) and (A), (b) and (B), or (c) and (C)

Cited from Reference 4

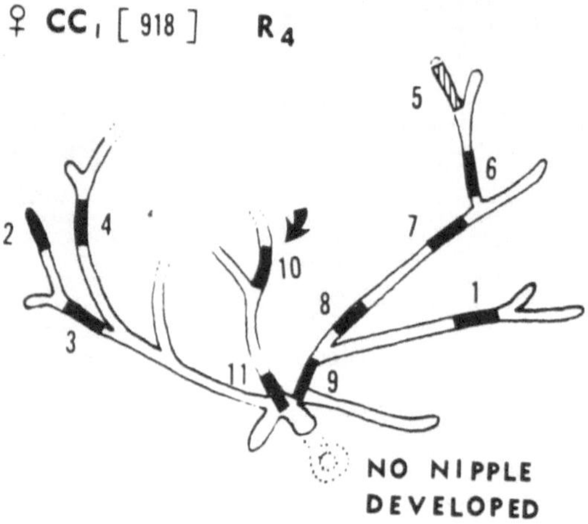

Figure 7.8

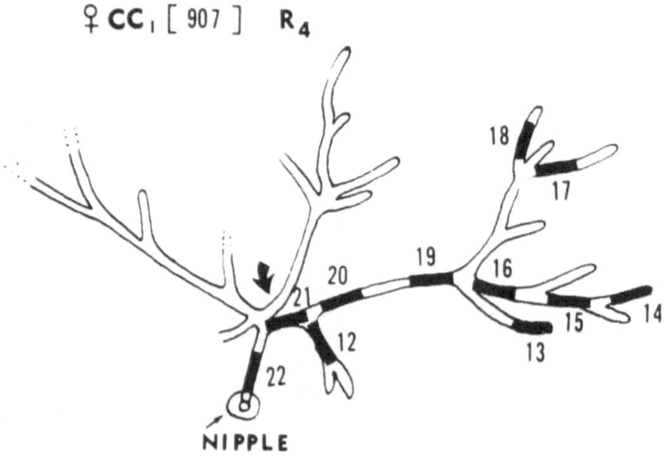

Figure 7.9

: SUCCESSFUL

: UNSUCCESSFUL

CC_1 : CBA × C_{57}

0.6 MM : TRANSPLANT LENGTH

Normal-looking and milk-secreting mammary glands developed from different portions of donor's mammary duct system in the isologous recipient mice (Figures 7.8, 7.9, 7.10 and 7.11). The sizes of the regenerating grafts of both groups were also comparable. A duct-segment (No. 10 in Figure 7.8) from a pretreated donor and another duct-segment (No. 21 in Figure 7.9) from a control donor grew in the same recipient (Figure 7.12) to comparable size and both underwent milk secretion.

Raynaud[3] observed no sexual dimorphism in the developing fetal mammary anlagen from day 12 to day 14 of gestation in mice. Sexual differences in mammary organogenesis appeared from day 15. In our experiments, days 12 and 15 of gestation were most critical in the fetal development of mammary anlagen in mice with special consideration of hormonal influence of androgen.

INFLUENCES OF ANDROGEN AND OESTROGEN UPON MAMMARY DEVELOPMENT AND GROWTH

In this series of experiments, the influence of prenatally administered TP or OB, with or without postnatal treatments with OB, upon the fetal development and postnatal growth of mammary glands were investigated.

Experimental set-up

Single injections of 1.0 μg OB or 5 mg TP were given subcutaneously to CC-F_1 hybrid pregnant mice on day 12 of gestation as prenatal treatment of the offspring. As a postnatal treatment, 0.3 μg OG was injected subcutaneously into CC-F_2 young daily for 7 days from 14 to 20 days of age. Animals were killed at 21 days of age and their mammary glands were examined. There were 216 control and 440 treated mammary glands examined in male mice, and 344 control and 496 treated mammary glands examined in female young.

Observations and discussion

The nipples were all absent in male mice regardless of the hormonal treatment. However, in female mice, the nipples were present in all the controls and in the mice prenatally or postnatally treated with OB. When TP was given

Figure 7.8 Sketch of the right 4th (inguinal) mammary gland obtained from a 68-day old CC-F_1 female mouse, 7 h after an intraperitoneal injection of 0.3 ml of 0.5% trypan blue aqueous solution in order to visualize the mammary duct system. The mother of this mouse received 5 mg of testosterone propionate subcutaneously on day 12 of pregnancy and therefore the nipple was absent. The 11 duct segments, each 0.6 mm, were excised as transplants at the location indicated on the sketch and were transplanted into the right 4th mammary gland-free fat pads of the hosts of the same CC-F_1 hybrid group. One out of 11 grafts, No. 5, was rejected by the host. A segment, No. 10 grew and secreted milk at parturition of the host (See Figure 7.12)

Figure 7.9 Sketch of the right 4th mammary gland obtained from 68-day old untreated intact female CC-F_1 mouse after the same trypan blue treatment. The 11 duct segments as indicated on the sketch were excised from this control mammary gland and transplanted into the left 4th mammary gland-free fat pads of the same host mice as described in Figure 7.8. All mammary grafts grew and secreted milk. One of them, No. 21, grew in the left 4th mammary gland-free fat pad and is shown in Figure 7.12. Cited from Reference 4

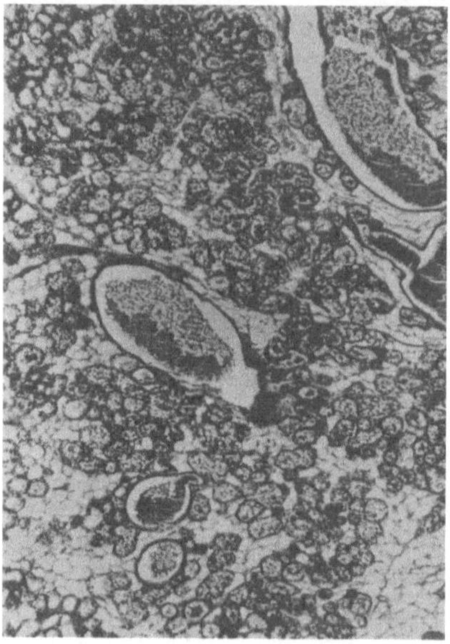

Figure 7.10 Photomicrograph of mammary graft in the left 4th mammary gland-free fat pad of the host, which was transplanted from a control donor mouse, shown with milk secretion. Haematoxylin and eosin. × 80

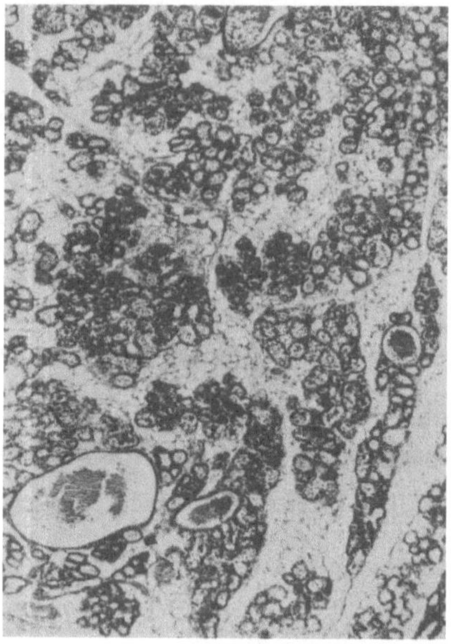

Figure 7.11 Photomicrograph of mammary graft in the right mammary gland-free fat pad of the same host mouse as in Figure 7.10, which was transplanted from a donor treated with androgen on day 12 of fetal life, also shown with milk secretion. Cited from Reference 4. Haematoxylin and eosin. × 80

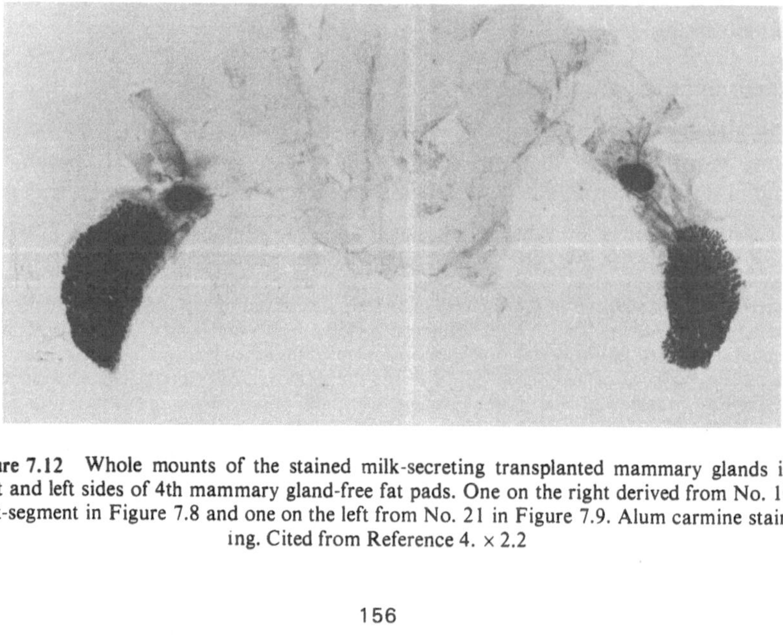

Figure 7.12 Whole mounts of the stained milk-secreting transplanted mammary glands in right and left sides of 4th mammary gland-free fat pads. One on the right derived from No. 10 duct-segment in Figure 7.8 and one on the left from No. 21 in Figure 7.9. Alum carmine staining. Cited from Reference 4. × 2.2

Table 7.14 Frequency of the existence of nipples in mice treated with testosterone propionate (TP) and oestradiol benzoate (OB) in mice

Treatments					
prenatally	*none*	*none*	*OB*	*TP*	*TP*
postnatally	*none*	*OB*	*none*	*none*	*OB*
Male young	0	0	0	0	0
Female young	100%	100%	100%	3.7% (3/81)	0 (0/53)

Prenatal treatments:
Single injections with either 1 μg OB or 5 mg TP were injected subcutaneously into CC-F_1 pregnant mice on day 12 of pregnancy
Postnatal treatments:
Daily injections of 0.3 μg OB were given to CC-F_2 young for 7 days from 14 to 20 days of age and they were killed at 3 weeks of age

on day 12 of pregnancy, the development of the nipple was arrested and postnatal treatment with OB did not have any influence upon the survival of the nipples (Table 7.14).

The development of the mammary duct system was also impaired by pre-

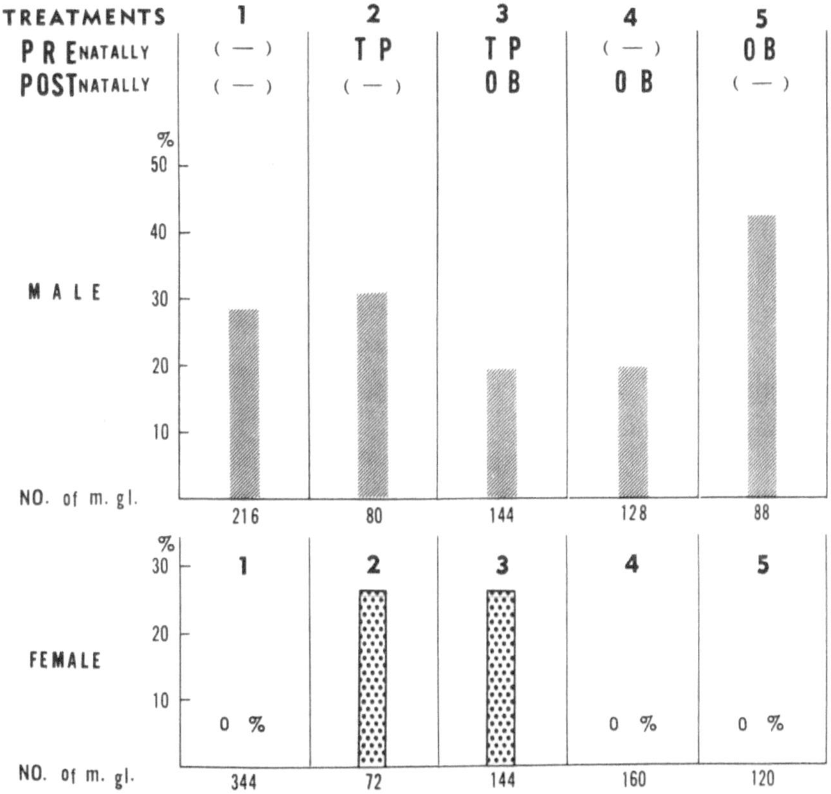

Figure 7.13 Incidence of the absence of mammary glands. After prenatal treatments of either testosterone (TP) or 17β-oestradiol benzoate (OB) followed by postnatal treatment with OB

natal treatment with TP in female young, and postnatal treatment with OB did not increase the chance of finding the persisting rudimentary mammary glands in prenatally TP treated females (Figure 7.13). In the experimental male mice, however, postnatal treatment with OB reduced the frequency of the absence of mammary parenchyma. This might be due to postnatal stimulation of mammary growth in rudimentary glands by oestrogenic action. However, prenatal exposure of mammary anlagen to OB was detrimental to fetal development of mammary glands in males (Figure 7.13).

The extent of mammary growth, determined by measurement of the greatest dimension of the persisting mammary glands, was significantly smaller in mammary glands of female young at 3 weeks of age when they were prenatally treated with TP on day 12 of gestation, whereas that of male mice was not influenced by prenatal TP exposure (Figure 7.14). Postnatal treatment with OB yielded beneficial effects on the growth of persisting glands in both male and female young. Mammary glands which had been exposed to the

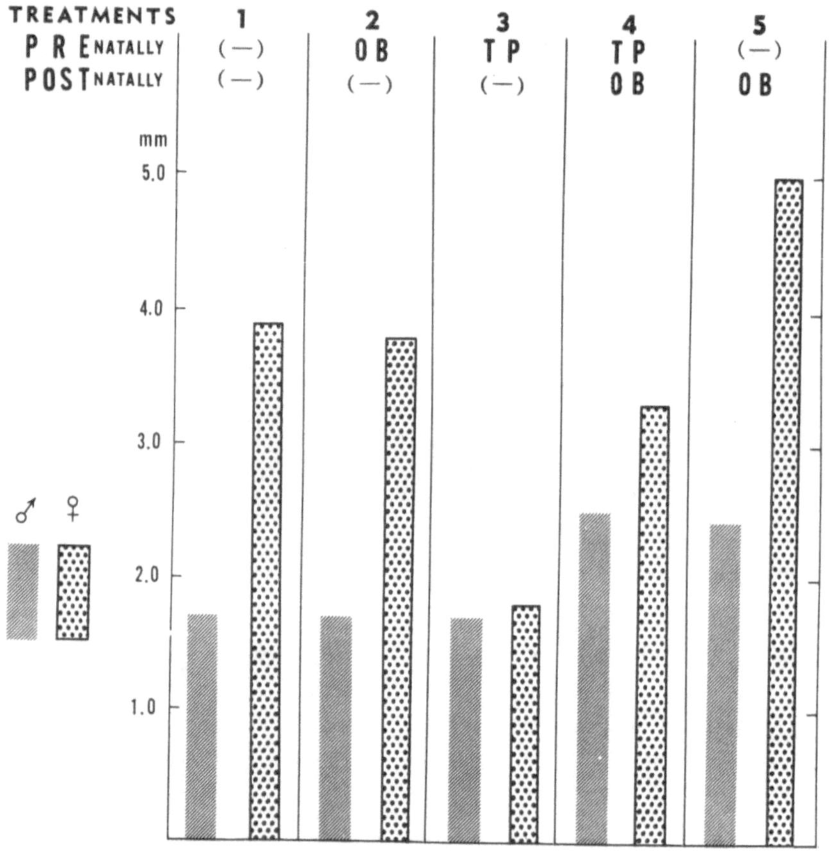

Figure 7.14 Growth extent of mammary glands of mice after prenatal treatments of either testosterone propionate (TP) or 17β-oestradiol benzoate (OB) followed by postnatal treatment with OB

detrimental effects of TP administered prenatally also responded favourably to postnatal oestrogenic stimulation (Figure 7.14).

The influences of prenatal TP or OB treatment and postnatal OB treatment upon fetal development of the nipples and the mammary anlagen, as well as on the growth of the persisting mammary duct systems, are summarized in Table 7.15.

Table 7.15 Influence of prenatal treatments with testosterone propionate (TP) or oestradiol benzoate (OB) and postnatal treatments with OB upon mammary development and growth in mice (prenatal treatments on day 12 of pregnancy)

	Prenatal treatment				*Postnatal treatment*	
	TP		*OB*		*OB*	
Parameter	*Male*	*Female*	*Male*	*Female*	*Male*	*Female*
No. of persisting nipples	no*	decreased	no	no	no	no
No. of persisting glands	no	decreased	decreased	no	increased	no
Extent of mammary growth	no	decreased	no	no	increased	increased

* 'no' means no significant changes caused by the treatment indicated

ACKNOWLEDGEMENTS

These investigations began in the Department of Anatomy, Health Sciences Centre, University of Western Ontario, London, Ontario, Canada, and were completed in the Cell Biology Research Laboratory, Department of Anatomy, Faculties of Medicine and Dentistry, University of Manitoba, Winnipeg, Manitoba, Canada. This work was supported by an operating grant as well as a Development Grant awarded to the author by the Medical Research Council of Canada.

The author is grateful to Mrs Sharon Griffin and Mrs Linda Nahnybida for their excellent technical assistance, and also to Miss Kyoko Hirai for her diligent assistance in the literature search and in the preparation of the manuscript.

References

1. Turner, C. W. (1939). The mammary glands. In: E. Allen (ed.). *Sex and Internal Secretions*, pp. 740–803. (London: Bailiere, Tindall and Cox)
2. Balinsky, B. I. (1950). On the prenatal growth of the mammary gland rudiment in the mouse. *J. Anat.*, **84**, 227
3. Raynaud, A. (1961). Morphogenesis of the mammary gland. In: S. K. Kon and A. T. Cowie (eds.). *Milk: The Mammary Gland and Its Secretion*, Vol. I, pp. 3–46. (New York and London: Academic Press)

4. Hoshino, K. (1965). Development and function of mammary glands of mice prenatally exposed to testosterone propionate. *Endocrinology*, **76**, 789

5. Hoshino, K. (1963). Morphogenesis and growth potentiality of mammary glands in mice. II. Quantitative transplantation of mammary glands of normal male mice. *J. Nat. Cancer Inst.*, **30**, 585

8
Reproductive and teratological studies with prostaglandins

T. V. N. PERSAUD

INTRODUCTION

Few groups of biological substances have stimulated such interest and intensive research as the prostaglandins. During the past few years there has been an exponential growth in the number of studies on the chemistry, metabolism, physiological role, and therapeutic applications of these substances.

Structurally, the prostaglandins are a family of closely related naturally-occurring, lipid soluble, unsaturated hydroxy acids. The prostaglandin molecule is considered to be derived from prostanoic acid which contains a cyclopentane ring with an alkyl and a carboxylic acid side chain. Depending on the functional groups present in the cyclopentane moiety (i.e. the arrangement of hydroxyl and/or ketone groups), the prostaglandins are subdivided into five major groups: E, F, A, B, and C. The two isomeric forms (α and β) of prostaglandin F depend on the arrangement of the two hydroxyl radicals in the cyclopentane molecule. The number of unsaturated bonds in the side chain is indicated by a numerical subscript[1–4].

The distribution of prostaglandins is ubiquitous. They are present in virtually all mammalian tissues and are noted for their extreme high potency and wide range of biological activity. Prostaglandins are formed enzymatically at low concentrations following a variety of stimuli and act locally within cells and between cells. These modulators of physiological processes are rapidly metabolized not far from their sites of production. For details of the biosynthesis and metabolism of the prostaglandins, see König[2], Horton[3] and Losert[4].

The involvement of prostaglandins in inflammatory response, pain, gastric secretion, reproduction, and other physiological and pathological processes appears certain[3,4]. As far as the clinical application of prostaglandins is concerned, many of the early promises have yet to be fulfilled. The one real use is in the field of human reproduction: for the purposes of induction of labour, mid-trimester abortion, and post-coital contraception[5–9].

Surprisingly, despite the widespread use of the prostaglandins in pregnant women, few studies have been undertaken to determine their safety. It would be no exaggeration to say that the clinical use of the prostaglandins as a whole, has preceded their full preclinical toxicological evaluation. The mode of action of these substances during pregnancy and their involvement in reproductive processes are not clearly understood. Furthermore, there is relatively little information on the influence of prostaglandins on the development of the conceptus[5,10]. This chapter reviews some of the effects of PGE_2 and $PFG_{2\alpha}$, two clinically important prostaglandins, in pregnant laboratory animals.

PROSTAGLANDIN E_2

Rat

Subcutaneous treatment of pregnant Sprague-Dawley rats with 100 μg PGE_2 on days 1 through 4 of gestation led to a retention of blastocysts in the uterine tube. The blastocysts that were recovered on gestational day 6 showed varying degrees of degeneration or retardation in development[11].

Whether this adverse action of PGE_2 in early rat pregnancy is due to a derangement in tubal transport, alterations in oviductal fluid, a direct action on the conceptus itself, changes at the site of attachment to the uterine wall, or a combination of these, remains to be determined. The luteolytic properties of certain prostaglandins in laboratory rodents have been reported, and it is conceivable that degeneration of the corpus luteum and the resulting rapid and marked decrease in progesterone level could account for these changes[5,12]. However, microscopic examination of the ovaries of the treated animals showed an intact ovarian stroma and no signs of venous congestion nor luteal degeneration. Also, serum progesterone levels were not affected[11].

At a relatively high dose level (300 μg), PGE_2 did not interrupt pregnancy in the rat when administered intraperitoneally from gestational days 12 through 15. The incidence of fetal resorptions was not significantly increased than that of the control and the number of viable offspring recovered near term was not affected by the treatment. No apparent malformations were seen in the offspring (Table 8.1), but 18.2% revealed extensive oedema and haemor-

Table 8.1 PGE_2 effects in pregnant rats following intraperitoneal injection

Treatment	*No. of litters*	*Implantations*	*Resorptions* (%)	*Abnormal offspring* (%)	*Mean fetal weight* (g)
300 μg PGE_2 on gestational days 12–15	8	93	5 (5.4)* NS	16† (18.2)	3.8
Controls	8	86	3 (3.5)	0	3.7
300μg PGE_2 on gestational days 9–12	8	85	18 (2.12)* ($p<0.01$)	0	3.6
Controls	8	88	2 (2.3)	0	3.3

* Significance of difference from control was evaluated by χ^2 test

† Fetuses showed extensive oedema and haemorrhagic lesions but no structural defects (see Figure 8.1)

After Persaud[13] with additional data

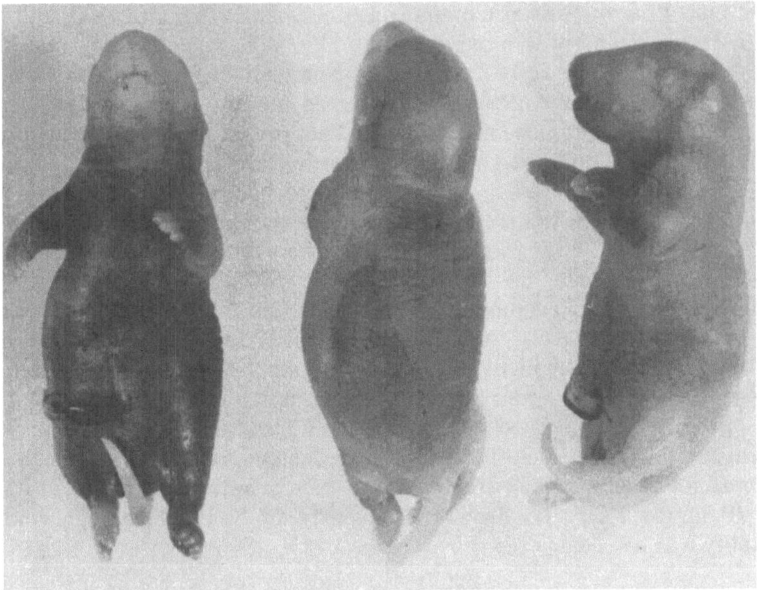

Figure 8.1 Rat fetuses (day 20), recovered from mothers that were treated with 300 μg PGE_2, administered intraperitoneally from day 12 through 15 of gestation. Note extensive oedema and haemorrhagic lesions. (From Persaud[13])

rhagic lesions (Figure 8.1). The E-type of prostaglandins are potent vasodilators and probably these lesions may have resulted from increased vascular permeability or vascular damage. In contrast, the same dose of PGE_2 administered intraperitoneally to pregnant rats from day 9 through 12 produced a higher incidence of resorptions than in the controls, but no fetal abnormalities[13] (Table 8.1).

In another series of experiments, PGE_2 (100 μg) was administered to rat fetuses on day 15 of gestation by the intra-amniotic route (Table 8.2). Fetal death and resorptions accounted for more than 16% of the control fetuses which was probably due to the technique itself. However, all of the prostaglandin-treated fetuses showed a significant decrease in mean weight, and microscopic examination revealed marked inhibition of decidua formation, venous congestion, inhibition of trophoblastic proliferation, and extensive vacuolization[13, 14]. Similar observations were made at the Research Institute

Table 8.2 PGE_2 effects on embryonic development in the rat when administered intra-amniotically

Treatment	*No. of embryos treated*	*Resorptions (%)*	*Abnormal fetuses*
100 μg PGE_2 on day 15 of pregnancy	62	62 (100) $p < 0.001$	0
Controls	67	11 (16.4)	0

After Persaud[14]

of the Ono Pharmaceutical Company Ltd., Japan[15]. PGE_2 (6, 120, and 240 mg/kg body weight) was administered orally to Sprague-Dawley rats from days 9 through 14 of gestation. The total number of implantation sites between the controls and prostaglandin-treated animals was not significantly different. External and skeletal defects were not present in any of the offspring, and the mean fetal body weight was not influenced by the treatment. However, compared with all the other groups, the incidence of fetal resorptions and death was significantly increased following treatment of the mothers at the highest dose level (240 mg/kg). All embryos were completely resorbed in four of these animals.

Mercier-Parot and Tuchmann-Duplessis[16] found that the administration of very high doses of prostaglandin E_2 to pregnant Wistar rats resulted in a high incidence of fetal resorptions and abnormal offsprings. Following treatment at dose levels of 15, 20, and 30 mg/kg from days 6 through 10 of gestation, PGE_2 produced a high incidence of gross malformations (eye, central nervous system and the face) which showed a dose-dependent relationship. The resorption rate followed a similar trend (Figure 8.2). In rats treated with 5, 10, 15 and 20 mg/kg PGE_2 on days 9 through 14 of gestation, a high embryo mortality was detected.

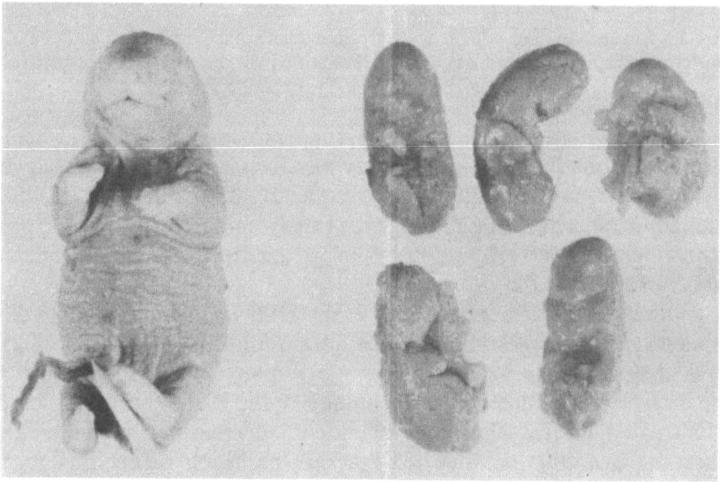

Figure 8.2 Left: Control rat fetus (day 20). Right: Resorbed embryo with recognizable features, induced by 100 μg PGE_2, administered intra-amniotically on day 15 of gestation. (From Persaud[13])

Mouse

PGE_2 was found to be highly teratogenic in Swiss-Webster mice following subcutaneous administration of 25 μg on either day 9, 12, or 15 of gestation. The results are summarized in Table 8.3. A significantly higher incidence of malformed offspring was detected in all three treatment groups, compared to the corresponding controls. Major skeletal defects were not present in any of the fetuses recovered, but in both control and experimental groups, a rela-

Table 8.3 Teratogenic effects of PGE_2 in pregnant mice

*Treatment**	*No. of litters*	*Implantations*	*Resorptions (%)*	*Abnormal offspring (%)*
25 μg PGE_2 on gestational day 9	12	142	12 (8.5)†	53 (40.8)‡
Controls	11	124	5 (4.0)	2 (1.7)
25 μg PGE_2 on gestational day 12	12	138	3 (2.2)†	53 (43)‡
Controls	12	127	5 (3.9)	0
25 μg PGE_2 on gestational day 15	13	126	3 (2.4)†	26 (21.1)‡
Controls	11	120	7 (5.8)	1 (0.9)

* Subcutaneous injections
† NS (χ^2 test)
‡ $p < 0.001$
After Persaud[17]

tively high frequency of minor skeletal variations were induced. These included incomplete sternebrae, missing phalanges, rudimentary ribs, and poor ossification of the skull, vertebrae and ribs. Compared to the controls, the incidence of fetal resorptions was not significantly altered. Intra-uterine growth of the offspring was impaired as a result of treatment on either gestational day 12 or 15; both the mean fetal body weights and mean crown–rump lengths differed significantly from those of the controls. Such an effect was not observed in the offspring treated on day 9 of gestation[17].

In the IRC-JCL strain of mice, PGE_2 administered at relatively high doses (6, 120 and 240 mg/kg body weight) by means of a gastric tube from day 7 through 12 of gestation was found not to be teratogenic. Cleft palate and polyphalangism were detected in three of the treated fetuses, and cleft palate was present in one of the control fetuses. However, the incidence of fetal death and resorptions was dose-dependent and significantly increased in all treatment groups compared to the controls. Also, the incidence of skeletal anomalies was significantly increased after treatment with the prostaglandin. These included delayed ossification and anomalies of the vertebrae and ribs[18].

Rabbit

Treatment of rabbit embryos with 50 and 100 μg PGE_2 administered intra-amniotically on day 14 of gestation, caused a high embryolethality. An increased incidence of congenital defects was observed following treatment with 25 μg (Figure 8.3). Furthermore, intra-uterine growth of fetuses treated with the prostaglandin was impaired (Table 8.4). The mean placental weight was significantly reduced only after treatment with 50 μg PGE_2, but extensive morphological changes were present at all dose levels. The decidua basalis showed marked thinning and a depletion of glycogen. Maternal blood spaces were congested and severely dilated, and extensive atrophic and degenerative changes were present in the labyrinthine trophoblast[19].

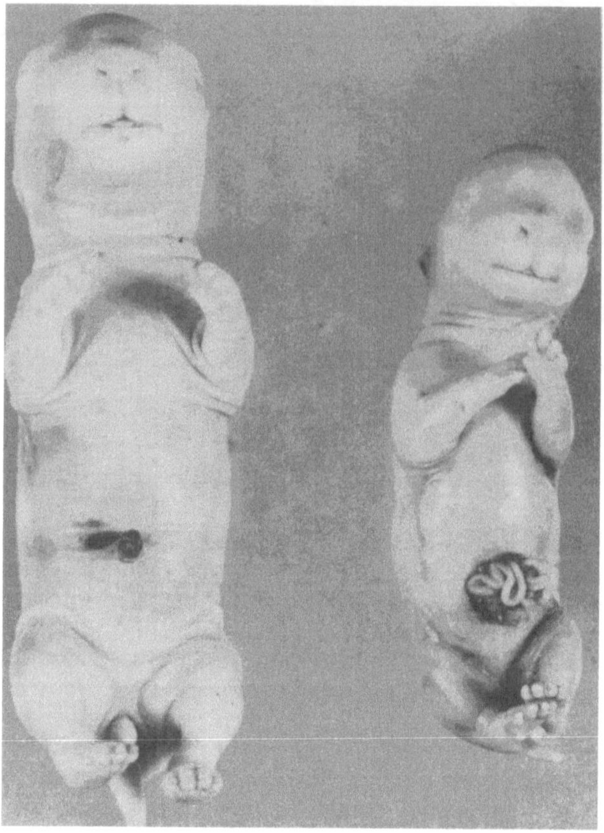

Figure 8.3 Left: Control rabbit fetus (day 24). Right: Stunted rabbit fetus with umbilical hernia and limb anomalies, induced by PGE_2 (25 μg), administered intra-amniotically on day 14 of gestation. (From Persaud[19])

Table 8.4 PGE_2 effects on embryonic development in the rabbit following intra-amniotic treatment

		PGE_2-treated		
	Controls	*25 μg*	*50 μg*	*100 μg*
No. of embryos treated	33	16	9	5
Resorptions (%)	9 (21)	1 (6.3)	4 (44)	5 (100)
Abnormal fetuses (%)	4 (16.7)	10 (66.7)	2 (40)	—
Fetal weight (g) (mean ± SEM)	10.4 ± 0.54	7.7 ± 0.83	6.4 ± 0.87	—
Fetal length (mm) (mean ± SEM)	57.8 ± 1.53	48.6 ± 1.91	48.6 ± 1.46	—
Placental weight (g) (mean ± SEM)	4.6 ± 0.31	4.5 ± 0.39	3.2 ± 0.92	—

After Persaud[19]

Chick

The avian embryo represents a convenient and reliable model for the teratological screening of drugs, providing the experimental conditions are properly controlled. It is a highly sensitive system and the response of the embryo to the treatment can often be evaluated within days. Furthermore, both maternal and placental influences are eliminated[20].

The incidence of embryonic mortality and defects was relatively higher than in the controls after treatment of 48 h old chick embryos with PGE_2 at doses of 50 and 100 μg, but this difference was significant only at the higher dose level. There was no evidence of an embryolethal or teratogenic effect following treatment at 72 h incubation. The frequency of the abnormal embryos revealed a dose-dependent relationship, and multiple malformations were present in many of these embryos. However, the embryos showed no evidence of retarded growth at any of the dose levels and incubation periods[21].

PROSTAGLANDIN $F_{2\alpha}$

Rat

The luteolytic properties and abortifacient action of $PGF_{2\alpha}$ were studied by several investigators. Fuchs and Mok[22] reported that $PGF_{2\alpha}$, at a dose of 125 μg per rat, caused 100% fetal resorption within 24 h following treatment on days 10–12 of gestation, and premature delivery of living offspring within an interval of 44 and 33 h following intravenous infusion on days 18–20 of gestation. At other stages of gestation, days 1 to 7 and 13, there was no embryolethality or it was minimal. Intra-uterine infusion of $PGF_{2\alpha}$ on days 7, 10 or 14 of gestation was less effective than the intravenous route. The abortifacient effect of $PGF_{2\alpha}$ on day 10 of gestation was counteracted with progesterone treatment.

Chatterjee[23] also found that the single injection of $PGF_{2\alpha}$ (2 mg/kg) on day 10 of pregnancy caused luteal regression followed by 100% resorption of the conceptuses. However, a similar treatment on gestational day 18 resulted in premature delivery of the offspring within 72 h (see also Strauss, *et al.*[24]). Simultaneous treatment of the pregnant animals with clomiphene citrate, an antioestrogenic agent, prevented termination of pregnancy or premature delivery of the fetuses in all cases. Previous studies[25] by these investigators have shown that treatment of the animals from day 5 through 8 of gestation with $PGF_{2\alpha}$ is ineffective as a luteolytic and antifertility agent. In a related study, Stcinctz *et al.*[26] found that a single subcutaneous injection of prostaglandin ($PGF_{2\alpha}$, PGE_2, PGE_1) on day 5 of pregnancy did not cause abortion. However, treatment twice daily on days 4 through 7 of gestation was always followed by resorption of the embryos although implantation had occurred normally. PGE_2 and PGE_1 were about half as effective as $PGF_{2\alpha}$. The abortifacient action of $PGF_{2\alpha}$ was prevented by daily prolactin injections. Because of this observation, the suggestion was made that the corpora lutea in rats are not affected by prostaglandin. This is often considered to be responsible for PG-induced abortions in rodents. Furthermore, because the abortifacient activity of $PGF_{2\alpha}$ in the rat was not increased by theophylline, it was considered not to be mediated by cyclic AMP, thereby refuting the suggestion of a common interceptive mechanism for oestrogens and prostaglandins.

Saksena *et al.*[27] found that the insertion of a silastic-PVP implant containing 2 mg, 400μg, or 200μg $PGF_{2\alpha}$ into one of the uterine horns of rats on day 10 of gestation induced abortion in all cases within 72 h. A similar treatment in rabbits on day 15 of pregnancy and hamsters on day 8, resulted in the same findings. The abortifacient action of $PGF_{2\alpha}$ in the three species was prevented by daily injections of progesterone. To account for its luteolytic action, it was concluded that $PGF_{2\alpha}$ either interferes with ovarian progesterone secretion or stimulates luteinizing hormone release, or both.

The effects of prostaglandin $F_{2\alpha}$ on embryonic development in both rats and mice were investigated by Matsuoka *et al.*[28]. Administered intravenously at doses of 0.1, 1.0, 1.5, and 2.0 mg/kg body weight to Sprague-Dawley rats from gestational day 9 through 14, $PGF_{2\alpha}$ caused significant fetal resorptions and minor abnormalities (cleft palate, fused ribs, short tail and oligodactylia) in some of the offspring. The incidence of malformed fetuses was significantly more only in the group treated at the highest dose level. In contrast, $PGF_{2\alpha}$ had no effect on the development of the conceptuses when administered intraperitoneally to mice from day 7 through 12 of gestation at doses of 0.05, 0.1, and 0.25 mg/kg body weight.

A significantly high incidence of fetal resorptions was reported by Tuchmann-Duplessis and Mercier-Parot[29] following intraperitoneal treatment with $PGF_{2\alpha}$ at doses of 0.3 to 7.5 mg/rat daily on gestational days 7 to 9. The embryotoxicity of the prostaglandin was reduced following simultaneous treatment of the mothers with progesterone. Chatterjee[25] also reported that the single subcutaneous injection of $PGF_{2\alpha}$ (2 mg/kg body weight) produced total fetal resorption on gestational days 10 or 12. The same treatment on gestational day 13 was found not to be embryolethal and the offspring recovered near term were apparently normal.

In our laboratory, 50 μg of $PGF_{2\alpha}$ dissolved in physiological saline was administered subcutaneously to two groups of pregnant rats on either gestational days 9 to 11, or 12 to 15[30]. Our findings are summarized in Tables 8.5 and 8.6. The incidence of fetal resorption and malformed offspring was not significantly different from that of the controls. However, intra-uterine growth was impaired; the mean fetal body weight and mean crown–rump length of

Table 8.5 Pregnancy and progeny in rats treated with $PGF_{2\alpha}$

No. of animals / implantations	*Treatment (s.c. injection)*	*Resorptions (%)*	*Fetuses alive*	*Abnormal offspring (%)*
7/93	50 μg/days 9–11 of gestation	0	93	4 (4.3)*
10/126	Control	0	126	0
7/83	50 μg/days 12–15 of gestation	2 (2.4)*	81	2 (2.5)*
11/144	Control	2 (1.4)	142	0

* NS (Significance of difference from control was evaluated by χ^2 test)

After Persaud and Thliveris[30]

Table 8.6 Influence of $PGF_{2\alpha}$ on intra-uterine growth in the rat

*Treatment**	*Fetal weight*† (g)	*Fetal CR length*† (mm)	*Placental weight*† (mg)
Days 9–11 experimental	2.1 ± 0.08‡	25.0 ± 0.45‡	433.8 ± 6.31†
Control	2.5 ± 0.03	26.2 ± 0.26	500.2 ± 7.63
Days 12–15 experimental	2.5 ± 0.04‡	28.3 ± 0.63‡	4441.7 ± 5.88‡
Control	3.0 ± 0.11	52.7 ± 26.20	559.2 ± 10.7

* 50 μg $PGF_{2\alpha}$ administered subcutaneously
† Mean ± SEM
‡ $p < 0.001$ (Student's *t* test)
After Persaud and Thliveris[30]

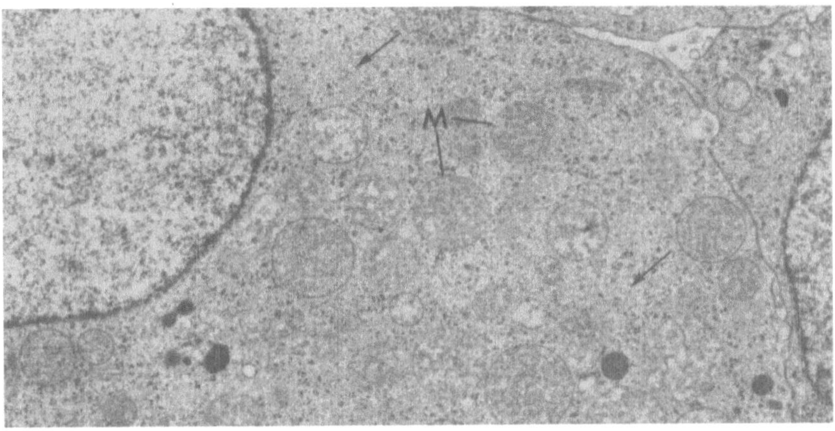

Figure 8.4 Fetal adrenal cortex from day 20 control animal. The most prominent organelles present are spherical mitochondria (M) with vesicular cristae and numerous tubular profile of smooth endoplasmic reticulum (arrows). × 5850. (From Persaud and Thliveris[30])

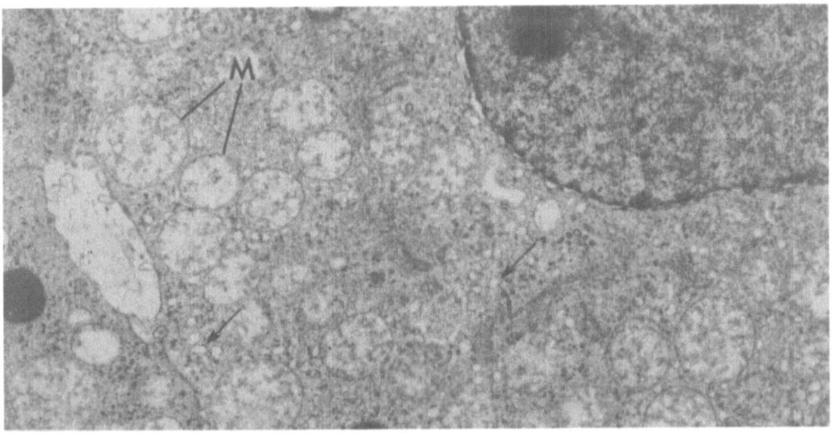

Figure 8.5 Fetal adrenal cortex from day 20 experimental animal. Note the numerous dilated profiles of smooth endoplasmic reticulum (arrows) and mitochondria (M) with clumped cristae. × 5850. (From Persaud and Thliveris[30])

the prostaglandin treated fetuses were significantly reduced than those of the controls. In addition, placental growth was impaired in those mothers treated with the prostaglandin from days 12 through 15 of gestation. Skeletal malformations were not present in any of the fetuses. Histological examination of maternal overies and fetal liver and adrenal glands, showed no significant morphological changes compared to the control. Other than a slight decrease in the thickness of the decidua basalis, the placenta appeared unaffected. However, at the ultrastructural level, extensive areas of degeneration, characteristic of placental insufficiency were detected. Also, the fetal adrenal cortex revealed changes indicative of intra-uterine distress (Figure 8.4–8.7). This

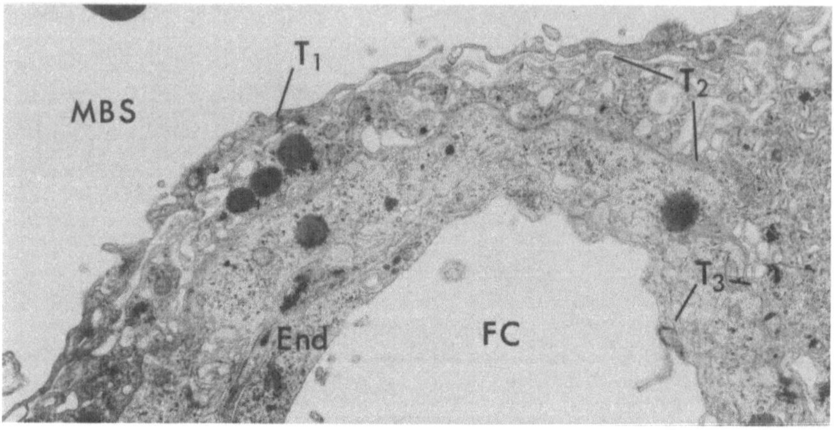

Figure 8.6 Placental labyrinth from day 20 control animal. Maternal blood space (MBS) and fetal capillary (FC) are separated by three layers of trophoblast tissue; an outer (T_1), middle (T_2), and inner (T_3) layer. Fetal capillary endothelium (End). × 7425. (From Persaud and Thliveris[30])

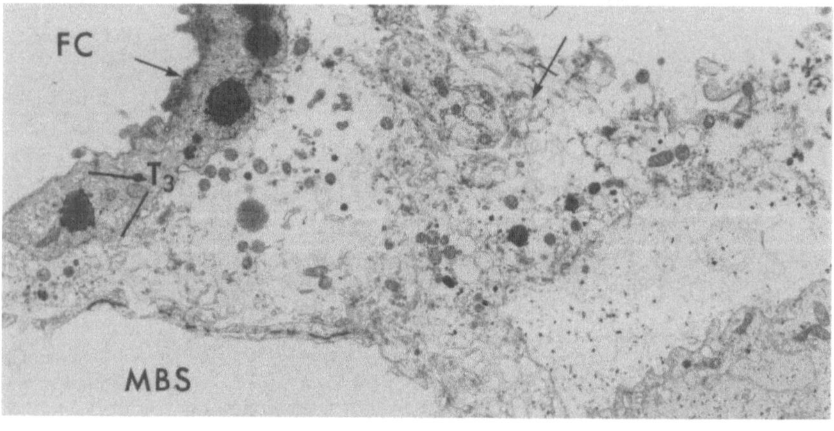

Figure 8.7 Placental labyrinth from day 20 experimental animal revealing disruption of the outer and middle trophoblast layers (crossed arrows). The inner trophoblast layer (T_3) and fetal capillary endothelium (arrow) appear normal. Maternal blood space (MBS); fetal capillary (FC). × 4875. (From Persaud and Thliveris[30])

study showed that $PGF_{2\alpha}$ caused fetal growth retardation unaccompanied by fetal loss or abnormalities[30]. Because of the changes found in the placenta and adrenal gland, it is possible that the observed intra-uterine growth retardation might be due to the direct effect of the prostaglandin on the embryo and/or developing placenta, or alternatively might involve a mechanism whereby the uterine and fetal placental blood circulation is redistributed[31].

Mouse

Embryonic development was impaired in mice following intravenous administration of $PGF_{2\alpha}$ on gestational day 12 or 13. When 100 μg was injected on gestational day 12, the incidence of resorption was 78%, but only 2.2% on day 13. A dose of 200 μg of the prostaglandin resulted in more than 90% of fetal resorption on gestational day 13. No abnormal offspring was detected in any of the groups, and the mean fetal weight showed no difference from that of the corresponding controls (Table 8.7). Treatment of the animals on either day 14 or 15 of pregnancy with 200 μg $PGF_{2\alpha}$ did not affect fetal development[30].

There was an insignificant increase in the incidence of fetal resorptions following treatment with 100 μg $PGF_{2\alpha}$, administered subcutaneously from day 12 to 15 of gestation. In 5.5% of the treated fetuses, reduction abnormalities in the hindlimb and tail (Figure 8.8) were detected[32]. As $PGF_{2\alpha}$ is a potent vasoconstrictor, the development of these abnormalities could have been due to an impairment of the blood supply to the tail and limbs caused by the prostaglandin. These observations clearly indicate that exogenous $PGF_{2\alpha}$ is embryotoxic in mice not only during very early gestation as reported by other workers[33,34], but even when administered after mid-gestation.

Table 8.7 Embryotoxic effects of $PGF_{2\alpha}$ in mice

Treatment with $PGF_{2\alpha}$*	*No. of litters*	*Implantations*	*Resorptions (%)*
100 μg on day 12 of gestation	6	73	57 (78)†
Controls	6	68	0
100 μg on day 13 of gestation	8	93	2 (2.2)
200 μg on day 13 of gestation	6	58	54 (93)†
Controls	8	86	4 (4.7)
200 μg on day 14 of gestation	6	68	6 (8.8)
Controls	6	71	3 (4.2)
200 μg on day 15 of gestation	8	89	3 (3.4)
Controls	8	64	2 (3.1)

* Administered intravenously
† $p<0.01$ (Significance of difference from control values was evaluated by χ^2 test; other values were not significantly different)

After Persaud[32]. Fetuses recovered showed no apparent anomalies

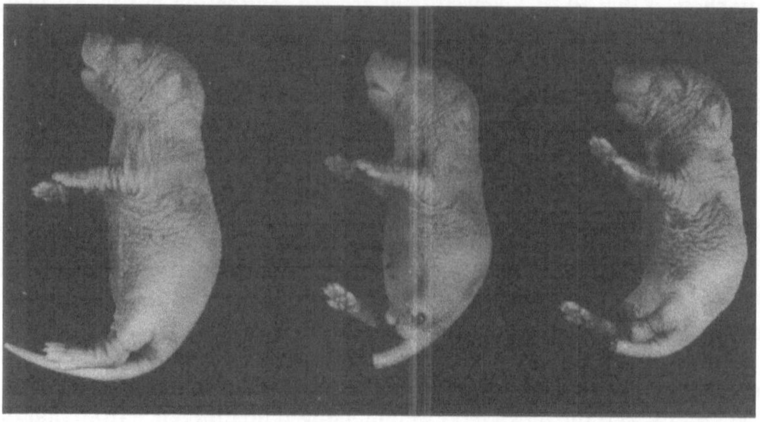

Figure 8.8 Twenty-day old mouse fetuses, all showing reduction anomalies of the tail, and also of the hindlimb in two cases. The mother was treated with 100 μg $PGF_{2\alpha}$, administered subcutaneously from day 12 through 15 of gestation (From Persaud[32])

In contrast to our findings, Matsuoka *et al.*[28] found that $PGF_{2\alpha}$ administered intraperitoneally to ICR–JCL strain of mice from day 7 through 12 of gestation at doses of 0.05, 0.1, 0.25 mg/kg produced no significant adverse effect on their fetuses and newborns.

Rabbit

Studies in the rabbit have shown that $PGF_{2\alpha}$ blocks implantation and induces luteolysis[35], accelerates ovum transport[36,37], causes premature expulsion of the eggs from the uterus and induces abortion[38–40].

Spilman[44] described blastocysts (2- and 4-cell stages) flushed from the oviducts of rabbits treated subcutaneously with prostaglandins ($PGF_{2\alpha}$ or PGE_1, PGE_2 at doses of 3 mg/kg body weight, 12 and 24 h after HCG) and subsequently cultured *in vitro*. Compared to the control cultures, PGE_1 and $PGF_{2\alpha}$ produced no significant effects, but PGE_2 treatment inhibited development of the early embryos. Whether this prostaglandin treatment altered the viability of spermatozoa, or of the oviductal fluid, or acted directly on the blastocyst by the oviductal transudate, is not known.

Chang and Hunt[38] also reported that the treatment of pregnant rabbits subcutaneously with $PGF_{2\alpha}$, at doses of 1 and 2 mg/kg body weight on gestational days 3, 4, 5, 6, and 7, significantly inhibited embryonic development, and at a dose of 2 mg/kg body weight on gestational days 8, 9, and 10, resulted in complete degeneration of the embryos. A dose of 5 mg/kg body weight administered on gestational days 3 to 7 completely inhibited embryonic development. Live and apparently normal fetuses were prematurely delivered following daily treatment with $PGF_{2\alpha}$ at a dose of 2 mg/kg body weight from gestational day 20 through 23.

Lau *et al.*[40,42] have investigated the abortifacient effect of a silastic polyvinyl pyrrolidone (silastic PVP) implant containing $PGF_{2\alpha}$. An intra-uterine implant with 2 mg $PGF_{2\alpha}$ induced abortion in 100% of the animals. An implant with a similar dose inserted intraperitoneally, subcutaneously, and intravaginally, induced abortion in 75%, 50% and 50% of the animals, respectively. Compared to parenteral injection of the prostaglandin, administration by means of a silastic PVP implant was more effective, particularly when inserted locally in the uterine lumen.

Chick

Chick embryos treated with $PGF_{2\alpha}$ (10, 40, 100 μg) at 48 and 96 h incubation revealed significantly higher mortality rates than the corresponding unopened controls, but the frequency of embryonic death between the groups treated with the solvent (saline) and $PGF_{2\alpha}$ showed no significant differences. In addition, treatment of the embryos with different dose levels of prostaglandin produced no significant differences in mortality rates. Although a small number of embryos treated with $PGF_{2\alpha}$ at 72 and 96 h incubation were malformed, the frequency of occurrence revealed no significant difference from that of the corresponding controls with respect to the time of treatment or the dose levels of the prostaglandin[43].

PGE_2 AND $PGF_{2\alpha}$ IN RHESUS MONKEYS

Observations on fetal morphology in the Rhesus monkey following maternal treatment with prostaglandins (Table 8.8) were made by Cole and Nutting[44]. The sexually mature female Rhesus monkeys were mated with sexually mature male animals around the time of estimated ovulation during the experimental menstrual cycle. Each female was subsequently treated twice a day with either the vehicle (controls) or with PGE_2 (1.5 or 15.0 mg) or $PGF_{2\alpha}$ (1.5 mg) for either 3 or 5 consecutive days during a period ranging from day 9 to day 20 of gestation. These pregnant animals were subsequently monitored for delivery of their young and the resulting infants were observed for gross morphological abnormalities. Although the mean gestational length for each group was somewhat different, all three groups of animals delivered spontaneously very close to the expected time of 165 days for this species. Likewise, the crown–rump length and body weight of the infants of each group fell well into the accepted range for this species of 180–200 mm and 400–500 g, respectively. There were no gross morphological abnormalities found in any of the offspring from either of the three experimental groups.

CONCLUSIONS

It is well recognized that the mammalian embryo is equally exposed to most drugs which are administered to the mother. The placenta readily transmits these substances or their metabolites to the fetus[45]. The influence on fetal activity of prostaglandins administered to pregnant women is undoubtedly related to the kinetics of transfer across the placenta. This problem has received relatively little attention.

It is important to determine the effects of prostaglandins on the placenta

Table 8.8 Pregnancy and progeny in Rhesus monkeys treated with PGE_2 and $PGF_{2\alpha}$

Compound	*No. of animals*	*Dose range* (mg/twice daily)	*Treatment* range* (gestation days)	*Day of parturition* (mean)	*Sex*	*Mean CR length* (mm)	*Mean fetal weight* (g)	*Observed abnormalities*
Vehicle	9	—	9–19	169.3	5M, 4F	192.2	489.8	None
PGE_2	4	1.5–15.0	9–20	164.2	2M, 2F	204.2	477.5	None
$PGF_{2\alpha}$	5	1.5	9–17	159.4	2M, 3F	185.0	436.4	None

* Subcutaneous injection for either three or five consecutive days

After Cole and Nutting[44]

itself, whether the substances are metabolized in the placenta, the extent of placental transfer, the disposition in the fetus, and how long they remain in the fetus. Further studies are most urgently needed in these areas.

Studies in our laboratory have shown that several prostaglandins are readily transferred across the rodent placenta and significant levels were detected over extended periods in both the fetus and amniotic fluid[45–48]. Recently, Green *et al.*[49] reported the kinetics of transfer of [^{3}H]PGF$_{2\alpha}$, administered intra-amniotically, in five patients for whom legal abortion was approved. High levels of the radioactivity were present in fetal tissues, including the liver, kidney, lung, ventricle and contents, suprarenal gland, skin, blood and in the placenta. Three metabolites were recovered from the amniotic fluid: 15-keto-PGF$_{2\alpha}$, 15-keto-13,14-dihydro-PGF$_{2\alpha}$ and 13,14-dihydro-PGF$_{2\alpha}$. Because several of these metabolites were isolated from the placenta and fetal tissues, it was suggested that both the fetus and placenta are capable of metabolizing PGF$_{2\alpha}$.

Although there is substantial data that PGE$_2$ and PGF$_{2\alpha}$ interrupt pregnancy and impair embryonic development in laboratory animals, the findings cannot conclusively indicate whether these substances will be teratogenic in humans. In the animal studies, the expression of embryopathic activity was variable and depended on the species of animals involved, the dose levels of the prostaglandins, and the route of administration. Further studies are clearly indicated here. However, in view of the high incidence of fetal resorption, fetal growth retardation and developmental defects induced by prostaglandins in laboratory animals, a similar deleterious action on the human embryo should be taken into consideration when prostaglandins are administered to pregnant women.

ACKNOWLEDGEMENTS

Our studies were supported by grants from the Medical Research Council of Canada, the University of Manitoba Research Board, Seller's Foundation, Winnipeg, and the Rh-Institute, Winnipeg. I am indebted to Dr John Pike, Upjohn Co., Kalamazoo, Michigan and to Mr Kazuo Sano, Ono Pharmaceutical Co. Ltd., Osaka, Japan for supplying the prostaglandins.

References

1. Schneider, W. P. (1976). The chemistry of prostaglandin. In: S. M. M. Karim (ed.) *Prostaglandins: Chemical and Biochemical Aspects*. pp. 1–23 (Baltimore: University Park Press)
2. König, H. (1975). Zur Chemie der Prostaglandine – Biogenese, Stoffwechsel, Totalsynthese. *Klin. Wschr.*, 53, 1041
3. Horton, E. W. (1976). Prostaglandins. *Sci. Prog., Oxf.*, 63, 335
4. Losert, W. (1975). Biologie der Prostaglandine unter Berücksichtigung therapeutischer Aspekte. *Arzneim. Forsch.* (*Drug Res.*), 25, 135
5. Goldberg, V. J. and Ramwell, P. W. (1975). Role of prostaglandins in reproduction. *Physiol. Rev.*, 55, 325
6. Gillespie, A. (1976). Clinical use of prostaglandins in obstetrics. *Drugs*, 11, 113
7. Brenner, W. E. (1975). The current status of prostaglandins as abortifacients. *Am. J. Obstet. Gynecol.*, 123, 306

8. Tsakok, F. H. M., Grudzinskas, J. G., Karim, S. M. M. and Ratnam, S. S. (1975). The routine use of oral prostaglandin E_2 in induction of labour. *Br. J. Obstet. Gynaecol.*, **82**, 894
9. Mackenzie, I. Z., Davies, A. J., Embrey, M. P. and Guillebaud, J. (1978). Very early abortion by prostaglandins. *Lancet*, i, 1223
10. Persaud, T. V. N. (1974). Embryonic and fetal development. In: P. W. Ramwell (ed.). *Prostaglandins*, Vol. 2, pp. 175–203. (New York: Plenum Press)
11. Persaud, T. V. N. (1979). Effects of prostaglandin E_2 on very early embryonic development in the rat. (In preparation)
12. Horton, E. W. and Poyser, N. L. (1976). Uterine luteolytic hormone: a physiological role for prostaglandin $F_{2\alpha}$. *Physiol. Rev.*, **56**, 595
13. Persaud, T. V. N. (1973). Pregnancy and progeny in rats treated with prostaglandin E_2. *Prostaglandins*, **3**, 299
14. Persaud, T. V. N. (1973). Prostaglandin E_2 effects on placenta and fetus following intra-amniotic administration in rats. *Int. Res. Commun. Syst.*, **(73–9)**, 8-5-4
15. Ono Pharmaceutical Company Ltd., Japan. (Personal communication)
16. Mercier-Parot, L. and Tuchmann-Duplessis, H. (1977). Action of prostaglandin E_2 on pregnancy and embryonic development of the rat. *Toxicol. Lett.*, **1**, 3
17. Persaud, T. V. N. (1975). The effects of prostaglandin E_2 on pregnancy and embryonic development in mice. *Toxicology*, **5**, 97
18. Ono Pharmaceutical Company Ltd., Japan. (Personal communication)
19. Persaud, T. V. N. (1974). Fetal development in rabbits following intra-amniotic administration of PGE_2. *Exp. Pathol.*, **9**, 336
20. Matthews, G. B. P., Persaud, T. V. N. and Mann, R. (1975). The chick embryo as an experimental model in teratological studies. In: P. G. Boehm, T. V. N. Persaud, K. Hoshino and C. V. Greenway (eds.). *Laboratory Animals in Biomedical Research*, pp. 211–221 (Ottawa: Canadian Association for Laboratory Animal Science)
21. Persaud, T. V. N., Mann, R. A. and Moore, K. L. (1973). Teratological studies with prostaglandin E_2 in chick embryos. *Prostaglandins*, **4**, 343
22. Fuchs, A. R. and Mok, E. (1973). Prostaglandin effects on rat pregnancy. II. Interruption of pregnancy. *Fertil. Steril.*, **24**, 275
23. Chatterjee, A. (1976). The possible mode of action of prostaglandins: XI. Clomiphene citrate and the reversal of luteolytic or abortifacient efficiency of prostaglandin $F_{2\alpha}$ in rats. *Contraception*, **14**, 447
24. Strauss III, J. F., Sokoloski, J., Caploe, P., Duffy, P., Mintz, G. and Stambaugh, R. L. (1975). On the role of prostaglandins in parturition in the rat. *Endocrinology*, **96**, 1040
25. Chatterjee, A. (1973). Possible mode of action of prostaglandins: Differential effects of prostaglandin $F_{2\alpha}$ before and after the establishment of placental physiology in pregnant rats. *Prostaglandins*, **3**, 189
26. Steinetz, B. G., Butler, M. C., Sawyer, W. K., O'Byrne, E. M. and Giannina, T. (1976). Post-implantation termination of pregnancy in rats. *Contraception*, **14**, 487
27. Saksena, S. K., Lau, I. F. and Chang, M. C. (1974). Prostaglandin $F_{2\alpha}$ implant-induced abortion: effect of progestin and luteinizing hormone concentration and its reversal by progesterone in rabbits, rats, and hamsters. *Fertil. Steril.*, **25**, 845
28. Matsuoka, Y., Fujita, T., Nozato, T., Yokohama, H., Onishi, Y. and Ohta, K. (1971). Toxicity and teratogenicity of prostaglandin $F_{2\alpha}$. *Iyakuhin Kenkyu (Japan)*, **2**, 403
29. Tuchmann-Duplessis, H. and Mercier-Parot, L. (1972). Action de la prostaglandine $F_{2\alpha}$ sur le corps jaune de la Ratte gestante. *C.R. Acad. Sci. (D) (Paris)*, **275**, 2033
30. Persaud, T. V. N. and Thliveris, J. A. (1978). Pregnancy and progeny in rats treated with prostaglandin $F_{2\alpha}$. *Prostaglandins and Medicine*, **1**, 23
31. Wigglesworth, J. S. (1974). Animal model of human disease: fetal growth retardation. *Am. J. Pathol.*, **77**, 347
32. Persaud, T. V. N. (1974). The effects of prostaglandin $F_{2\alpha}$ on pregnancy and fetal development in mice. *Toxicology*, **2**, 25
33. Bartke, A., Merrill, A. P. and Baker, C. F. (1972). Effects of prostaglandin $F_{2\alpha}$ on pseudopregnancy and pregnancy in mice. *Fertil. Steril.*, **23**, 543
34. Marley, P. B. (1972). Effects of prostaglandins $F_{2\alpha}$, E_2 and E_1 on fertility in mice. *Nature (New Biol.)*, **235**, 213

35. Gutknecht, G. D., Cornette, J. C. and Pharriss, B. B. (1969). Antifertility properties of prostaglandin $F_{2\alpha}$. *Biol. Reprod.*, 1, 367
36. Ellinger, J. V. and Kirton, K. T. (1972). Ovum transport in rabbits injected with prostaglandin E_1 or $F_{2\alpha}$. *Biol. Reprod.*, 7, 106
37. Chang, M. C., Hunt, D. M. and Polge, C. (1973). Effects of prostaglandins (PGs) on sperm and egg transport in the rabbit. *Adv. Biosci.*, 9, 805
38. Chang, M. C. and Hunt, D. M. (1972). Effect of prostaglandin $F_{2\alpha}$ on the early pregnancy of rabbits. *Nature (Lond.)*, 236, 120
39. Chang, M. C., Saksena, S. K. and Hunt, D. M. (1974). Effects of prostaglandin $F_{2\alpha}$ on ovulation and fertilization in rabbit. *Prostaglandins*, 5, 341
40. Lau, I. F., Saksena, S. K. and Chang, M. C. (1974). Intravaginal insertion of a dimethylpolysiloxane-polyvinyl pyrrolidone–prostaglandin $F_{2\alpha}$ tube for midterm abortion in rabbits. *Am. J. Obstet. Gynecol.*, 120, 837
41. Spilman, C. H. (1974). Effect of prostaglandins on the development of rabbit embryos. *J. Reprod. Fertil.*, 39, 403
42. Lau, I. F., Saksena, S. K. and Chang, M. C. (1975). Prostaglandin $F_{2\alpha}$ for induction of midterm abortion: A comparative study. *Fertil. Steril.*, 26, 74
43. Matthews, G. B. P. and Persaud, T. V. N. (1976). Influence of prostaglandin $F_{2\alpha}$ on the development of the chick embryo. Morphological and biochemical aspects. *Exp. Pathol.*, 12, 100
44. Cole, M. P. and Nutting, E. F. (1974). Observations on fetal morphology after maternal treatment with prostaglandins during early pregnancy in the Rhesus monkey (*Macaca mulatta*). (Personal communication)
45. Mirkin, B. L. and Singh, S. (1976). Placental transfer of pharmacologically active molecules. In: B. L. Mirkin (ed.). *Perinatal Pharmacology and Therapeutics*, pp. 1–69. (New York: Academic Press)
46. Persaud, T. V. N. and Clancy, Jr., J. (1974). Placental transfer of ^{3}H-prostaglandin E_2 in mice. *Exp. Pathol.*, 9, 247
47. Persaud, T. V. N. and Jackson, C. W. (1975). Placental transfer of tritium-labelled prostaglandin A_1 in the rat. *Exp. Pathol.*, 10, 353
48. Persaud, T. V. N. (1978). Prostaglandin $F_{2\alpha}$ distribution in pregnant rats. *Exp. Pathol.*, 15, 46
49. Gréen, K., Bygdeman, M. and Wiqvist, N. (1974). Kinetic and metabolic studies of prostaglandin $F_{2\alpha}$ administered intraamniotically for induction of abortion. *Life Sci.*, 14, 2285

9
Percutaneous embryotoxicity testing of chemicals

E. F. STULA

INTRODUCTION

The skin is an important interface between humans and the environment as well as animals and the environment. The skin is also an important portal of entry for various chemicals. The most common exposure of humans or animals to foreign chemicals of all types is by exposure through accidental or intentional contact of the chemical with the skin[1]. The incidence of occupational skin disease as determined from workmen's compensation statistics has ranged from 25 to 80% of all occupational diseases[2]. It is reasonable to expect that at least trace amounts of the causative toxic agents are absorbed internally from the affected skin. Many chemicals have been found to be absorbed percutaneously in humans and animals[3–17]. The incidence where significant amounts of toxic chemicals are absorbed internally from the skin of pregnant women is not known. Factors that determine the rate of percutaneous absorption of chemicals include: concentration and rate of application, pH, water/lipid solubility, solvents/vehicles, chemical structure, molecular size, state of ionization, temperature, number of hair follicles, thickness, state of hydration and other changes to the stratum corneum[18]. The mechanisms of various body defences against toxic chemicals are not well understood. Likewise, the specific body defence differences that exist in the major portals of entry (skin, oral and inhalation) are not well understood. It is for these reasons that a good toxicological test design should include exposure of experimental animals by routes where humans are most apt to be exposed. The experimental exposure should mimic as closely as possible the potential human exposure.

Embryotoxic effects include all embryonal and fetal abnormalities (structural and functional) but are not necessarily detectable at birth. Teratogenic effects are those embryotoxic abnormalities that relate to development. The causes of embryotoxicity may be genetic and/or environmental in origin.

EMBRYOTOXIC POTENTIAL OF CHEMICALS ABSORBED PERCUTANEOUSLY

The application of chemicals to the skin of animals to test for skin irritation, absorption, and sensitization has been used for many years, whereas the application of chemicals to the skin of pregnant animals for the evaluation of embryotoxicity is a more recent development[3]. Theirsch[18], in 1962, induced embryotoxic effects when *N*-methylacetamide (MMAC) or *N,N*-dimethylacetamide (DMAC) was applied to the skin of pregnant rats. Oettel and Froberg[19] found teratogenic effects in mice exposed to *N*-Methylforamide (MMF) or foramide (F) by skin application. Tuchmann–Duplessis and Mercier–Parot[20] found MMF to be teratogenic in rats when applied to the skin in amounts that did not appear to affect the mother. MMAC, DMAC and MMF are used as industrial solvents and may readily penetrate the skin.

Palazzolo *et al.*[7] used the percutaneous route in rabbits to test for the teratogenic potential of a phenylmethylcyclosiloxane; an appropriate route since one of the many applications of organopolysiloxanes has been use as ingredients of topical cosmetic formulations.

Palmer *et al.*[13] compared the embryotoxic potential of the surfactant linear alkylbenzene sulphonate and soap in mice, rats, and rabbits after percutaneous application. Surfactants are used commonly as cleaning agents and dermal contact is considerable.

A percutaneous teratogenic rabbit study of pyrithione, an antidandruff agent, was reported by Nolen *et al.*[14] Various hair dye formulations were tested for percutaneous teratogenicity in rats by Burnett *et al.*[17]

Stula and Krauss[3], using the percutaneous route in pregnant rats, tested the embryotoxic potential of 11 chemicals, some of which are used commercially as solvents. This information was considered to be essential in the formulation of safe industrial hygienic practices. The 11 chemicals tested are shown in Table 9.1: six 'amide-type' solvents, tetramethylurea (TMU) dimethylsulphoxide (DMSO), and three thioureas (in DMSO). In addition, four 'amide-type' solvents were applied percutaneously to pregnant rabbits. Tables 9.1 to 9.11 give a summary of the experimental design together with the results obtained by Stula and Krauss[3] on the 11 chemicals noted above. Table 9.12, from Stula and Krauss[3], ranks these chemicals in order of degree of effect on the following parameters of toxicity in rats: (1) maternal mortality (skin ALD), (2) embryomortality, and (3) teratogenicity. As the number of tests performed with each chemical was not entirely comparable, this table is an approximation of the comparative toxicity. Test chemicals in the ALD range of 1000 to 1500 mg/kg were designated as having a marked effect on maternal mortality, those with an ALD in the 2500 to 7500 range as moderate, and those 11000 and above as slight. In ranking of chemicals as to the degree of embryomortality, the fraction of the ALD required to produce embryomortality was considered. Chemicals that produced over 50% embryomortality with a daily dose of less than one-eighth of the ALD were designated as having marked embryomortality; those requiring more than one-eighth of the ALD for 50% embryomortality were classed as moderate. Chemicals having a slight degree of embryomortality potential produced less than 50%

Table 9.1 Skin absorption approximate lethal dose (ALD) for pregnant rats and rabbits*

Test chemical	*Rat ALD* (mg/kg)	*Rabbit ALD* (mg/kg)
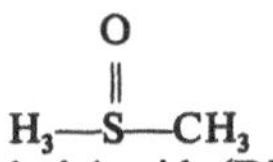 Dimethylsulphoxide (DMSO)	36000	—†
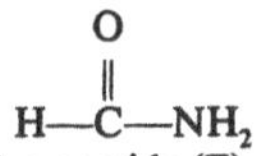 Formamide (F)	>17000	—
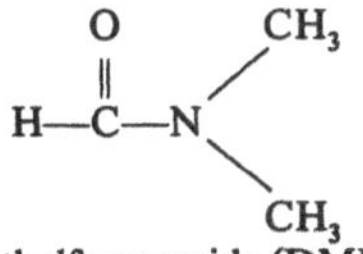 *N,N*-Dimethylformamide (DMF)	17000	3400
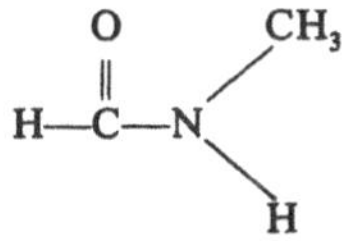 *N*-Methylformamide (MMF)	11000	1500
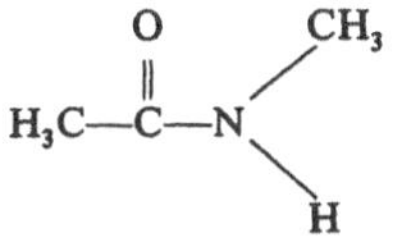 *N*-Methylacetamide (MMAC)	11000	—
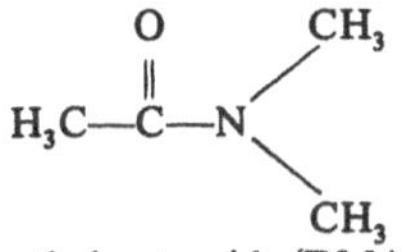 *N,N*-Dimethylacetamide (DMAC)	7500	5000
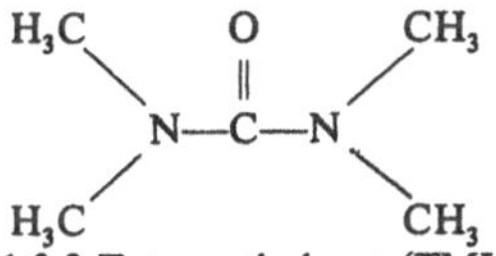 1,1,3,3-Tetramethylurea (TMU)	5000 2250 (in DMSO)‡	3400
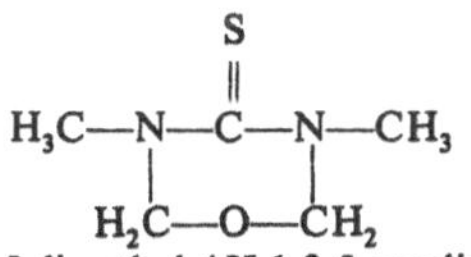 Tetrahydro-3,5,dimethyl-4*H*-1,3,5-oxadiazine-4-thione (TDOT)	3400 (in DMSO)‡	—

* In determining the ALD, one animal was used per dose level with a factor of 1.5 between dose levels. Rats were treated on Day 11 and rabbits on Day 15 of gestation. Rats were sacrificed on Day 21 and rabbits on Day 30 of gestation

† — = not determined

‡ 1.0 g in DMSO to make 5.0 ml of DMSO solution

(*continued*)

Table 9.1 (*continued*)

Test chemical	*Rat ALD* (mg/kg)	*Rabbit ALD* (mg/kg)
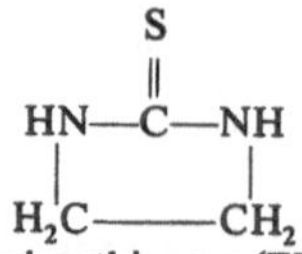 Ethylenethiourea (ETU)	2250 (in DMSO)‡	—
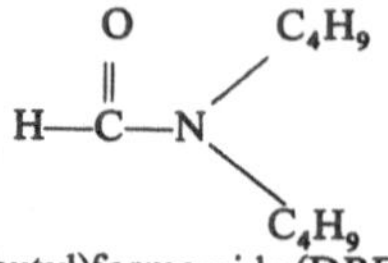 *N,N*-Di(n-butyl)formamide (DBF)	1500	—
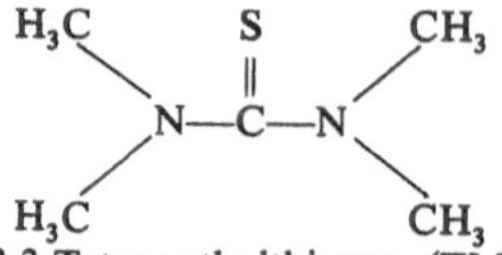 1,1,3,3-Tetramethylthiourea (TMTU)	1000 (in DMSO)‡	—

‡ 1.0 g in DMSO to make 5.0 ml of DMSO solution

embryomortality and/or required a dose that approximated the ALD together with maternal mortality. Those chemicals showing 10% or less embryomortality were considered as having no effect, as up to 10% has been found in historical controls. Teratogenicity was considered marked when fetal malformation could be produced consistently with involvement of multiple organs together with no apparent clinical effect on the mother and on embryomortality. Moderate teratogenicity was described when malformation could be produced consistently, but clinical effects on the mother together with embryomortality were also found. Chemicals with a slight teratogenic potential

Table 9.2 Embryotoxic effects of formamide applied to skin of pregnant rats

Number of pregnant rats in group	*Daily dose [mg/kg] (fraction of ALD)*	*Gestation days applied*	*48-h mother body weight change (%)*	*Embryo-mortality (%)*	*Average fetal weight (g)*	*Fetal abnormalities**
6	600 H_2O Control	10 + 11	+4	2	2.4	—
7	600(<1/28)	9	0	5	2.4	—
7	600(<1/28)	10 + 11	+1	5	2.1	One of 53: distorted face
7	600(<1/28)	11 + 12	+2	13	2.3	—
6	600(<1/28)	12 + 13	+2	5	2.1	Four of 60: subcutaneous haemorrhage

* No entry under 'Fetal abnormalities' indicates 'none'

had a sporadic incidence of malformation.

With the four solvents (TMU, DMF, DMAC, MMF) tested in both rats and rabbits, the skin ALD was lower in rabbits than in rats (on a milligram per kilogram body weight basis)[3]. This species difference may be due to a lower skin permeability in rats than in rabbits as reported by Bartek *et al.*[6] for chemicals other than the ones used in these experiments.

The rates of application and absorption of the chemical through the skin appear to be important factors in the evaluation of maternal and embryotoxicity[3]. Mixing TMU with mineral oil increased the maternal toxicity of TMU in rats. Six applications of MMF, each amounting to one-sixth of the single daily dose, applied during an 8 h period were more toxic to the mother and embryo than a single daily dose in rats. It appears that for a toxic change to occur, a specific tissue level of the chemical and/or its metabolites must be maintained for a specific length of time.

Table 9.3 Embryotoxic effects of *N*-methylformamide applied to skin of pregnant rats and rabbits

Number of pregnant animals in group	*Daily dose [mg/kg] (fraction of ALD)*	*Gestation days applied*	*48-h mother body weight change (%)*	*Embryo-mortality (%)*	*Average fetal weight (g)*	*Fetal abnormalities**
Rats						
7*	600 H_2O Control	11 + 12	+7	1	2.4	—
8	600 (1/18)	9	—†	95	1.6	1 of 4: umbilical hernia
8	600 (1/18)	10 + 11	−9	72	2.4	1 of 23: umbilical hernia
7	600 (1/18)	11 + 12	−7	25	2.2	18 of 51: encephalocele
8	600 (1/18)	12 + 13	−9	64	2.1	32 of 32: encephalocele; 12 of 32: umbilical hernia
Rats						
8	0 Control	—‡	+4	1	2.4	—
8	600 (1/18)	12 + 13	−11	63	2.1	5 of 25: encephalocele
10	1200 (1/9)	12 + 13	−14	64	2.2	13 of 37: encephalocele
8	2400 (1/4)	12 + 13	−13	100	1.9	14 of 21: encephalocele
Rats						
4	0 Control	—‡	+2	3	2.3	—
6	200 (1/55)	12 + 13	−4	2	2.2	12 of 60: encephalocele; 1 of 60: umbilical hernia; 5 of 60: subcutaneous haemorrhage .
6	600 (1/18)	12 + 13	−9	24	2.2	46 of 52: encephalocele; 12: umbilical hernia, all from one mother
6	100 × 6 (1/18)	12 + 13	−15	77	1.8	12 of 12: encephalocele
Rabbits						
4	200 H_2O Control	8–16	—†	3	28.4	—
2§	200 (1/18)	8–16	—†	100	—‖	—

* No entry under 'Fetal abnormalities' indicates 'none'
† Not weighed
‡ Untreated
§ Not included were four of eight adults that died in 3 to 9 days during the exposure
‖ No live fetuses

Table 9.4 Embryotoxic effects of *N,N*-dimethylformamide applied to skin of pregnant rats and rabbits

Number of pregnant animals in group	*Daily dose [mg/kg] (fraction of ALD)*	*Gestation days applied*	*48-h mother body weight change (%)*	*Embryo-mortality (%)*	*Average fetal weight (g)*	*Fetal abnormalities**
Rats						
6	600 H_2O Control	11 + 12	+4	3	2.7	—
7	600 (1/28)	9	—†	3	2.7	—
3	600 (1/28)	10 + 11	0	2	2.5	—
6	600 (1/28)	11 + 12	+2	4	2.9	—
8	600 (1/28)	12 + 13	0	4	2.5	—
Rats						
6	600 H_2O Control	10 + 11	+4	3	2.3	—
6	600 (1/28)	10 + 11	+2	8	2.5	—
8	1200 (1/14)	10 + 11	+2	4	2.6	—
7	2400 (1/7)	10 + 11	0	26‡	2.4	—
Rats						
6	600 H_2O Control	10 + 11	+3	9	2.1	—
9	2400 (1/7)	10 + 11	0	9	2.2	—
Rats						
8	600 H_2O Control	12 + 13	+4	1	2.5	—
8	600 (1/28)	12 + 13	+4	5	2.4	9 of 84: subcutaneous haemorrhage
7	1200 (1/14)	12 + 13	+2	4	2.5	5 of 73: subcutaneous haemorrhage
8	2400 (1/7)	12 + 13	+2	4	2.4	4 of 83: subcutaneous haemorrhage
Rats						
4	0 Control	—§	+3	4	2.3	—
7‖	200 × 6 (1/14)	11 + 12 + 13	0	43	2.2	3 of 40: subcutaneous haemorrhage
8¶	400 × 6 (1/7)	11 + 12 + 13	−2	30	1.9	—
Rabbits						
4	200 H_2O Control	8–16	—†	3	28.4	—
5	200 (1/17)	8–16	—†	6	32.9	—

* No entry under 'Fetal abnormalities' indicates 'none'
† Not weighed
‡ Eight of 14 resorptions in one mother
§ Untreated
‖ Not included were three rats that died
¶ Not included was one rat that died

Table 9.5 Embryotoxic effects of *N,N*-di(n-butyl) formamide applied to skin of pregnant rats

Number of pregnant rats in group	*Daily dose [mg/kg] (fraction of ALD)*	*Gestation days applied*	*48-h mother body weight change (%)*	*Embryo-mortality (%)*	*Average fetal weight (g)*	*Fetal abnormalities**
9	600 H_2O Control	10	+2†	7	2.6	—
8	600 (1/2)	10	0†	7	2.5	—
4‡	1200 (1/1)	10	−5†	33	2.7	—
6	600 H_2O Control	10 + 11	+3	9	2.1	—
5	600 (1/2)	10 + 11	−4	8	1.6	—

* No entry under 'Fetal abnormalities' indicates 'none'
† 24-h mother body weight change
‡ Not included were one of five pregnant mothers that died; three of three nonpregnant females survived this dosage

Table 9.6 Embryotoxic effects of *N*-methylacetamide applied to skin of pregnant rats

Number of pregnant rats in group	*Daily dose [mg/kg] (fraction of ALD)*	*Gestation days applied*	*48-h mother body weight change (%)*	*Embryo-mortality (%)*	*Average fetal weight (g)*	*Fetal abnormalities**
8	600 H_2O Control	11 + 12	+5	3	2.4	—
8	600 (1/18)	9	—†	5	2.4	—
8	600 (1/18)	10 + 11	+2	14	2.3	—
7	600 (1/18)	11 + 12	+7	4	3.4	—
8	600 (1/18)	12 + 13	+3	7	2.2	1 of 71: diffuse subcutaneous oedema

* No entry under 'Fetal abnormalities' indicates 'none'
† Not weighed

Table 9.7 Embryotoxic effects of *N,N*-dimethylacetamide applied to skin of pregnant rats and rabbits

Number of pregnant animals in group	*Daily dose [mg/kg] (fraction of ALD)*	*Gestation days applied*	*48-h mother body weight change (%)*	*Embryo-mortality (%)*	*Average fetal weight (g)*	*Fetal abnormalities**
Rats						
8	600 H_2O Control	10 + 11	+6	6	2.7	—
7	600 (1/12)	9	—†	15	2.5	—
7	600 (1/12)	10 + 11	+4	12	2.2	—
8	600 (1/12)	11 + 12	+4	16	2.1	—
8	600 (1/12)	12 + 13	+2	4	2.1	—
7	1200 (1/6)	9	—†	10	2.4	—
5	1200 (1/6)	10 + 11	−1	45	2.2	3 of 34: encephalocele; all from one mother
8	1200 (1/6)	11 + 12	+1	19	2.2	—
Rats						
7	600 H_2O Control	10 + 11	+4	3	2.3	—
8	600 (1/12)	10 + 11	+4	14	2.3	—
8	1200 (1/6)	10 + 11	+1	89	1.9	1 of 8: diffuse subcutaneous oedema
6	2400 (1/3)	10 + 11	−1	100		
Rabbits						
4	200 H_2O Control	8–16	—†	3	28.4	—
5	200 (1/25)	8–16	—†	0	32.6	—

* No entry under 'Fetal abnormalities' indicates 'none'
† Not weighed

Table 9.8 Embryotoxic effects of 1,1,3,3-tetramethylurea applied to skin of pregnant rats and rabbits

Number of pregnant animals in group	Daily dose [mg/kg] (fraction of ALD)	Gestation days applied	48-h mother body weight change (%)	Embryo-mortality (%)	Average fetal weight (g)	Fetal abnormalities*
Rats						
6	600 H_2O Control	12 + 13	+6	0	2.6	—
8	600 (1/18)	9	—†	27	1.7	1 of 48: malformed head
8	600 (1/8)	10 + 11	–5	100	—‡	—
7	600 (1/8)	11 + 12	–4	78	1.7	—
7	600 (1/8)	12 + 13	0	79	2.0	4 of 19: diffuse subcutaneous oedema
Rats						
8	600 H_2O Control	10 + 11	+3	8	2.5	—
7	600 (1/8)	10 + 11	–1	25	2.1	—
5	400 (1/12)	10 + 11	–1	11	2.0	—
6	200 (1/25)	10 + 11	+1	4	2.2	—
8	100 (1/50)	10 + 11	+2	19§	2.3	—
Rats						
7	600 H_2O Control	11 + 12	+5	0	2.5	—
7	600 (1/8)	9	—†	11	2.1	—
8	600 (1/8)	10 + 11	–2	71	2.0	—
8	600 (1/8)	11 + 12	–1	2	1.9	—
5	600 (1/8)	12 + 13	–1	14	2.1	8 of 67: diffuse subcutaneous oedema
Rats						
8	1600 Mineral oil	9 + 10	—†	5	3.7	—
6	600	8 + 9	—†	49	2.6	—
7	600 (1/8)	9 + 10	—†	72	4.1	1 of 21: short tail
5‖	600 (1/8)¶	9 + 10	—†	76	3.4	—
7	400 (1/12)	8–13	—†	57	3.2	—
Rats						
8	1600 Mineral oil	10 + 11	+8	5	2.3	—
8	600 (1/8)	10 + 11	–4	81	2.7	—
Rats						
8	6000 H_2O Control	10 + 11	0	5	3.9	—
8	6000 (1/6)	10 + 11	+3	9	3.9	1 of 40: encephalocele
7	500 (1/4)**	10 + 11	–4	100	—‡	—
Rats						
6	6000 H_2O Control	12 + 13	–5	3	3.8	—
8	6000 (1/6)	12 + 13	–5	10	3.8	—
8	500 (1/4)**	12 + 13	–8	7	3.0	—
Rabbits						
4	200 H_2O Control	8–16	—†	3	28.4	—
6	200 (1/17)	8–16	—†	25	28.3	—

* No entry under 'Fetal abnormalities' indicates 'none'
† Not weighed ‡ No live fetuses
§ Six of 12 resorptions from one mother
‖ Not included were three pregnant rats that died
¶ In 1000 mg/kg of mineral oil
** 1.0 g of TMU in DMSO to make 5.0 ml of DMSO solution

Table 9.9 Embryotoxic effects of 1,1,3,3-tetramethylthiourea applied to skin of pregnant rats

Vumber of 'regnant rats in group	Daily dose [mg/kg] (fraction of ALD)	Gestation days applied	48-h mother body weight change (%)	Embryo-mortality (%)	Average fetal weight (g)	Fetal abnormalities*
8	6000 H_2O Control	10 + 11	0	5	3.9	—
8	6000 DMSO (1/6)	10 + 11	+3	9	3.9	1 of 40: encephalocele
7	250 (1/4)†	10 + 11	−5	52	3.2	—
7	250 (1/4)†	12 + 13	−10	20	2.9	—
8	6000 H_2O Control	12 + 13	−5	3	3.8	—
8	6000 DMSO (1/6)	12 + 13	−5	10	3.8	—
5‡	500 (1/2)†	12 + 13	−13	74	3.0	2 of 13: subcutaneous haemorrhage

No entry under 'Fetal abnormalities' indicates 'none'
1.0 g of TMTU in DMSO to make 5.0 ml of DMSO solution
Not included were three pregnant rats that died at this level of exposure

Table 9.10 Embryotoxic effects of ethylenethiourea applied to skin of pregnant rats

Vumber of regnant rats in group	Daily dose [mg/kg] (fraction of ALD)	Gestation days applied	48-h mother body weight change (%)	Embryo-mortality (%)	Average fetal weight (g)	Fetal abnormalities*
8	6000 H_2O Control	10 + 11	0	5	3.9	—
8	6000 DMSO (1/6)	10 + 11	+3	9	3.9	1 of 40: encephalocele
8	50 (1/45)†	10 + 11	+4	7	4.1	3 of 83: fetuses with short tail; 2 of 83: fused ribs
5	25 (1/90)†	10 + 11	+4	7	4.2	—
6	6000 H_2O Control	12 + 13	−5	3	3.8	—
8	6000 DMSO (1/6)	12 + 13	−5	10	3.8	—
6	50 (1/45)†	12 + 13	−5	4	3.3	All 73 fetuses malformed‡

No entry under 'Fetal abnormalities' indicates 'none'
1.0 g of ETU in DMSO to make 5.0 ml of DMSO solution
One or more of the following malformations were found: encephalocele, a part or entire tail missing, missing leg nes, hunchback curvature to spine, short mandible, fusion of ribs, fusion of sternabrae

able 9.11 Embryotoxic effects of tetrahydro-3,5-dimethyl-4*H*-1,3,5-oxadiazine-4,-thione applied to skin of pregnant rats

Vumber of regnant rats in group	Daily dose [mg/kg] (fraction of ALD)	Gestation days applied	48-h mother body weight change (%)	Embryo-mortality (%)	Average fetal weight (g)	Fetal abnormalities*
8	6000 H_2O Control	10 + 11	0	5	3.9	—
8	6000 DMSO (1/6)	10 + 11	+3	9	3.9	1 of 40: encephalocele
6	500 (1/7)†	10 + 11	0	36	3.3	—
8	500 (1/7)†	12 + 13	0	10	3.3	1 of 70: focal subcutaneous haemorrhage
6	6000 H_2O Control	12 + 13	−5	3	3.8	—
8	6000 DMSO (1/6)	12 + 13	−5	10	3.8	—
7‡	1500 (1/2)†	12	−6	35	2.9	—

No entry under 'Fetal abnormalities' indicates 'none'
1.0 g of TDOT in DMSO to make 5.0 ml of DMSO solution
Not included was one pregnant rat that died at this dose

Table 9.12 Summary of comparative toxicity of various chemicals in pregnant rats after skin application

Degree of effect	*Maternal mortality*	*Embryomortality*	*Teratogenicity*
Marked	TMTU* DBF	MMF TMU	ETU*
Moderate	ETU* TDOT* TMU DMAC	DMAC TMTU* TDOT*	MMF
Slight	MMAC MMF DMF F DMSO	F DBF MMAC DMF	TMU DMAC F MMAC
No effect demonstrated		ETU* DMSO	TMTU* DMF DMSO TDOT* DBF

* In DMSO

In making a species comparison of the potential hazard of percutaneous absorption of chemicals, humans appear to have a cutaneous defence mechanism that is superior to that found in rats or rabbits. Bartek *et al.*[6] reported that skin permeability to six chemicals tested decreased in the following order: rabbit, rat, pig, and human.

SUMMARY

The need to include percutaneous teratological studies in a complete toxicological evaluation of certain chemicals has been presented. Selected references to percutaneous teratological studies using a variety of chemicals have been presented. Tabular results of the work of Stula and Krauss[3] have been presented. For further details of these experiments, the original paper should be referred to.

ACKNOWLEDGEMENT

We thank Academic Press, Inc. for permitting the reprinting of Tables 9.1 to 9.12 from *Toxicol. Appl. Pharmacol.*, 1977, **41**, 35

References

1. Lomis, T. A. (1974). *Essentials of Toxicology*. 2nd ed. p. 63. (Philadelphia, PA: Lea and Febiger)
2. Johnson, M. L. (1969). *Diability*. pp. 46 and 48. (Evanston, Ill: American Academy of Dermatology)

3. Stula, E. F. and Krauss, W. C. (1977). Embryotoxicity in rats and rabbits from cutaneous application of amide-type solvents and substituted ureas. *Toxicol. Appl. Pharmacol.*, **41**, 35
4. Feldmann, R. J. and Maibach, H. I. (1970). Absorption of some organic compounds through the skin in man. *Invest. Dermatol.*, **54**, 399
5. Wester, R. C. and Maibach, H. I. (1975). Percutaneous absorption in the Rhesus monkey compared to man. *Toxicol. Appl. Pharmacol.*, 32, 394
6. Bartek, M. J., LaBudde, J. A. and Maibach, H. I. (1972). Skin permeability *in vivo*: Comparison in rat, rabbit, pig, and man. *J. Invest. Dermatol.*, **58**, 114
7. Palazzolo, R. J., McHard, J. A., Hobbs, E. J., Fancher, O. E. and Calandra, J. C. (1972). Investigation of the toxicologic properties of a phenylmethylcyclosiloxane. *Toxicol. Appl. Pharmacol.*, **21**, 15
8. Malkinson, F. D. (1960). Percutaneous absorption of toxic substances in industry. *Arch. Indust. Health.*, **21**, 87
9. Suskind, R. R. (1977). Environment and the skin. *Envir. Health Perspectives*, **20**, 27
10. Chow, C. P., Buttar, H. S. and Downie, R. H. (1977). Percutaneous absorption of chlorhexidine in rats. *Toxicol. Lett.*, 1, 213
11. Maibach, H. I., Feldmann, R. J., Milby, T. H. and Serat, W. F. (1971). *Arch. Envir. Health*, **23**, 208
12. Scheuplein, R. J. (1976). Permeability of the skin: A review of major concepts and some new developments. *J. Invest. Dermatol.*, **67**, 672
13. Palmer, A. K., Readshaw, M. A. and Neuff, A. M. (1975). Assessment of the teratogenic potential of surfactants. Part III. Dermal application of LAS and soap. *Toxicology*, **4**, 171
14. Nolen, G. A., Patrick, L. F. and Dierckman, T. A. (1975). A percutaneous teratology study of zinc pyrithione in rabbits. *Toxicol. Appl. Pharmacol.*, **31**, 430
15. Shah, H. C. and Lal, H. (1976). Effects of 1,1,1-trichloroethane administered by different routes and in different solvents on barbiturate hypnosis and metabolism in mice. *J. Toxicol. Envir. Health*, **1**, 807
16. Jacob, L. (1977). *Survey and Evaluation of Techniques Used in Testing Chemical Substances For Teratogenic Effects*. p. 95. (Washington, DC: Environment Protection Agency)
17. Burnett, C., Goldenthal, E. I., Harris, S. B., Wazeter, F. X., Strausburg, J., Kapp, R. and Voelker, R. (1976). Teratology and percutaneous toxicity studies on hair dyes. *J. Toxicol. Envir. Health*, **1**, 1027
18. Theirsch, J. B. (1962). Effects of acetamides and formamides on the rat litter *in utero*. *J. Reprod. Fertil.*, **4**, 219
19. Oettel, H. and Frohberg, H. (1965). *Rapport au 4e Congrés International de Medécine Preventive*. p. 331. (Vienne, Austriche)
20. Tuchmann-Duplessis, M. H. and Mercier-Parot, L. (1965). Production chez le rat d'anomalies aprés applications cutanèes d'un solvant industriel: la mono-methyl-formamide. *C. R. Acad. Sci.*, **261**, 241
21. Grasso, P. and Lansdown, A. B. G. (1972). Methods of measuring and factors affecting percutaneous absorption. *J. Soc. Cosmet. Chem.*, 23, 481

10
Effects of mycotoxins on development

R. D. HOOD

MYCOTOXINS AS A HEALTH PROBLEM

Epidemics of 'St Anthony's fire' (ergotism) were not uncommon in the Western World during the Middle Ages. These outbreaks of dry gangrene and neurological disorders are now known to have been caused by consumption of rye heavily contaminated with a group of related alkaloids from the sclerotia of the fungus *Claviceps purpurea*[1]. Upon consuming ergot-infected cereal grains, domestic animals have been affected in a similar fashion by ergot alkaloids[2]. In the 1930s and '40s, reports of other mycotoxic diseases, such as alimentary toxic aleukia and 'yellow rice disease' in man and stachybotryotoxicosis of horses, began to appear[3–5]. Such reports were followed by those of acute haemorrhagic hepatitis of swine and cattle and deaths of dogs fed 'mouldy' corn[6,7]. Despite such evidence, it was not until 1960, when the death of 100000 turkey poults in Great Britain due to consumption of aflatoxin-contaminated peanut meal was reported[8], that the significance of mycotoxins as an aetiological factor in diseases of man or domestic animals became widely appreciated. Interest in toxic fungal metabolites has been sustained by evidence linking aflatoxins with human carcinogenesis[9] and by reports[10,11], such as those that associate ochratoxin A and citrinin with endemic porcine nephropathy and T-2 toxin with a lethal toxicosis of dairy cattle. Intake of mouldy grain was suggested as the culprit in both cases.

The effects of prenatal exposure to mycotoxins had not been investigated until the 1960s, when Le Breton *et al.*[12] reported the fetocidal effect of transplacental exposure to aflatoxin B_1. This report on aflatoxin was soon followed by others[13–15], but the first paper[16] describing the teratogenic effects of any other mycotoxins was not presented until 1973, and the vast majority of known mycotoxins have never been screened for teratogenicity.

OCCURRENCE OF MYCOTOXINS IN FOODS

It has been stated that 'Mycotoxins may well be among the world's most significant food contaminants. The sources of many of the unexplained diseases of man and animals may lie in their exposure, through food, to these chemicals'[17]. A large number of mycotoxins have been identified, produced by a variety of moulds growing under different conditions and on many different substrates. Since most fungal contamination occurs under high moisture conditions in stored materials, high humidity together with poor harvesting or storage practices is particularly conducive to mycotoxin production. In most cases where significant mycotoxin levels have occurred, mould contamination is relatively obvious, but this is not always the case. The fungal growth may have regressed, leaving certain stable toxins behind, or the food may have been processed, thus destroying the visual evidence. Also, consumption of toxin-containing feeds by livestock, as well as use of toxigenic fungal strains in production of mould-ripened foods, can result in contaminated meat as well as milk and other dairy products[18]. Certain mycotoxins, such as ochra-

Table 10.1 Occurrence of toxigenic fungi* in agricultural products

Toxin	*Organism*	*Sources*	*Reference*
Aflatoxins	various Aspergilli, e.g. *A. flavus*, *A. parasiticus*, *A. ochraceus*	corn, cottonseed, peanuts, pistachio nuts, pecans, dried beans, various food products	23–28
'*Alterneria* toxins'	various *Alterneria*, e.g. *A. mali*, *A. tenuis*	grains, peanuts, alfalfa, grass hay	29, 30
Citrinin	*Penicillium viridicatum*, *P. citrinum* and various *Penicillia* and *Aspergilli*	wheat, rye, barley, dried beans	10, 31
Cytochalasins	*Phoma exigua*, *Hormiscium* spp., *Phomopsis paspalli*, various *Helminthosporia*	kodo millet, tomatoes, potatoes	32–34
Ergot alkaloids	various *Claviceps*, e.g. *C. purpurea*	rye, wheat, pearl millet	35, 36
Ochratoxin A	*Aspergillus ochraceus*, *Penicillium viridicatum*	various food products, oats, corn, barley, wheat, cured ham, dried beans, peanuts, pecans	10, 23, 25, 28, 31, 37
Patulin	various *Penicillia* and *Aspergilli*, e.g. *P. expansum*, *P. urticae*, *A. flavus*	apples, grapes, flour, dried beans	25, 38, 39, 40
Penicillic acid	various *Penicillia* and *Aspergilli*, e.g. *P. martensii*, *A. ochraceus*	corn, beans, tobacco	25, 41–42
Penitrem A	*Penicillium cyclopium*	corn	44
Rubratoxin B	*Penicillium rubrum*, *P. purpurogenum*	corn	45
Stachybotrys toxins	*Stachybotrys alternans*	oats, barley, straw, hay	5
T-2 toxin	various *Fusaria*, e.g. *F. tricinctum*	corn	11, 19
Zearalenone (F-2)	various *Fusaria*, e.g. *F. roseum*, *F. graminearum*	corn	19, 46–48

* This list is not meant to be exhaustive

toxin A, zearalenone, T-2 toxin, and aflatoxins, are relatively heat-stable and thus not certain to be destroyed during processing and cooking[19,20].

Although current food storage, handling and grading practices in the more technologically advanced nations are thought to keep levels of most mycotoxin contamination relatively low, livestock are still subjected to significant amounts of mould-contaminated feeds. The situation in less developed countries, particularly where the climate is warm and humid, is often less favourable. War, drought or other causes of short food supplies in such areas result in human consumption of mycotoxin-contaminated foods[17,21], followed by mycotoxicoses and sometimes deaths[22].

Toxigenic fungi have been isolated from a wide variety of agricultural products throughout the world, in both tropical and temperate climates. Examples of known toxin-producing moulds are presented in Table 10.1.

TERATOGENIC MYCOTOXINS

Mycotoxins that have been screened for teratogenicity and/or prenatal toxicity are listed in the following portion of this chapter. Each type will be discussed separately, because each is a unique compound, making broad generalizations impossible.

In the discussions which follow, gestation days have been converted to conform with designating the day of finding sperm or a copulation plug (rats and mice) or the day following mating the evening before (hamsters) as day one. This was done in order to avoid the necessity to state in each case whether the first day was called 'day one' or 'day zero.'

Aflatoxins

The aflatoxins consist primarily of four compounds, designated B_1, B_2, G_1 and G_2 on the basis of their fluorescence ('B' for blue, 'G' for green) and their relative mobilities when chromatographed. Aflatoxins invariably contain a coumarin nucleus linked to a bifuran. The B_1 and B_2 molecules also contain a pentanone ring, while in G_1 and G_2 there is a six-carbon lactone instead. Similarly toxic 4-hydroxylated derivatives, designated M_1, M_2 and GM_1, are seen in milk of animals fed diets containing B_1, B_2 and G_1, respectively[24].

Aflatoxins are highly toxic, with day-old ducklings being particularly sensitive (oral $LD_{50}=0.36$ and 0.78 mg/kg for aflatoxins B_1 and G_1, respectively)[49], while the rat is more resistant (oral $LD_{50}=5$ mg/kg)[50]. Aflatoxins are also among the most potent known carcinogens[19]. This fact, combined with their widespread occurrence in agricultural products, has served to generate more interest in the aflatoxins than in any other mycotoxins.

LeBreton *et al.*[12] reported that in rats, a maternal intraperitoneal dose of 300 μg aflatoxin B_1 caused fetal deaths, with haemorrhage at the utero-placental junction. Repeated small doses resulted in fetal growth retardation. Butler and Wigglesworth[13] reported decreased fetal weights and a slight reduction in placental weight when the dams were subjected to a single oral dose of 1 mg/rat on gestation day 17. This was said to be one-quarter of the LD_{50} for non-pregnant female rats. Litters treated earlier (days 8–13) exhibited no growth inhibition. No placental pathology other than a possible slight weight decrease was noted, nor did fetal organs show histological evidence of damage, though the maternal livers exhibited the expected lesions of aflatoxicosis. Chromatography of fetal liver extracts obtained after toxin administration on day 21 indicated that, at least at that point in gestation, aflatoxin crossed the 'placental barrier'. The authors concluded that the fetal growth retardation associated with aflatoxin treatment may have been an indirect effect brought about as a consequence of maternal liver damage, as fetal effects were correlated with visible effects on the maternal liver. The conflict between these results and those of the previous study may have been due to the difference in administration route employed (oral versus intraperitoneal).

The developing golden hamster (*Cricetus auratus*) appears to be more susceptible than the rat to damage from aflatoxin exposure. Elis and DiPaolo[14] injected pregnant hamsters i.p. with single doses of 4 mg/kg aflatoxin B_1 on day 8 of pregnancy, after having determined an LD_{50} of 6 mg/kg. Results were assessed on days 9, 11, 13 or 15. A variety of malformations were noted, but number and severity were negatively correlated with examination at progressively later stages of development. Conversely, the number of dead or resorbed fetuses was positively correlated with stage at examination. Such results are not unusual, and are assumed to be due to failure of the more severely malformed fetuses to survive until late in gestation. Since aflatoxin B_1 is known to bind to DNA[51], Elis and DiPaolo[14] attempted to use calf thymus DNA to alleviate the teratogenicity of aflatoxin B_1. In these experiments, mixing aflatoxin (doses of 4 or 6 mg/kg) with DNA in solution for 24 h prior to injection into the pregnant hamsters appeared to reduce the teratogenic effects of the mycotoxin, as well as reducing maternal mortality. At an aflatoxin dose of 8 mg/kg, with or without DNA, no mothers survived. Injection of aflatoxin and DNA separately was not generally beneficial, nor was injection of fractionated doses of DNA at intervals following aflatoxin treatment. Concurrent or subsequent DNA injection did, however, prevent or decrease the fetal liver damage noted in the hamster; DNA prophylaxis was less protective of the maternal liver. Such fetal liver damage was not noted in fetuses of the rat[13] or the mouse[15]. The authors[14] speculated that the protective effect of DNA for the hamster fetus may have been related to decreased ability of the toxin–DNA complex to cross the placenta or to slow release of aflatoxin from the complex.

In a subsequent report, DiPaolo *et al.*[15] reported that following a low i.p. dose (2 mg/kg) of aflatoxin B_1 on gestation day 8 or a higher dose (4 mg/kg) on day 13 – after the time of major organogenesis – significant effects on mouse fetuses were not seen. When C3H/He mice were given 8 or 12 mg/kg on gestation day 9, 90% prenatal mortality resulted, while the survivors were normal in appearance. Repeated small doses had a lesser effect on fetal survival and again resulted in no malformations or other detectable adverse effects. These differences in susceptibility of different species of laboratory animals to teratogenesis are not unusual findings. Such results are thought to be due to differences in metabolism or excretion, or to differences in inherited susceptibility to alteration of specific developmental processes.

The precise mechanism by which aflatoxins influence developing systems remains to be elucidated. Upon binding to DNA, aflatoxins have been found to inhibit both DNA and RNA synthesis[24]. Numerous reports have also described direct effects on protein synthesis, due to mechanisms such as polyribosome disaggregation[52]. Aflatoxins have also been implicated in inhibition of mitochondrial respiration, pathological alterations of lipid metabolism, and binding to nuclear histones, resulting in alteration of the chromatin[52]. The increased aflatoxin susceptibility of the rat compared with that of the mouse may be related to the ability of the latter to metabolize the toxin more rapidly[53]. In view of the many effects of aflatoxins, their effects on the conceptus are likely to be the result of a variety of mechanisms.

'*Alterneria* toxins'

Toxic metabolites of *Alterneria* have not been extensively studied. *Alterneria* are phytopathogens and cause a number of important crop diseases, e.g. early blight of potato and tomato and seedling chlorosis of cotton and citrus[30]. They also have been found to elaborate a number of compounds of significant toxicity to animals, including dibenzopyrones, tetramic acids and a third class of compounds of unspecified structure[29].

Two of these dibenzopyrones, alternariol (AOH) and alternariol monomethyl ether (AME), were isolated from *Alternaria tenuis* and administered to pregnant DBA mice on gestation days 9–12 by Pero *et al.*[29] A combination of the two toxins in DMSO, administered subcutaneously at a dose level of 25 mg/kg each, was found to be fetocidal and to result in underweight fetuses. The incidence of fetal malformations appeared to increase, but the increase only approached statistical significance. AOH alone (100 mg/kg) had similar effects, but was not notably teratogenic; similar treatment with AME had no effect. AOH at a dose level of 100 mg/kg on days 13–16 was teratogenic, producing a variety of defects, while AME at 50 mg/kg on the same days was not. These results were reported to suggest a synergistic effect, although when there are small numbers of litters in treatment groups, such effects are at times difficult to assess. Synergistic effects for lethality in *Bacillus mycoides*, but not for HeLa cells or mice, were also reported, but the mechanism is unknown[9].

Citrinin

Citrinin is a cyclic, low molecular weight metabolite of various *Penicillium* and *Aspergillus*[54] species that has been known since 1931[55]. Citrinin has been found to be a potent nephrotoxin in mammals[54], with a 14-day LD_{50} (in mg/kg) of 35 (s.c. or i.p.) in mice, 67 (s.c.) in rats, and 19 (i.v.) in rabbits[56,57].

The effects of prenatal exposure of mice to citrinin were investigated by Hood *et al.*,[58] who administered citrinin in propylene glycol at doses of 10, 20, 30 or 40 mg/kg on one of days 7–10 of pregnancy. The highest dose resulted in some maternal deaths. It also resulted in significant decreases in fetal survival to gestation day 18 and slight decreases in fetal weight. A few malformations were seen, but the incidence was not significant. Since there appears to be little information on the molecular basis for citrinin toxicity, other than the toxin's ability to inhibit proteolysis in phagolysosomes[59], and to adversely affect hepatic lipid metabolism[52], it is difficult to speculate on the cause of its fetotoxicity.

Cytochalasins

Cytochalasins are a group of mycotoxins that are of particular interest because of their effects on microfilaments[60, 61]. Such agents with known effects at the molecular or organelle level are potentially useful as tools for investigating mechanisms of teratogenicity, as well as normal developmental processes.

The cytochalasins are produced by moulds of the genera *Phoma*[34], *Hormiscium*[3], *Helminthosporium*[62], and others. These compounds are highly toxic, with an LD_{50} of 2.1–2.2 mg/kg in three mouse strains for single i.p. doses[63].

The initial experiments with cytochalasins involved the use of chick embryos. Linville and Shepard[64] treated explanted embryos of 1–4 somites with cytochalasin B dissolved in DMSO and added to the thin albumin culture medium, comparing their neural tube development with that of embryos exposed to DMSO alone. Each embryo received a total dose of 0.5, 1.0 or 2.0 μg. Treatment resulted in neural tube closure defects, generally associated with missing optic vesicles. The highest doses also resulted in general developmental retardation and increased mortality. No direct evidence was presented, however, as to whether disruption of microfilaments was involved.

In another study involving the chick embryo, Karfunkel[65] also used explanted embryos of the 4-somite stage and exposed them to cytochalasin B in DMSO mixed in a thin albumin solution. The results obtained were similar to those in the previously discussed study[64]. The neural tissue flattened, rather than rising in folds, and already-formed portions of the neural folds separated at the midline and regressed to a flattened state. The neural cells remained columnar but lost their wedge shape and were lacking in microfilaments. Treatment also prevented the formation of the wedge shape typically attained by elongating neural cells posterior to the initially formed portions of the neural folds. These results appear to implicate microfilaments in the normal apical constriction of neural cells which is correlated with the elevation of the neural folds, and implicate disruption of microfilaments as a possible teratogenic mechanism.

Greenaway *et al.*[66] compared the relative potencies of cytochalasins A, B, D and E on explanted chick embryos. They found that all four toxins were teratogenic *in vitro*. Cytochalasin D caused open neural tubes at doses lower than was required for A or B, which were intermediate in potency; cytochalasin E required the highest doses for effect. These results, however, differed from the effects of the same mycotoxins in a binucleation assay of a mouse embryo cell line, an assay thought to test for interference with cytokinesis. Cytochalasin E proved most effective in inducing binucleation, with A and D intermediate, and B least effective.

To determine if species differences in response of cells to the cytochalasins were responsible for the apparently contradictory results, the authors[66] tested chick fibroblasts in the binucleation assay. In this case, the results were similar to those obtained with mouse cells. Greenaway *et al.*[66] suggested that the discrepancy between the results of the binucleation assays – which are believed to be measures of microfilament disruption – and the assay of neural tube disruption may be due to other effects of the cytochalasins. They give examples of the ability to interfere with such processes as hexose transport and cell surface attributes (which may also, however, be influenced by microfilaments). If one or more such processes are crucial to neural tube closure, cytochalasin effects on these may be more significant than those measured in the binucleation assay.

In experiments involving the rat, Ruddick *et al.*[67] observed no effects on development following maternal exposure to cytochalasin B dispersed in water and administered daily by gavage on gestation days 6–15. Dose levels were 0.25 or 1.0 mg/kg. Such negative results may indicate a resistance of the developing rat to cytochalasins (or at least to cytochalasin B) or may merely be due to use of a resistant strain. In a more recent study, Shepard and Greenaway[63] produced exencephaly, hypognathia and skeletal defects in C57BL/6J and BALB/c mice (but not in a Swiss-Webster strain) by maternal treatment with cytochalasin D. In these experiments, the toxin was dissolved in DMSO and injected i.p. at doses of 0.4 to 0.9 mg/kg on days 8–12 of pregnancy. An increased resorption rate was noted in all three mouse strains. Additional BALB/c females were dosed orally with cytochalasin D in corn oil; their litters also exhibited an enhanced incidence of exencephaly. The production of exencephalies is consistent with the hypothesis that such defects result from prevention of neural tube closure[68] and the evidence from chick studies that cytochalasins can have such effects[64, 65].

Ergot alkaloids

Ergot is the term applied to members of the genus *Claviceps*. Ergot fungi primarily parasitize grasses, of which cultivated grains are examples. Perhaps the best known species is *C. purpurea*, one of the first described producers of mycotoxins[35]. Ergot sclerotia (the dormant stage in which the fungus overwinters) contain a large number of physiologically (and sometimes behaviourally[69]) active chemicals[2]. Many alkaloids produced by ergot are toxic, and their consumption in ergot-infested grain results in two types of mycotoxicoses: so-called 'gangrenous' and 'convulsive' ergotism.

Only a few ergot derivatives have been studied for possible developmental effects. Grauwiler and Schon[70] exposed pregnant mice, rats, and rabbits p.o. to ergotamine tartrate suspended in 2% gelatin. Doses (in mg/kg/day) were 0, 30, 100, and 300 on gestations days 7–16 for mice, 1, 3, 10, and 100 on days 7–16 for rats, and 1, 3, 10, and 30 on days 7–19 for rabbits. Apart from increased prenatal mortality in the rat, there was little evidence of a significant effect on the developing young. The authors speculate that such results could derive from the known vasoconstrictor effects of ergotamine on the uterine and placental vasculature. Use of labelled ergotamine in the rat showed that little of the compound reached the fetus, indicating that any effects were probably not direct.

Witters *et al.*[71] administered elymoclavine, an ergot toxin that is also produced by higher plants, such as morning glories and other members of the genus *Ipomoea*, to pregnant mice. When dose levels of 3, 30 and 60 mg/kg i.p. were employed, skeletal defects were noted in fetuses treated at the higher dose levels. Exposure of the pregnant dams to elevated environmental temperatures of 35 or 40 °C for 24 h following elymoclavine injection tended to be associated with enhanced prenatal and maternal mortality, indicating an interaction between temperature stress and the effect of the toxic alkaloid.

Ochratoxin A

Ochratoxin A is the most toxic of the isocoumarin derivatives produced by various storage fungi, such as *A. ochraceus* and *P. viridicatum*[72]. Ochratoxin A has been identified as the 7-carboxy-5-chloro-8-hydroxy-3,4-dihydro-3-methyl isocoumarin amide of L-β-phenylalanine[73]. It has been reported to cause toxic effects in a number of species, particularly nephropathy, lymphoid necrosis, enteritis, and liver damage[74, 75].

Still *et al.*[76] described increased prenatal mortality of rats whose mothers were gavaged with ochratoxin A. The same report implicated ochratoxin A in field outbreaks of abortion in swine and cattle; no malformations were mentioned, however, and the first report of the teratogenicity of a mycotoxin other than aflatoxin was a paper[16] presented to the Society of Toxicology by Hayes and Hood in 1973 which reported on the prenatal effects of ochratoxin A and rubratoxin B. This preliminary report was followed by a more detailed account[77], which described a variety of developmental defects in CD-1 mice following i.p. injection of 5 mg/kg ochratoxin A in propylene glycol and water (1 : 1). Treatment on one of days 7–12 of gestation resulted in increased prenatal mortality and a wide variety of gross and skeletal malformations. The most striking malformation was a median facial cleft accompanied by exencephaly, a syndrome we have found to be consistently produced with varying degrees of severity by ochratoxin A treatment in the mouse. Such median clefts may arise from damage to the prechordal mesoderm[78] and are generally less common than lateral clefts (which are thought to be produced by failure of coalescence of the facial swellings)[79]. These characteristic ochratoxin-induced median clefts are similar to those produced in mice by hypervitaminosis A[80]. Two facial halves, each consisting of nasal and upper maxillary elements, are spread apart to varying degrees, while the tongue and maxilla remain in the midline. Giroud *et al.*[80] suggest that such malformations may be caused by deficiency in the inductive ability of prechordal mesoderm or by lack of competence on the part of the anterior portion of the neural plate.

Moré and Galtier[81] confirmed the teratogenicity of ochratoxin A, reporting decreased fetal weights and an increased resorption rate for offspring of rats treated during pregnancy. Total doses of 4 or 5 mg/kg by the i.p. or p.o. routes (respectively) were employed. Treatments were given on from 2 to 8 consecutive days. The only malformation noted was described as 'coelosomy'. Adverse effects observed were attributed to possible effects on carbohydrate metabolism, and it was noted that ochratoxin A crossed the placenta and was detectable in the fetus.

Brown *et al.*[82] also treated pregnant Sprague-Dawley-derived rats p.o. with ochratoxin A dissolved in 0.1 N $NaHCO_3$. They chose doses of 0.25, 0.50, 0.75, 1, or 2 mg/kg/day on gestation days 7–16, 4 mg/kg/day on days 7–13, or 8 mg/kg/day on days 7–11 (treatment at the 4 and 8 mg/kg dose levels was discontinued early due to signs of overt toxicity in the dams and maternal deaths; only one rat in 10 survived to day 21 at the high dose). At doses below 4 mg/kg/day, there were no overt signs of maternal toxicosis. In the groups given doses of 0.75 mg/kg/day or above, fetal resorptions increased dramatically and fetal weights decreased. Doses above 2 mg/kg/day were incompatible with fetal survival. Ochratoxin A treatment was also associated with increased malformation rates, with open eyes, snout malformations, and skeletal defects predominating.

The golden hamster is also affected by *in utero* exposure to ochratoxin A, according to Hood *et al.*[83] Doses employed ranged from 2.5 to 20.0 mg/kg dissolved in 0.1 N sodium bicarbonate given i.p. on one of days 7–10 of pregnancy. Ochratoxin A tolerance of the developing hamster was lowest early in gestation. Although no effect was seen at the lowest dose, a large increase in dead or resorbed fetuses was seen following treatment at 5.0 mg/kg on day 7, and at 7.5 mg/kg prenatal mortality reached 99%. Treatment at later times was associated with decreasing mortality, even though the dose levels were increased; on day 10, treatment with 20 mg/kg failed to result in increased prenatal mortality. Fetal growth, as reflected in weights at sacrifice (day 15), was generally unaffected at any dose allowing fetal survival. The only exception was the 20 mg/kg dose given on day 9. Malformations were observed in fetuses exposed to ochratoxin A on day 7, 8 or 9. Defects reported were mainly shortened snouts, hydrocephalus, short tails, and a few limb malformations. Some cleft lips were also seen, but these were lateral clefts, which are presumed to differ in aetiology from the median clefts seen in mice by Hayes *et al.*[77]

As is the case with most mycotoxins, the molecular basis for the prenatal effects of ochratoxin A are not clear. Although the toxin is capable of damaging the maternal kidney and other organs critical to maintenance of a normal internal environment, it is likely that the teratogenic mechanism(s) involve a direct effect on the conceptus. This contention is supported by the ability of ochratoxin A in the mouse to cause teratogenic effects at dose levels well below those toxic to the mother, by the tendency of ochratoxin A to produce widely different effects in different species, and by the specific nature of some of the defects observed, e.g. the median facial clefts.

Ochratoxin A in the rat is said to enter the blood quickly – within 30 minutes to one hour – and at relatively high levels, whether given i.p.[84] or p.o.[85] Ochratoxin A levels remain higher in the blood than in the tissues, presumably due to binding of the toxin by serum albumin[84], and the mycotoxin is excreted largely in the urine and faeces, either intact or after metabolism to the less toxic ochratoxin-α[85]. There is also evidence that ochratoxin A may be less toxic to ruminants because it is metabolized to innocuous metabolites by microorganisms in the stomach[86].

Ochratoxin A has been said to affect mitochondrial activity[87,88] and enzymes involved in protein and carbohydrate metabolism[89–91], although reports of effects on carbohydrate metabolism are contradictory[52].

Of the known effects of ochratoxin A, interference with mitochondrial function must be considered a likely candidate for a teratogenic mechanism. Indeed, Wilson[93] among others has suggested that interference with energy metabolism may well be a common mechanism of teratogenicity. In the case of ochratoxin A it is likely that other effects may be involved, however, such as the aforementioned interference with enzyme function.

Patulin

Patulin, a mutagenic and carcinogenic lactone with the formula 4-hydroxy-4H-furo[3,2c]pyran-2(6H)-one, is produced by a number of *Penicillium* and *Aspergillus* species. Patulin has also been known by other names, such as clavacin, claviformin, and penicidin[93].

Patulin was tested for teratogenicity by Hayes and Hood[94], who administered the toxin to mice. A dose of 4 mg/kg dissolved in propylene glycol and administered on gestation days 8, 9, or 10 had no effect. A 6 mg/kg dose was generally lethal to the dam, but one survivor who was treated on gestation day 10 produced a litter with decreased fetal weights. Since the high dose level was fatal to 5 out of 6 females, the fetal weight reduction observed is likely to have been a result of maternal toxicity. Only small numbers of animals were tested, however, and further evidence is needed to determine the teratogenic potential of patulin in the mouse.

A study of patulin teratogenicity involving the chick was published by Ciegler *et al.*[95], who administered 3 to 75 μg per egg to unincubated eggs and 0.5 to 9 μg to 4-day-old embryos. Teratogenic effects were seen at doses of 10 μg/egg in the preincubation treated group and 1–2 μg/embryo in the group treated on day 4, with limb defects predominating. Additional 4-day embryos were treated with patulin that had been reacted with cysteine in an attempt to verify a proposed method of patulin detoxification (reaction with sulphydryl-containing molecules). Although the reaction product was not lethal at doses of up to 50 times those of patulin alone, teratogenic effects similar to those attributed to patulin alone were noted at doses of only 15 μg/embryo. Such results are further evidence that embryotoxic and teratogenic effects may at times be caused by different mechanisms. Presumably the decreased toxicity and teratogenicity of patulin following injection into unincubated eggs compared with effects in the 4-day chicken embryo was due to reaction of the mycotoxin with egg constituents (sulphydryl-containing amino acids or proteins) prior to development of sensitivity in the embryo.

Patulin has been found to cause a variety of toxic effects in the adult, and has been said to inhibit cellular respiration as well as a variety of other enzyme-mediated processes. Patulin seems to particularly affect enzymes, such as lactate dehydrogenase and aldolase, which contain essential thiol groups[52]. The data of Ciegler *et al.*[95] indicate that the lethal effects of patulin in the chicken embryo are likely to be due to its reaction with SH-containing molecules, while the teratogenic effects have other causes.

Penicillic acid

Penicillic acid, another carcinogenic mycotoxin, can exist in either of two forms, as the substituted γ-keto acid (γ-keto-β-methoxy-δ-methylene-A^{a}-hexenoic acid) or as the γ-hydroxylactone[93]. It is produced by a wide variety of *Penicillium* and *Aspergillus* species[96].

The only prenatal study of penicillic acid known to this author was done by Hayes and Hood[97], who treated mice with the mycotoxin on one of days 7–10 of pregnancy. Dose levels of 30 or 50 mg/kg i.p. failed to produce detectable effects. A 90 mg/kg dose was also administered on gestation day 10, and resulted in decreased fetal survival and weight gain. This effect, however, may be related to maternal toxicity, as only two out of six treated females survived to sacrifice. At such toxic dose levels, effects on the maternal organism should be considered of more practical significance than those on the offspring.

The maternal mortality observed after a 90 mg/kg i.p. dose was not unexpected, as a subcutaneous LD_{50} of 100 mg/kg in mice has been reported[98]. Penicillic acid is mutagenic as well as carcinogenic, and like patulin, reacts with a variety of enzymes, particularly those containing SH groups. In view of such potentially unspecific toxic effects, it is not surprising that penicillic acid did not exhibit differential effects between mother and conceptus.

Penitrem A

Penitrem A is a toxic metabolite of various *Penicillium* species. The toxin is capable of causing tremors and convulsions in a number of mammalian species, but its molecular structure is not yet known[99].

The literature on prenatal effects of penitrem A is sparse, with again only one reference known to this author. Hayes and Hood[97] injected mice i.p. with 1–3 mg/kg penitrem A on one of gestation days 7–10. No dose-related effect on resorptions and deaths or weight gain was seen, nor were gross malformations increased significantly. There was an increase, however, in skeletal malformations in fetuses exposed to the higher doses on days 8, 9 or 10. Defects observed consisted mainly of vertebral malformations and fused ribs.

Penitrem A is recognized as a potent neurotoxin and is also capable of causing diuresis with loss of glucose and electrolytes[44]. It has been proposed that penitrem A adversely affects inhibitory interneurons[100] or alters presynaptic transmitter release[101]. It is doubtful that such effects were responsible for the skeletal defects observed in prenatally exposed mice[97], but no biochemical actions of the mycotoxin are currently known that offer a likely explanation for such effects.

Rubratoxin B

Both *P. rubrum* and the closely related *P. purpurogenum* elaborate rubratoxin B, a hepatotoxin with two disubstituted maleic anhydride groups and an α,β-unsaturated δ-lactone attached to a nine member ring[102]. Rubratoxin B is a relatively potent toxin, with an LD_{50} of approximately 3 mg/kg i.p. in the mouse[102], although oral rubratoxin B is much less toxic[103], possibly due to instability of the toxin in stomach acid[104].

The first report[16] of the teratogenicity of rubratoxin B was presented at the 12th annual meeting of the Society of Toxicology, followed by a more detailed report[105] and soon confirmed by others[106].

According to Hood *et al.*[105], i.p. injections of rubratoxin B in propylene glycol and water (1 : 1) at dose levels of 0.6, 0.9 or 1.2 mg/kg given to mice on one of gestation days 6–12 resulted in increased prenatal mortality. A 0.4 mg/kg dose was given to additional mothers on day 8 in order to allow survival of fetuses treated on that day. Rubratoxin B treatment was also associated with decreased fetal weight following administration on all days, with the exception of day 9. A number of gross malformations (mainly exencephaly, ear and jaw defects, umbilical hernias, and open eyes) were also associated with toxin treatment, particularly following treatment on days 6, 7 or 8, but no skeletal abnormalities were seen. Similar results were reported by Koshakji *et al.*[106]

These studies were followed by a report[107] which indicated that rubratoxin B was fetotoxic and teratogenic when given i.p. to pregnant female mice at a dose of 0.5 mg/kg. When the hydrogenated analogue (with the lactone ring saturated) of rubratoxin B was similarly tested, prenatal toxicity and teratogenicity were abolished at dose levels up to 10 times the highest dose of the parent compound.

The previous results indicate that the prenatal effects of rubratoxin B are associated with the reactivity of the α,β-unsaturated lactone ring, a result anticipated by the evidence[102] of decreased adult toxicity in mice seen when the lactone was either saturated or removed prior to i.p. injection. Biological effects, including carcinogenic properties, have previously been associated with α,β-unsaturated lactones[45], although there is no evidence that rubratoxin B is carcinogenic.

Subcellular effects of rubratoxin B have been variously attributed to inhibition of mitochondrial function[108, 109], inhibition of adenosine triphospatase[10], and alteration of informational macromolecules, resulting in polyribosomal disaggregation[111]. It is not yet known, however, if any of the above mentioned effects are related to the teratogenesis or embryotoxicity induced by rubratoxin B.

Stachybotrys toxins

Consumption of fodder contaminated with toxin-producing strains of *Stachybotrys alternans* has been associated with acute toxicity and a high mortality rate in horses[5] and with equine and porcine abortion[112]. The mycotoxic disease has been termed stachybotryotoxicosis and was one of the earlier recognized mycotoxicoses[113]. Several trichothecenes, including roridins and verrucarins, have been isolated from toxic extracts of *S. alternans*, and such results are in agreement with the known toxic effects of the trichothecenes and those reported for '*Stachybotrys* toxins'[113].

Stachybotrys toxins have been tested for teratogenicity in pregnant mice by Korpinen[112], who administered grain infested with toxic strains of *Stachy-*

botrys alternans in the diet, or a toxin-containing liquid growth medium or 'partly purified preparation' by gavage. Treatments were given on single days or on five consecutive days. Various treatments resulted in either implantation failure, prenatal mortality, fetal stunting or some combination thereof. Histopathological examination revealed little, with the exception of uteroplacental haemorrhage; no malformations were noted.

The author suggested that the effects of *Stachybotrys* toxin on the conceptus were likely to have been a result of the observed uteroplacental damage and that a direct effect on the fetus was unlikely due to the presence of apparently normal fetuses in the same litters with dead or stunted individuals. It seems likely, however, that any direct histologically discernable effects on the embryo or fetus could have undergone repair prior to examination. Also, finding apparently normal fetuses in the same litter with stunted or grossly abnormal fetuses is a well known phenomenon to teratologists. It is believed to merely indicate mechanisms such as differing susceptibility among individuals due to differences in genetic background or development stage, or perhaps to receipt by the fetuses of differing dose levels (due to variations in blood supply to each placenta).

T-2 toxin

Mouldy corn toxicosis associated with consumption by livestock of corn stored at low temperatures has been primarily associated with contamination by *Fusarium tricinctum*[114]. The causative agent involved is thought to be T-2 toxin (3-hydroxy-4,15-diacetoxy-8-[3-methylbutyryloxy]-12,13-epoxy-Δ-trichothecene), one of a number of toxic 12,13-epoxytricothecenes[11]. T-2 toxin is readily absorbed through the skin, as well as from the gut. It has an acute oral LD_{50} of only 4 mg/kg in swine and rats and can produce intestinal, liver, and kidney lesions in mammals, as well as skin necrosis[11], damage to the vascular system, and cellular damage in proliferating tissues[15]. T-2 toxin has been implicated as a teratogen by the work of Stanford *et al.*[116], who administered the mycotoxin i.p. to mice on one of days 7–11 of gestation. Dose levels of 0.5, 1.0 and 1.5 mg/kg were used, with the higher dose levels frequently resulting in maternal deaths. Effects on the conceptus included decreased survival and growth. Limb and tail malformations, exencephaly, open eye and retarded jaw development were associated with T-2 toxin treatment on days 8, 9, 10 or 11, and defective skeletal development followed treatment on days 9 or 10 of pregnancy.

Another developmental effect attributed to T-2 toxin is the alteration of feather development in growing chicks[117]. T-2 toxin was fed in the diet at levels of 1, 2, 4, 8 or 16 mg/g of feed for a period of three weeks post-hatching. At the three highest levels, the mycotoxin caused growth inhibition and sparse feathering. The feathers grew out at angles different from the angle of growth of control feathers, and the feathers often exhibited bizarre shapes.

Toxic effects of T-2 toxin and related 12,13-epoxytrichothecenes have generally been attributed to their inhibition of protein synthesis. This effect has been demonstrated in a number of systems, including tumour cells, liver, reticulocytes, tonsil tissue and others[52]. According to Cundliffe *et al.*[118], T-2 toxin is a specific and potent inhibitor of the initiation of eukaryote polypeptide chain synthesis. They also contend that while some other 12,13-epoxytrichothecene mycotoxins (e.g. nivalenol and varrucarin A) share the same mechanism of action, additional related compounds (e.g. trichodermin) act by inhibiting polypeptide elongation or termination. These activities are said to be related to the ability of such compounds to bind to ribosomes, with such binding possible due to opening of the epoxide ring. The groups attached to carbon-15 are then postulated to affect the precise mode of inhibition.

Although general protein synthesis inhibitors are often not very effective producers of malformations, T-2 toxin seems to act preferentially on dividing cells[115]. This may offer an explanation for the mycotoxin's teratogenicity, as it would allow selective damage of certain tissues. Selective, as opposed to generalized, damage appears necessary to initiate teratogenesis rather than merely cause death of the developing organism.

T-2 toxin has also been said by Lary and Hood[119] to interact with the brachyury (T) gene of mice. When the mycotoxin was administered at a dose of 1.0 mg/kg to females whose litters contained heterozygous for *T*, the incidence of taillessness was significantly increased. Also, although untreated *T*/+ fetuses tend to have abnormal caudal and sacral vertebrae, T-2 treatment increased both the incidence and severity of such defects as well as other skeletal anomalies. Such results indicate that T-2 toxin can enhance the expressivity of *T* with regard to tail length, while the presence of the brachyury gene increases the effect of the toxin on skeletal development.

These results have been said[120] to be consistent with the hypothesis of Bennett[121] that *T*-locus alleles control normal development by sequentially and briefly directing the production of specific proteins needed for processes essential to cell behaviour during embryogenesis. The role of T-2 toxin in increasing the expressivity of short tail length in *T*/+ fetuses could then be to further decrease essential proteins (cell surface antigens?) already in short supply (due to the presence of only one wild-type allele) in the cells of the developing caudal end of the notochord.

Zearalenone (F-2)

Zearalenone, an oestrogenic mycotoxin elaborated by *Fusarium graminearum* and related *Fusarium* species during periods of low temperatures, has been identified in livestock feeds and is known to cause mycotoxicoses in swine. An oestrogenic syndrome of swine is known to be caused by consumption of zearalenone-contaminated feed[122]. Zearalenone is also believed to be a relatively frequent cause of feed refusal and emesis in swine[123]. The mycotoxin has been chemically characterized as 2,4-dihydroxy-6-(10-hydroxy-6-oxo-trans-1-undecenyl) benzoic acid μ-lactone and is found in a significant percentage of corn samples in the United States[124].

Miller *et al.*[125] fed *Fusarium* cultures to pregnant sows and reported increased prenatal mortality and limb defects, but the failure to use purified zearalenone leaves its role in producing such results unclear.

Ruddick *et al.*[126] treated rats p.o. on gestation days 6–15 with zearalenone doses of 1, 5, or 10 mg/kg/day. The high dose resulted in decreased fetal weights and all doses were associated with an increased incidence of skeletal anomalies, although the majority of these were apparently delayed ossification, malpositioned sternebrae or extra ribs.

Effects of combined exposure

Hood *et al.*[127] carried out a recent study to determine if concurrent exposure of the conceptus to two different mycotoxins would result in a synergistic effect. They administered ochratoxin A and T-2 toxin, singly or in combination, to pregnant mice by i.p. injection on either day 8 or 10 of gestation. T-2 toxin was invariably given at a dose level of 0.5 mg/kg, a marginally teratogenic dose, with a greater effect when given on day 10. Ochratoxin A was given at a dose level of 2 or 4 mg/kg, with both levels being significantly teratogenic when given on day 8, but not day 10.

When the effects of the combined treatments were evaluated, it was seen that they did not result in an increased malformation rate if given on day 8 but an apparent synergistic effect was seen when the high dose combination (4 mg/kg ochratoxin A + 0.5 mg/kg T-2 toxin) was used on day 10. The malformations observed were those typical of exposure to T-2 toxin, with none of the type exclusively attributable to ochratoxin A. This indicated that simultaneous exposure to two teratogenic mycotoxins may result in exacerbation of the effects of only one. The cause of this effect is not clear but is likely to be dependent on the particular combination of teratogenic agents and the developmental stage during exposure. The effects of the combination treatments on fetal weight and prenatal mortality were never more than additive.

Ochratoxin A treatment on days 15–17 (but not days 8–10) is associated with brain necrosis observable gestation day 18. Early cellular damage to tissues of the conceptus tends to be repaired by late in gestation, and this case was no exception. Such ochratoxin A-induced necrosis has been reported by Szczech and Hood[128, 129].

References

1. Christensen, C. M. (1975). *Molds, Mushrooms and Mycotoxins*..34. (Minneapolis: University of Minnesota Press)
2. Burfening, P. J. (1973). Ergotism. *J. Am. Vet. Med. Ass.*, **163**, 1288
3. Joffe, A. Z. (1971). Alimentary toxic aleukia. In: S. Kadis, A. Ciegler and S. J. Ajl (eds.). *Microbial Toxins*, Vol. VII, pp. 139–189. (New York: Academic Press)
4. Saito, M., Enomoto, M., Tatsuno, T. and Uragucki, K. (1971). Yellowed rice toxins. In: A. Ciegler, S. Kadis and S. J. Ajl (eds.). *Microbial Toxins*, Vol. VI, pp. 299–380. (New York: Academic Press)
5. Forgacs, J. (1972). Stachybotryotoxicosis. In: S. Kadis, A. Ciegler and S. J. Ajl (eds.). *Microbial Toxins*, Vol. VIII, pp. 95–128. (New York: Academic Press)
6. Sippel, W. L., Burnside, J. E. and Atwood, M. B. (1953). A disease of swine and cattle caused by eating moldy corn. *Proc. 90th Annu. Meet. Am. Vet. Med. Ass.*, Toronto, p. 174
7. Newberne, J. W., Bailey, W. S. and Seibold, H. R. (1955). Notes on a recent outbreak and experimental reproduction of hepatitis X in dogs. *J. Am. Vet. Med. Ass.*, **127**, 59
8. Blount, W. P. (1961). Turkey X disease. *Turkeys* (*J. Br. Turkey Fed.*), **9**, 52
9. Shank, R. C., Bourgeios, C. H., Keschamras, N. and Chandavimol, P. (1971). Aflatoxins in autopsy specimens from Thai children with an acute disease of unknown etiology. *Food Cosmet. Toxicol.*, **9**, 501
10. Krogh, P., Hald, B. and Pedersen, E. J. (1973). Occurrence of ochratoxin A and citrinin in

cereals associated with mycotoxic porcine nephropathy. *Acta Pathol. Microbiol. Scand., Sec. B*, **81**, 689

11. Hsu, I.-C., Smalley, E. B., Strong, F. M. and Ribelin, W. E. (1972). Identification of T-2 toxin in moldy corn associated with a lethal toxicosis in dairy cattle. *Appl. Microbiol.*, **24**, 684
12. LeBreton, E., Frayssinet, C., Lafarge, C. and de Recondo, A. M. (1964). *Food Cosmet. Toxicol.*, 2, 675
13. Butler, W. H. and Wigglesworth, J. S. (1966). The effects of aflatoxin B_1 on the pregnant rat. *Br. J. Exp. Pathol.*, **47**, 242
14. Elis, J. and DiPaolo, J. A., (1967). Aflatoxin B_1 induction of malformations. *Arch. Pathol.*, **83**, 53
15. DiPaolo, J. A., Elis, J. and Erwin, H. (1967). Teratogenic response by hamsters, rats and mice to aflatoxin B_1. *Nature (Lond.)*, **215**, 638
16. Hayes, A. W. and Hood, R. D. (1973). Mycotoxin induced developmental abnormalities in mice. *Toxicol. Appl. Pharmacol.*, **25**, 457
17. Fischbach, H. and Rodricks, J. V. (1973). Current efforts of the Food and Drug Administration to control mycotoxins in food. *J. Ass. Off. Analyt. Chem.*, **56**. 767
18. Jarvis, B. (1976). Mycotoxins in food. In: F. A. Skinner and J. G. Carr (eds.). *Microbiology in Agriculture, Fisheries and Food*, pp. 251–267. (London: Academic Press)
19. Harwig, J. and Munro, I. C. (1975). Mycotoxins of possible importance in diseases of Canadian farm animals. *Can. Vet. J.*, **16**, 125
20. Chu, F. S., Chang, C. C., Ashoor, S. H. and Prentice, N. (1975). Stability of aflatoxin B_1 and ochratoxin A in brewing. *Appl. Microbiol.*, **29**, 313
21. Campbell, T. C. and Stoloff, L. (1974). Implications of mycotoxins for human health. *J. Agric. Food Chem.*, 22, 1006
22. van Rensburg, S. J. (1977). Role of epidemiology in the elucidation of mycotoxin health risks. In: J. V. Rodricks, C. W. Hesseltine and M. A. Mehlman (eds.). *Mycotoxins in Human and Animal Health*, pp. 699–711. (Park Forest South: Pathotox)
23. van Walbeek, W., Scott, P. M. and Thatcher, F. S. (1968). Mycotoxins from food-borne fungi. *Can. J. Microbiol.*, **14**, 131
24. Detroy, R. W., Lillehoj, E. B. and Ciegler, A. (1971). Aflatoxin and related compounds. In: A. Ciegler, S. Kadis and S. J. Ajl (eds.). *Microbial. Toxins*, **Vol. VI**, pp. 3–178. (New York: Academic Press)
25. Mislivec, P. B., Dieter, C. T. and Bruce, V. R. (1975). Mycotoxin-producing potential of mold flora of dried beans. *Appl. Microbiol.*, **29**, 522
26. Shotwell, O. L., Goulden, M. L. and Hesseltine, C. W. (1976). Aflatoxin M_1. Occurrence in stored and freshly harvested corn. *Agric. Food Chem.*, **24**, 683
27. Sommer, N. F., Buchanan, J. R. and Fortlage, R. J. (1976). Aflatoxin and sterigmatocystin contamination of pistachio nuts in orchards. *Appl. Environ. Microbiol.*, **32**, 64
28. Schindler, A. F., Abadie, A. N., Gecan, J. S., Mislivec, P. B. and Brickey, P. M. (1974). Mycotoxins produced by fungi isolated from inshell pecans. *J. Food Sci.*, **39**, 213
29. Pero, R. W., Posner, H., Blois, M., Harvan, D. and Spalding, J. W. (1973). Toxicity of metabolites produced by the '*Alterneria*.' *Environ. Hlth. Perspect.*, **6**, 87
30. Templeton, G. E. (1972). *Alternaria* toxins related to pathogenesis in plants. In: S. Kadis, A. Ciegler and S. J. Ajl (eds.). *Microbial Toxins*, **Vol. VIII**, pp. 169–192. (New York: Academic Press)
31. Scott, P. M., van Walbeek, W., Kennedy, B. and Anyeti, D. (1972). Mycotoxins (ochratoxin A, citrinin, and sterigmatocystin) and toxigenic fungi in grains and other agricultural products. *Agric. Food Chem.*, **20**, 1103
32. Patwardhan, S. A., Pandey, R. C. and Pendse, G. S. (1974). Toxic cytochalasins of *Phomopsis paspalli*, a pathogen of kodo millet. *Phytochemistry*, **13**, 1985
33. Pribela, A., Tomko, J. and Dolejš, L. (1975). Cytochalasin B from tomatoes contaminated by *Hormiscium* sp. *Phytochemistry*, **14**, 285
34. Scott, P. M., Harwig, J., Chen, Y-K. and Kennedy, B. P. C. (1975). Cytochalasins A and B from strains of *Phoma exigua* var. *exigua* and formation of cytochalasin B in potato gangrene. *J. Gen. Microbiol.*, **87**, 177
35. Groger, D. (1972). Ergot. In: S. Kadis, A. Ciegler and S. J. Ajl (eds.). *Microbiol. Toxins*, **Vol. VIII**, pp. 321–373. (New York: Academic Press)

36. Bhat, R. V., Roy, D. N. and Tulpule, P. G. (1976). The nature of alkaloids of ergoty pearl millet or bajra and its comparison with alkaloids of ergoty rye and ergoty wheat. *Toxicol. Appl. Pharmacol.*, **36**, 11
37. Escher, F. E., Koehler, P. E. and Ayres, J. C. (1973). Production of ochratoxins A and B on country cured ham. *Appl. Microbiol.*, **26**, 27
38. Hesseltine, C. W. and Graves, R. R. (1966). Microbiology of flours. *Econ. Bot.*, **20**, 156
39. Harwig, J., Chen, Y-K., Kennedy, B. P. C. and Scott, P. M. (1973). Occurrence of patulin and patulin-producing strains of *Penicillium expansum* in natural rots of apple in Canada. *Can. Inst. Food Sci. Technol. J.*, **6**, 22
40. Scott, P. M., Fuleki, T. and Harwig, J. (1977). Patulin content of juice and wine produced from moldy grapes. *Agric. Food Chem.*, **25**, 434
41. Ciegler, A. and Kurtzman, C. P. (1970). Penicillic acid production by blue-eye-fungi on various agricultural commodities. *Appl. Microbiol.*, **20**, 761
42. Snow, J. P., Lucas, G. B., Harvan, D., Pero, R. W. and Owens, R. G. (1972). Analysis of tobacco and smoke condensates for penicillic acid. *Appl. Microbiol.*, **24**, 34
43. Thorpe, C. W. and Johnson, R. L. (1974). Analysis of penicillic acid by gas–liquid chromatography. *J. Ass. Off. Anal. Chem.*, **57**, 861
44. Wilson, B. J. (1971). Miscellaneous *Penicillium* toxins. In: A. Ciegler, S. Kadis and S. J. Ajl (eds.). *Microbial Toxins*, **Vol. VI**, pp. 459–521. (New York: Academic Press)
45. Moss, M. O. (1971). The rubratoxins, toxic metabolites of *Penicillium rubrum* Stoll. In: A. Ciegler, S. Kadis and S. J. Ajl (eds.). *Microbial Toxins*, Vol. VI, pp. 381–407. (New York: Academic Press)
46. Eppley, R. M., Stoloff, L., Trucksess, M. W. and Chung, C. W. (1974). Survey of corn for *Fusarium* toxins. *J. Ass. Off. Analyt. Chem.*, 57, 632
47. Mirocha, C. J., Pathre, S. V. Schauerhamer, B. and Christensen, C. M. (1976). Natural occurrence of *Fusarium* toxins in feedstuff. *Appl. Environ. Microbiol.*, 32, 553
48. Bacon, C. W., Robbins, J. D. and Porter, J. K. (1977). Media for identification of *Gibberella zeae* and production of F-2 (zearalenone). *Appl. Environ. Microbiol.*, **33**, 445
49. Windholz, M., ed. (1976). *The Merck Index*, 9th Ed. (Rahway: Merck)
50. Wogan, G. N. and Newberne, P. M. (1967). Dose–response characteristics of aflatoxin B carcinogenesis in the rat [*Aspergillus flavus*]. *Cancer Res.*, **27**, 2370
51. Clifford, J. I. and Rees, K. R. (1966). Aflatoxin: A site of action in the rat liver cell. *Nature (Lond.)*, **209**, 312
52. Chu, F. S. (1977). Mode of action of mycotoxins and related compounds. In: D. Perlman (ed.). *Advances in Applied Microbiology*, Vol. 22, pp. 83–143. (New York: Academic Press)
53. Portman, R. S., Plowman, K. M. and Campbell, T. C. (1968). Aflatoxin metabolism by liver microsomal preparations of two different species. *Biochem. Biophys. Res. Commun.*, **33**, 711
54. Carlton, W. W., Sansing, G., Szczech, G. M. and Tuite, J. (1974). Citrinin mycotoxicosis in beagle dogs. *Food Cosmet. Toxicol.*, 12, 479
55. Heatherington, A. C. and Raistrick, H. (1931). Studies in the biochemistry of micro-organisms. Part XIV. On the production and chemical constitution of a new yellow colouring matter, citrinin, produced from glucose by *Penicillium citrinum*. Thom. *Phil. Trans. Roy. Soc. Ser. B.*, **220B**, 269
56. Ambrose, A. M. and DeEds, F. (1945). Acute and subacute toxicity of pure citrinin. *Proc. Soc. Exp. Biol. Med.*, **59**, 289
57. Ambrose, A. M. and DeEds, F. (1946). Some toxicological and pharmacological properties of citrinin. *J. Pharmacol. Exp. Ther.*, **88**, 173
58. Hood, R. D., Hayes, A. W. and Scammell, J. G. (1976). Effects of prenatal administration of citrinin and viriditoxin to mice. *Food Cosmet. Toxicol.*, **14**, 175
59. Farb, R. M., Mego, J. L. and Hayes, A. W. (1976). Effect of mycotoxins on uptake and degradation of [^{125}I]albumin in mouse liver and kidney lysosomes. *J. Toxicol. Environ. Health*, **1**, 985
60. Carter, S. B. (1967). Effects of cytochalasins on mammalian cells. *Nature (Lond.)*, **213**, 261
61. Wessels, W. K., Spooner, B. S., Ash, J. F., Bradley, M. O., Luduena, M. A., Tayler, E. L., Wrenn, J. T., and Yamada, K. M. (1971). Microfilaments in cellular and developmental processes. *Science*, **171**, 135

62. Aldridge, D. C., Armstrong, J. J., Speake, R. N. and Turner, W. B. (1967). Cytochalasins, a new class of biologically active mold metabolites. *Chem. Commun.*, **1**, 26
63. Shepard, T. H. and Greenaway, J. C. (1977). Teratogenicity of cytochalasin D in the mouse. *Teratology*, **16**, 131
64. Linville, G. P. and Shepard, T. H. (1972). Neural tube closure defects caused by cytochalasin B. *Nature New Biol.*, **236**, 247
65. Karfunkel, P. (1972). The activity of microtubules and microfilaments in neurulation in the chick. *J. Exp. Zool.*, **181**, 289
66. Greenaway, J. C., Shepard, T. H. and Kuc, J. (1977). Comparison of cytochalasins (A, B, D, and E) in chick explant teratogenicity and tissue culture systems. *Proc. Soc. Exp. Biol. Med.*, **155**, 239
67. Ruddick, J. A., Harwig, J. and Scott, P. M. (1974). Nonteratogenicity in rats of blighted potatoes and compounds contained in them. *Teratology*, **9**, 165
68. Giroud, A. (1960). Causes and morphogenesis of anencephaly. In: G. E. W. Wolstenholme and C. M. O'Connor (eds.). *Ciba Foundation Symposium on Congenital Malformations*, pp. 199–212. (London: J. and A. Churchill)
69. Hood, R. D., Melvin, K. B. and Starling, P. B. (1974). Effects of agroclavine on avoidance behavior in the hamster. *Bull. Psychon. Soc.*, **3**, 71
70. Grauwiler, J. and Schön, H. (1973). Teratological experiments with ergotamine in mice, rats, and rabbits. *Teratology*, **7**, 227
71. Witters, W. L., Wilms, R. A. and Hood, R. D. (1975). Prenatal effects of elymoclavine administration and temperature stress. *J. Anim. Sci.*, **41**, 1700
72. Scott, de B. (1965). Toxigenic fungi isolated from cereal and legume products. *Mycopath. Mycol. Appl.*, **25**, 213
73. Van der Merwe, K. J., Steyn, P. S. and Fourie, L. (1965). Mycotoxins. Part II. The constitution of ochratoxins A, B, and C, metabolites of *Aspergillus ochraceus* Wilh. *J. Chem. Soc.*, 7083
74. Szczech, G. M., Carlton, W. W. and Tuite, J. (1973). Ochratoxicosis in beagle dogs. II. Pathology. *Vet. Pathol.*, **10**, 219
75. Szczech, G. M., Carlton, W. W., Tuite, J. and Caldwell, R. (1973). Ochratoxin A toxicosis in swine. *Vet. Pathol.*, **10**, 347
76. Still, P. E., Macklin, A. W., Ribelin, W. E. and Smalley, E. B. (1971). Relationship of ochratoxin A to foetal death in laboratory and domestic animals. *Nature* (*Lond.*), **234**, 563
77. Hayes, A. W., Hood, R. D. and Lee, H. L. (1974). Teratogenic effects of ochratoxin A in mice. *Teratology*, **9**, 93
78. DeMyer, W. (1971). Median cleft lip. In: W. C. Grabb, S. W. Rosenstein and K. R. Bzoch (eds.). *Cleft Lip and Palate*, pp. 359–369. (Boston: Little, Brown and Co.)
79. Arey, L. B. (1965). *Developmental Anatomy*. (Philadelphia: Saunders)
80. Giroud, A., Martinet, M. and Deluchat, C. (1969). Fissuration faciale médiane. *Arch. Anat. Histol. Embryol.*, **52**, 207
81. Moré, J. and Galtier, P. (1974). Toxicité de l'ochratoxin A. I. Effet embryotoxique et tératogène cheze le rat. *Ann Rech. Veter.*, **5**, 167
82. Brown, M. H., Szczech, G. M. and Purmalis, B. P. (1976). Teratogenic and toxic effects of ochratoxin A in rats. *Toxicol. Appl. Pharmacol.*, **37**, 331
83. Hood, R. D., Naughton, M. J. and Hayes, A. W. (1976). Prenatal effects of ochratoxin A in hamsters. *Teratology*, **13**, 11
84. Chang, F. C. and Chu, F. S. (1976). The fate of ochratoxin A in rats. *Food Cosmet. Toxicol.*, **15**, 199
85. Suzuki, S., Satoh, T. and Yamazaki, M. (1971). The pharmacokinetics of ochratoxin A in rats. *Japan J. Pharmacol.*, **27**, 735
86. Hult, K., Teiling, A. and Gatenbeck, S. (1976). Degradation of ochratoxin A by a ruminant. *Appl. Environ. Microbiol.*, **32**, 443
87. Moore, J. H. and Truelove, B. (1970). Ochratoxin A: Inhibition of mitochondrial respiration. *Science*, **168**, 1102
88. Meisner, H. and Chan, S. (1974). Ochratoxin A, an inhibitor of mitochondrial transport systems. *Biochemistry*, **13**, 2795
89. Pitout, M. J. (1968). The effect of ochratoxin A on glycogen storage in the rat liver. *Toxicol. Appl. Pharmacol.*, 13, 299

90. Pitout, M. J. and Nel, W. (1969). The inhibitory effect of ochratoxin A on bovine carboxypeptidase A *in vitro*. *Chem. Pharmacol.*, **18**, 1837
91. Suzuki, S., Satoh, T. and Yamazaki, M. (1975). Effect of ochratoxin A on carbohydrate metabolism in rat liver. *Toxicol. Appl. Pharmacol.*, 32, 116
92. Wilson, J. G. (1973). *Environment and Birth Defects*. (New York: Academic Press)
93. Ciegler, A., Detroy, R. W. and Lillehoj, E. B. (1971). Patulin, penicillic acid, and other carcinogenic lactones. In: A. Ciegler, S. Kadis and S. J. Ajl (eds.). *Microbial Toxins*, Vol. VI, pp. 409–434. (New York: Academic Press)
94. Hayes, A. W. and Hood, R. D. (1975). Effect of prenatal exposure to mycotoxins. In: *The Prediction of Chronic Toxicity from Short Term Studies, Proceedings of the European Society of Toxicology*, Vol. XVII, p. 209. (Amsterdam: Excerpta Medica)
95. Ciegler, A., Beckwith, A. C. and Jackson, L. K. (1976). Teratogenicity of patulin and patulin adducts formed with cysteine. *Appl. Environ. Microbiol.*, 31, 664
96. Oxford, A. E., Raistrick, H. and Smith, G. (1942). Antibacterial substances from molds. II. Penicillic acid, a metabolic product of *Penicillium puberulum* Bainier and *Penicillium cyclopium* Westling. *Chem. Ind.* (*London*), **61**, 48
97. Hayes, A. W. and Hood, R. D. (1978). Effects of prenatal administration of penicillic acid and penitrem A to mice. *Toxicon*, **16**, 92
98. Spector, W. S. (1957). *Handbook of Toxicology*, **Vol. II.** (Philadelphia: Saunders)
99. Hayes, A. W., Phillips, R. D. and Wallace, L. C. (1977). Effect of penitrem A on mouse liver composition. *Toxicon*, 15, 293
100. Stern, P. (1971). Pharmacological analysis of the tremor induced by cyclopium toxin. *Jugoslav. Physiol. Pharmac. Acta*. 7 187
101. Wilson, B. J., Hoekman, T. and Dettbarn, W. D. (1972). Effects of a fungus tremorgenic toxin (penitrem A) on transmission in rat phrenic nerve-diaphragm preparations. *Brain Res.*, **40**, 540
102. Rose, H. M. and Moss, M. O. (1970). The effect of modifying the structure of rubratoxin B on the acute toxicity to mice. *Biochem. Pharmacol.*, **19**, 612
103. Wogan, G. N., Edwards, G. S. and Newberne, P. M. (1971). Acute and chronic toxicity of rubratoxin B. *Toxicol. Appl. Pharmacol.*, **19**, 712
104. Wogan, G. R. and Mateles, R. I. (1968). Mycotoxins. *Progr. Ind. Microbiol.*, **7**, 149
105. Hood, R. D., Innes, J. E. and Hayes, A. W. (1973). Effects of rubratoxin B on prenatal development in mice. *Bull. Environ. Contam. Toxicol.*, **10**, 200
106. Koshakji, R. P., Wilson, B. J. and Harbison, R. D. (1973). Effect of rubratoxin B on prenatal growth and development in mice. *Res. Comm. Chem. Path. Pharmacol.*, 5, 584
107. Evans, M. A. and Harbison, R. D. (1977). Prenatal toxicity of rubratoxin B and its hydrogenated analog. *Toxicol. Appl. Pharmacol.*, **39**, 13
108. Bernard, C. and Dumas, P. (1975). Action de la rubratoxine B sur la respiration des mitochondries hépatiques de rat. *Mycopathologia*, **55**, 53
109. Hayes, A. W. (1976). Action of rubratoxin B on mouse liver mitochondria. *Toxicology*, **6**, 253
110. Desaiah, D., Hayes, A. W. and Ho, I. K. (1977). Effect of rubratoxin B on adenosine triphosphatase activities in the mouse. *Toxicol. Appl. Pharmacol.*, **39**, 71
111. Watson, S. A. and Hayes, A. W. (1977). Evaluation of possible sites of action of rubratoxin B-induced polyribosomal disaggregation in mouse liver. *J. Toxicol. Environ. Health*, **2**, 639
112. Korpinen, E.-L. (1974). Studies on *Stachybotrys alternans*. IV. Effect of low doses of *Stachybotrys* toxins on pregnancy of mice. *Acta Pathol. Microbiol. Scand., Sec. B.* **82**, 457
113. Eppley, R. M. (1977). Chemistry of stachybotryotoxicosis. In: J. V. Rodricks, C. W. Hesseltine and M. A. Mehlman (eds.). *Mycotoxins in Human and Animal Health*, pp. 285–293. (Park Forest South: Pathotox)
114. Smalley, E. B. (1973). T-2 toxin. *J. Am. Vet. Med. Ass.*, **163**, 1278
115. Sato, N., Ueno, Y. and Enomoto, M. (1975). Toxicological approaches to the toxic metabolites of *Fusaria*. VIII. Acute and subacute toxicities of T-2 toxin in cats. *Jap. J. Pharmacol.*, **25**, 263
116. Stanford, G. K., Hood, R. D. and Hayes, A. W. (1975). Effect of prenatal administration of T-2 toxin to mice. *Res. Commun. Chem. Pathol. Pharmacol.*, **10**, 743

117. Wyatt, R. D., Hamilton, P. B. and Burmeister, H. R. (1975). Altered feathering of chicks caused by T-2 toxin. *Poultry Sci.*, **54**, 1042
118. Cundliffe, E., Cannon, M. and Davies, J. (1974). Mechanism of inhibition of eukaryotic protein synthesis by trichothecene fungal toxins. *Proc. Nat. Acad. Sci.* (*USA*), **71**, 30
119. Lary, J. M. and Hood, R. D. (1977). Developmental interactions between T-2 toxin and the brachyury (T) gene in mice. *Teratology*, **15**, 19A
120. Lary, J. M. (1978). *Developmental Interactions Between Selected Teratogens and T-locus Alleles in the Mouse*. Ph.D. Thesis, The University of Alabama, Tuscaloosa
121. Bennett, D. (1975). *T*-locus mutants: Suggestions for the control of early embryonic organization through cell surface components. In: M. Balls and A. E. Wild (eds.). *The Early Development of Mammals*, pp. 207–218. (London: Cambridge University Press)
122. Mirocha, C. J., Christensen, C. M. and Nelson, G. H. (1971). F-2 (zearalenone) estrogenic mycotoxin from *Fusarium*. In: S. Kadis, A. Ciegler and S. J. Ajl (eds.). *Microbial Toxins*, **Vol. VII**, pp. 107–138. (New York: Academic Press)
123. Curtin, T. M. and Tuite, J. (1966). Emesis and refusal of feed in swine associated with *Gibberella zeae*-infected corn. *Life Sci.*, **5**, 1937
124. Marasas, W. F. O., Kriek, N. P. J., van Rensburg, S. J., Steyn, M. and van Schalkwyk, G. C. (1977). Occurrence of zearalenone and deoxynivalenol, mycotoxins produced by *Fusarium grammearum* Schwabe, in maize in southern Africa. *S. African J. Sci.*, **73**, 346
125. Miller, J. K., Hacking, A., Harrison, J. and Gross, V. J. (1973). Stillbirths, neonatal mortality and small litters in pigs associated with the ingestion of *Fusarium* toxin by pregnant sows. *Vet. Rec.*, **93**, 555
126. Ruddick, J. A., Scott, P. M. and Harwig, J. (1976). Teratological evaluation of zearalenone administered orally to the rat. *Bull. Environ. Contam. Toxicol.*, **15**, 678
127. Hood, R. D., Kuczuk, M. H. and Szczech, G. M. (1978). Effects in mice of simultaneous prenatal exposure to ochratoxin A and T-2 toxin. *Teratology*, **17**, 25
128. Szczech, G. M. and Hood, R. D. (1978). Animal model: Ochratoxicosis in dogs, mice, pigs and rats. *Am. J. Pathol.* (In press)
129. Szczech, G. B. and Hood, R. D. (1978). Unpublished results.

11
Embryotoxicity of polybrominated biphenyls*

A. R. BEAUDOIN

INTRODUCTION

A mixture of polybrominated biphenyls (PBB) was accidentally mixed with livestock feed in Michigan, and resulted ultimately in prolonged periods of exposure to large numbers of farm animals. Several writers have reviewed the episode[1–4], and the following account is a brief summary of their chronologies. PBB was manufactured solely by Michigan Chemical Company as a fire retardant, under the trade name of Firemaster. The company also manufactured magnesium oxide as a feed supplement for lactating cows, under the trade name of Nutrimaster. Somehow, some PBB was packaged in the colour bags usually reserved for magnesium oxide and, although labelled Firemaster, was shipped to the Michigan Farm Bureau as part of a shipment of the Nutrimaster. Substitution of the toxic PBB for magnesium oxide in feed went unrecognized. Compounding the problem of feed contamination was the cross-contamination that occurred to all other feeds handled in the facilities that had been exposed to PBB, thereby resulting in contamination of other livestock and poultry. Widespread contamination of farms further resulted from the rendering of PBB contaminated animals for feed, and by feed swapping between individual farms and between feed mills. The Michigan PBB episode has resulted in thousands of farm animals killed and buried, and in the destruction of huge quantities of eggs and dairy products, at great economic loss to Michigan dairy farmers.

Polybrominated biphenyls are industrial compounds with a chemical structure similar to the polychlorinated biphenyls. They are used mainly as fire retardants. Fries and Marrow[5] reported that gas chromatography showed Firemaster contained two major components and at least six minor components. The two major components were identified as a hexabromobiphenyl and a mixture of heptabromobiphenyl and octabromobiphenyls. The minor compounds were not identified.

* Supported by NIH Research Grant HD 00400

The first description of the effects of PBB in a dairy herd was given in 1974 by Jackson and Halbert[6]. During the first two weeks after the first feeding of the contaminated feed, the cows' food consumption decreased dramatically to one-half the normal dry matter intake. Accompanying the anorexia was a dramatic fall in the herd's milk production, from 13000 lb/day to 7600 lb/day. Some cows exhibited increased frequency of urination and lacrimation. After 16 days, the feed ration was changed completely and the cows' appetites improved immediately, but milk production remained low. Cows in early pregnancy, having bred 4 to 6 weeks prior to eating the contaminated feed, came back into oestrus. Early embryonic resorption was suspected, but could not be proven. Some cows developed lameness. During the second month, a few cows developed haematomas and abscesses in various body locations. The herd's weight loss continued. In the third month, one-quarter of the cows developed abnormal growths on their hooves; some cows had matting of their hair and hair loss. The skin of the neck, thorax, and shoulder became thickened in many animals. Cows that had received PBB during the last trimester of pregnancy delivered 2–4 weeks late and produced many stillborn calves. Those that were born alive died soon after birth. Laboratory findings were inconclusive. Necropsy findings suggested the PBBs to be hepatotoxins: liver lesions were the most consistently reported pathological change. Also observed was kidney damage, and haematomas and abscesses in the thoracic and peritoneal cavities. PBB content of body fat ranged from 110 to 2500 ppm and in milk, 40–900 ppm.

The evidence for embryotoxicity is not clear. Cecil *et al.*[7] claimed 20 ppm PBB in the diet was teratogenic in the hen's egg. There was no effect on hatchability but 3% of the embryos were said to be abnormal; however, the malformations were not described. Chicken eggs from hens receiving more than 60 ppm failed to hatch[8]. Quail eggs were more resistant and only failed to hatch after a maternal dose of 100 ppm[9]. The bird embryos died during the first or second day of development. The no effect level on hatchability was less than 30 ppm[9]. At intermediate doses, eggs hatched but the offspring were less viable, exhibiting both growth retardation and higher mortality during the first 3 weeks of life. There were no malformations reported in any of the hatched quail or chicken young, but many newly hatched chicks were oedematous in the neck and shoulder region. Deposition of PBB in the hen's egg was equivalent to approximately 60% of the daily intake[9].

Aftosmis[10] reported octabromobiphenyl, fed to primigravid rats at 1000 or 10000 ppm in their diets from day 6 to day 15 caused massive generalized oedema in two fetuses and gastroschisis in two fetuses. No other abnormalities were reported. Dose related levels of bromine were found in the fetuses. Dent *et al.*[11] fed Sprague-Dawley rats a diet containing 500 ppm PBB from day 8 of gestation through day 14 postpartum. No mention was made of malformations nor postnatal distress in the pups. Repeated gavage doses of up to 10 mg PBB on each day of pregnancy from day 7 through day 15 had no effect on the incidence of resorption, nor were any malformations seen in the fetuses[12]. Corbett *et al.*[13] reported exencephaly in two fetuses and cleft palate in four fetuses in the 261 surviving fetuses from pregnant Swiss ICR mice fed 1000 ppm PBB in their diet from day 7 through day 18 of pregnancy.

Seven monkeys fed 0.3 ppm PBB in their diet continuously for six months were subsequently bred, and all conceived[14]. One animal aborted a mummified fetus and one gave birth to a stillborn infant. The remaining five animals delivered normal appearing but small infants.

The present experiment was conceived amidst great public concern of the threat to health posed by the PBB consumed (often in large quantities) by women on contaminated farms. The experiment was designed to determine the nature of PBB embryotoxicity in rats. Portions of the results have already been published[15].

MATERIALS AND METHODS

Virgin female Wistar-derived rats from my colony were used. The animals were maintained on Teklad Rat Diet (Teklad Mills, Winfield, Ia) *ad libitum* with supplemental feedings of lettuce once each week. Water was available at all times. The animal quarters were maintained at constant temperature and humidity with a light/dark cycle of 14/10 hours. Groups of five sexually mature females were placed overnight with an experienced male. The following morning vaginal smears were examined for the presence of sperm. The day of finding sperm was designated day 0 of pregnancy.

The polybrominated biphenyl mixture used in the experiment was the commercial preparation Firemaster BP-6 (Michigan Chemical Co., now Velsicol Chemical Corp., Chicago). Solutions of PBB were prepared in sesame oil. The lower dosages dissolved completely, whilst the highest dose formed a suspension. PBB solutions were administered by gavage as a single dose at one day of pregnancy, from day 3 through day 17, or by gavage as a chronic dose, commencing at day 0 and continuing on alternate days through day 14. The period of chronic dosing encompassed the preimplantation period and most of the period of organogenesis. For single doses, each animal received either 40, 200, 400 or 800 mg PBB/kg maternal body weight. Rats in the chronic dosing group received a total dose of PBB calculated to approximate the dose received by a 200 g rat in a single dose. The total chronic doses administered were 8, 40, 80, or 160 mg per animal. On an alternate day injection schedule, there were eight injections from day 0 to day 14. Consequently each gavage feeding contained one-eighth of the total dose. Control animals were untreated or received sesame oil by gavage. Since the higher doses of PBB caused a sharp drop in food and water consumption, pair fed controls were used to evaluate the effect of the nutritional state of the animals on the outcome of the experiment.

Pregnancy was terminated at day 20, resorption sites were counted, and the fetuses and placentae examined and weighed. Fetuses fixed in Bouin's fluid were subsequently free-hand sectioned with a razor blade and each slice was examined for malformations. Fetuses fixed in 95% alcohol were prepared for staining with alizarin red S for visualization of the skeleton.

Experiments were performed to analyse for the presence of PBB in maternal liver and fat, and to determine the transfer of PBB across the rat's placenta. In one experiment, pregnant rats were gavaged with 800 mg/kg PBB at day 12 and the embryos recovered 24 and 48 h later, together with samples of maternal liver and fat. All tissues were quick-frozen in liquid nitrogen and stored frozen for subsequent determination of PBB content. In a second experiment, pregnant rats received a chronic dosing with PBB from day 0 to day 14 (a total dose of 40 or 80 mg). At day 20 the fetuses were recovered, frozen, and subsequently analysed for the presence of PBB.

RESULTS

The results of administering low doses (40 or 200 mg/kg) of PBB are shown in Table 11.1. No malformations ascribable to PBB were induced at any treatment day; therefore, only selected days are included in the table. Two malformations were found, however, one in the control group and another in an experimental group. Both fetuses came from mothers treated at day 12 and both fetuses had identical abnormalities: the absence of all digits on all paws. It is concluded that this malformation is not due to PBB treatment.

The higher doses (400 and 800 mg/kg) of PBB induced malformations (Table 11.2). Except for treatment at day 7 with 800 mg/kg, PBB caused few maternal deaths (seven animals died for every 100 injected). For some unexplained reason five of the six animals injected at day 7 died. Up through day 12, doubling the dose caused a marked increase in the number of fetal resorptions. After day 12, the embryolethal effect declined abruptly. At day 17, PBB treatment resulted in the death of fetuses during the interval between the injection and the autopsy at day 20. The most susceptible period for the production of malformations consisted of days 11, 12, and 13. At each of these days doubling the dose induced a significant increase in the number of fetuses malformed.

The chronic administration of PBB from the day of fertilization until near

Table 11.1 Embryotoxicity following maternal ingestion of low doses of PBB

Day of treatment	*Dose* (mg/kg)	*No. of implantation sites*	*% Resorbed*	*% Survivors malformed*
6	40	48	6.2	0
	200	39	41.4	0
8	40	39	3.4	0
	200	47	14.1	0
10	40	46	15.4	0
	200	59	11.8	0
12	40	56	5.4	1.8*
	200	61	6.6	0
14	40	58	10.3	0
	200	47	13.5	0
11, 12, or 13 sesame oil				1.0†

* One embryo lacking digits on all four paws
† One embryo lacking digits on all four paws, treated at day 12

Table 11.2 Embryotoxicity following maternal administration of 400 or 800 mg/kg PBB

Day of treatment	*Dose* (mg/kg)	*No. of implantation sites*	*% Resorbed*	*% Survivors malformed*
3	400	52	48.1	7.4
	800	*	—	—
5	400	60	71.6	0
	800	49	100	—
6	400	46	100	—
	800	52	100	—
7	400	67	61.3	3.8
	800	7	100	—
8	400	52	28.6	0
	800	79	72.1	13.6
9	400	90	28.8	3.1
	800	87	87.4	0
10	400	63	23.8	4.1
	800	62	59.6	12.0
11	400	73	6.8	11.8
	800	79	81.0	60.0
12	400	122	18.0	0
	800	165	37.5	24.3
13	400	107	4.6	1.9
	800	101	8.9	24.7
14	400	73	5.4	1.4
	800	77	9.1	1.4
17	400	59	37.3†	0
	800	56	7.4‡	0
11, 12, or 13 sesame oil	0	109	10.0	1.0§

* No observable implantation sites in seven injected mothers
† Three resorbed, 19 recently dead
‡ Two resorbed, 2 recently dead
§ One embryo lacking digits on all four paws, treated at day 12

the end of organogenesis (day 14) did not produce significant numbers of malformed fetuses (Table 11.3). The chronic administration of the lower doses (8 and 40 mg) markedly increased embryonic death over that seen following the acute administration of an equivalent dose (compare Table 11.1 with Table 11.3). The higher doses (80 and 160 mg) not only caused embryonic death but also appeared to interfere with the process of implantation. It is possible that the rats exhibiting no visible signs of implantation were not pregnant, but it seems unlikely that non-pregnant animals would be found only in the groups receiving the two higher doses. The administration of a total dose of 160 mg over an extended period of time (day 0 to day 14) caused the death of one-half of the recipient rats, whereas approximately the same amount of PBB (800 mg/kg) administered as a single dose caused few maternal deaths (9% of all rats injected from day 8 through day 14).

Table 11.3 Embryotoxicity following chronic treatment with PBB*

Total dose (mg)	*No. of implantation sites*	*% Resorbed*	*% Survivors malformed*
8	87	6.9	0
40	61	62.3	0
80	49†	33.3	3.3‡
160	16§	100	—

* By gavage on alternate days, day 0 through day 14
† Sites found in only three rats. In three other injected, and presumed pregnant, rats there were no visible signs of implantation
‡ One fetus lacking all digits on all four paws
§ Sites found in only one rat. One-half of the injected, and presumed pregnant, rats died. In two others, there were no visible signs of implantation

It appears that only two malformations are attributable to PBB, cleft palate and diaphragmatic hernia (Table 11.4). One other malformation was seen, complete absence of the digits on all four paws, but it occurred in one sesame oil control fetus as well as in two fetuses from treated animals (Tables 11.1 and 11.2). Single gavage treatment with 800 mg/kg PBB at day 12 or 13 produced the greatest number of abnormal fetuses. The only other indication of aberrant development was scattered cases of oedema, subcutaneous haemorrhage, and apparent hydronephrosis, but these conditions may be transitory and, therefore, are not included in the table. There were no skeletal malformations following PBB ingestion at any day of treatment. Alizarin Red S preparations of the skeletons of 456 embryos were examined. There was considerable variation in the time of appearance of ossification centres in the sternum, metacarpals and metatarsals in both experimental and control fetuses. This is thought to reflect individual differences in growth and matu-

Table 11.4 Malformations following PBB treatment

*Day of treatment**	*Dose* (mg/kg)	*No. with cleft palate/no. survivors*	*No. with diaphragmatic hernia/no. survivors*
7	400	1/26	—
8	800	—	3/22
9	400	—	1/64
10	400	—	2/48
	800	1/25	2/25
11	400	—	7/68
	800	3/15	7/15
12	800	17/103	12/103
13	400	2/102	—
	800	21/92	2/92
14	400	—	1/69
	800	1/70	—

* Only treatment days with abnormal survivors are included in the table

ration, and the apparent absence of ossification centres is not considered a malformation.

Treatment with low doses of PBB had no effect on fetal weight or placental weight. The two higher doses, however, did cause significant reductions in both fetal and placental weight. During the period of greatest susceptibility (days 11–13), the degree of weight reduction was dependent on the dose administered (Table 11.5).

Daily weight changes were recorded for all animals treated acutely. During the 24 to 72 h immediately following PBB administration, all treated rats lost weight. The duration of the period of weight loss was directly related to the dose of PBB given. At the end of the experiment, those animals treated with the lower doses had regained the weight lost immediately following treatment, and their total weight gain was not significantly different from controls. Rats treated with 400 or 800 mg/kg PBB, however, gained less weight than controls or continued to lose weight throughout pregnancy (Table 11.5).

To determine the extent of anorexia, food and water consumption were measured from day 5 to day 20 of pregnancy in five animals chosen for similar initial weight. The average daily food intake from day 5 to day 12 was 30.0 ± 4.0 g per day; water drunk, 49.0 ± 0.8 ml per day. At 9 a.m. on day 12, 800 mg/kg PBB was administered. During the subsequent 48 h period the rats ate a daily average of 6.4 ± 2.6 g of food and drank a daily average of 27.7 ± 8.0 ml of water. Daily food consumption gradually increased during the remainder of pregnancy, but with a great deal of individual variation among the rats (range 12.8 g to 31.0 g). The average daily food consumption never did return to pretreatment levels, and at day 20 it was 24.3 ± 7.0 g. Water consumption subsequent to PBB treatment was highly variable and exceeded pretreatment levels in most animals. At day 20 the average amount of water drunk was 64.2 ± 22.2 ml. Three of the five treated rats continually lost weight

Table 11.5 **The effects of high doses of PBB on maternal weight gain and fetal and placental weight**

Day of treatment	*Dose* (mg/kg)	*Maternal weight gain**	*Fetal weight† mean ± SD*	*Placental weight‡ mean ± SD*
Untreated control		+77.0 ± 11.8	3.88 ± 0.45	0.51 ± 0.06
Sesame oil control		+88.0 ± 9.0	4.55 ± 0.71	0.58 ± 0.07
11	400	+20.0 ± 44.8	3.36 ± 0.27	0.42 ± 0.05
	800	−42.8 ± 43.2	2.00 ± 0.60	0.31 ± 0.10
12	400	+29.0 ± 37.2	3.66 ± 0.60	0.44 ± 0.04
	800	−30.6 ± 33.4	2.66 ± 0.60	0.35 ± 0.06
13	400	+30.8 ± 23.6	3.64 ± 0.23	0.45 ± 0.06
	800	−23.2 ± 16.7	2.80 ± 0.46	0.37 ± 0.03

* From day of treatment to day 20. Controls, day 11 to 20

† All fetuses from treated mothers weighed significantly less than fetuses from sesame oil control, $p = 0.01$ or less

‡ All placentae from treated mothers weighed significantly less than placentae from sesame oil controls, $p = 0.01$ or less. p values determined by analysis of independent samples (t independent)

following treatment and at day 20 weighed less than at day 12 (the remaining two animals gained only 6 and 10 g respectively. Fetal weights averaged 2.94 ± 0.35 g and placental weights 0.38 ± 0.03 g, both significantly less than fetal and placental weights in sesame oil controls (4.55 ± 0.71 and 0.58 ± 0.07 g, respectively).

For the determination of the effect of decreased food intake on the outcome of the experiment, five additional pregnant rats were paired by weight with the five above treated animals. Each rat received the same amount of food and water as its pair member from day 5 to day 20 of gestation. At day 12, each rat received 1.5 ml of sesame oil, but no PBB. At day 20, the average maternal weight gain was 47.5 ± 3.5 g. The average weight of the fetuses was 3.91 ± 0.40 g and the average weight of placentae, 0.52 ± 0.01 g. The weights compare closely with the average fetal weight (3.88 ± 0.45 g) and placental weight (0.51 ± 0.06 g) of untreated controls.

Subsequent to maternal treatment, PBB can be found in the maternal liver and fat, and in the embryo and fetus. Twenty-four hours after a single dose of 200 mg PBB on day 12 the maternal liver contained 267 ppm PBB, maternal fat 51 ppm, and a pooled sample of embryos from two litters 13 ppm. At 48 h post-treatment maternal liver contained 248 ppm, maternal fat 250 ppm, and a sample of pooled embryos from two litters 7 ppm PBB. A single dose of 280 mg PBB (800 mg/kg) administered at day 17 resulted in day 20 fetuses from a single litter containing an average of 0.26 ± 0.02 mg PBB per fetus. Chronic dosing with eight injections from day 0 to day 14 (total dose of 40 mg) resulted in day 20 fetuses with an average of 0.028 ± 0.001 mg PBB, and following a total dose of 80 mg the average PBB content in fetuses from two litters was 0.021 ± 0.004 mg.

DISCUSSION

It has been demonstrated repeatedly that PBB can be toxic to animals. In addition to the toxic syndrome first described in a Michigan dairy herd[6], toxicity has also been reported in birds[7,8], sheep[16], rodents[17] and monkeys[14]. PBB has been shown to induce changes in liver microsomal enzymes in fish[18], birds[8,9] and rodents[17,19,20]. The toxic effect, however, is directly related to the amount of PBB consumed, for low doses do not always produce overt clinical signs of toxicity[5,16,17,21,22], although liver damage may still occur[16,17,22,25]. All of the authors cited above report PBB to accumulate in adipose tissue of the respective animals. PBB accumulates in the eggs of birds[7–9], appears in the milk of lactating cows[5,24,25] and rats[17,26], and crosses the bovine[22,27] and rodent placenta[13,15].

Studies on the distribution and clearance of PBB in cows[24] revealed the faeces to be the major route of excretion during the exposure period. Approximately one-half of a single dose was excreted by 168 h post-dosing. Thereafter, however, the amount found in the faeces was low. Very little, if any, was excreted in the urine. In rats it has been estimated that over the lifetime of a rat only 10% of the ingested dose would be excreted in the faeces[26]. During lactation, however, a considerable amount is lost in the milk[24,26].

Tilson *et al.*[28] report PBB to have behavioural and neurological effects in rats and mice. The continued administration of low doses did not produce clinical signs of acute PBB toxicity but did, nonetheless, produce measurable changes. PBB depressed motor reflexes, impaired forelimb grip strength, and decreased motor activity. The effects were more marked in rats than in mice. Visual placement responses were decreased in some animals. Emotionality, as measured by the number of defaecations and urinations in the open field, was not affected.

The results of this experiment show that single oral doses of PBB administered at one day of pregnancy during the organogenetic period in the rat can have embryolethal and teratogenic effects. Prior to and subsequent to organogenesis the toxicity of PBB to the developing rat is much reduced. Low doses did not cause frank malformations nor affect fetal or placental weight. This agrees with observations of the effects of low-level PBB contamination in other animals[5,8,16,17,21,22].

Treatment with high doses of PBB is quite another matter. Following treatment at day 6 there were no surviving embryos. Implantation is in progress at this time and it may be that PBB interferes with this complex process, or simply kills the embryo outright. Following the period of implantation, the embryolethal effects of PBB decreased markedly with the 400 mg/kg dose and remained low for each subsequent day of treatment. The 800 mg/kg dose, however, continued to cause large numbers of embryos to die, until at day 13 the embryolethal effect diminished.

The chronic administration of PBB from the day of fertilization until near the end of organogenesis did not induce malformations ascribable to PBB. The digital anomalies present in one fetus were identical to those appearing in one fetus exposed to sesame oil alone. It is apparent from the results that a dose of PBB administered by repeated gavage feeding is more detrimental to the embryo and mother than the same dose administered as a single gavage feeding at one day in pregnancy. The reason for this is unclear but may be related to the ability and speed with which the intestine absorbs PBB. The large single dose may overwhelm the absorptive process in the intestine.

PBB caused only cleft palate and diaphragmatic hernia in the offspring of rats used in this experiment. Why only these developmental events are singled out remains unknown. Critical times in the development of the two structures, however, do coincide. The pleuroperitoneal folds complete the formation of the diaphragm around day 15–16 of gestation. The palatal shelves begin elevation around day 16 and fusion usually is completed during day 17–18. Thus, the two events are related in time, and both may involve the growth and migration of mesoderm. Why earlier developmental processes apparently show no visible effect of PBB treatment remains an enigma. During the period of organogenesis, at least one malformed fetus was found in litters from each day of treatment. Only following treatment at day 12 were there many fetuses with both defects. No significance is attached to the observation that a delay in the appearance of sternal, metacarpal, and metatarsal ossification centres often followed PBB treatment. There was also considerable variation among control embryos in the appearance of these ossification centres. The delay in the experimental rats probably only reflects the delayed growth of the fetus, as indicated by lower fetal weights.

PBB has been reported either to lack teratogenic activity in birds[8], rats[11,12], monkeys[14] and cattle[22] or to possess teratogenic action in birds[7], rats[10,15] and mice[13], albeit sometimes weak. There are no documented reports of births of malformed calves in dairy herds contaminated with PBB, although, hearsay evidence abounds.

There are several differences between the present teratological experiment and other teratological experiments using rats. Two differences stand out. First, in all but one investigation PBB was fed mixed with the diet and no attempt was made to measure the amount of PBB eaten[10–13]. No direct comparison of the amount of PBB consumed is therefore possible between this study and the others. A second difference is that the Sprague-Dawley rat was used in three of the four investigations, and in the fourth investigation an unspecified strain of rat was used. It is well-known in teratology that species differences exist in the responses to teratogens. There is one report[12] of gavage feeding of PBB to pregnant rats. The greatest amount of PBB administered was 90 mg over a 10-day period. No effect on embryonic development was reported. In the present experiment, chronic dosing with 80 mg PBB did not induce malformations, but did cause embryonic death. The same amount given as a single dose was teratogenic. The results of the present experiment suggest that the chronic administration of PBB is less likely to produce mal formations than the acute administration of the same dose. This may explain the lack of PBB teratogenicity reported by other authors.

The anorexia and weight loss seen in the rats following high doses of PBB is similar to that reported in other animals. The fact that pair-fed control rats had similar numbers of fetuses with comparable weight to sesame oil-fed controls indicates that inanition, while reducing weight gain, was not the cause of resorptions or malformations in the PBB-treated rats. It is, however, probably responsible for the marked reduction in fetal and placental weights.

The results of this and other studies[10,13,15,27] demonstrate that PBB can cross the mammalian placenta. It can traverse the newly formed rat chorio-allantoic placenta at day 12 and the mature rat placenta at day 17. PBB was found in the day 13 embryo 24 h after its administration and in the day 20 fetus six days after the last gavage treatment. It is interesting to note that the amount of PBB accumulated by the day 20 fetus did not depend on the dose administered chronically. After a PBB dose of either 40 or 80 mg the amount accumulated by the fetus was nearly the same, 0.28 and 0.21 mg respectively. The reason for this is unknown. That the fetus can accumulate greater amounts of PBB is obvious. Following a single dose of 280 mg given at day 17, the day 20 fetus contains 0.26 mg PBB. It may be that during the six-day interval following the last gavage with PBB an equilibrium is established in PBB concentration between the maternal tissues and plasma and between the maternal plasma and the fetus. This could account for the similarities in fetal PBB concentrations following two different maternal doses.

PBB appears to be a persistent environmental contaminant. Significant quantities of PBB were found in the river-water, fish and ducks downstream from the manufacturing plant[29]. Follow-up studies for three years subsequent to termination of PBB manufacture indicated a decline in river-water and sediment PBB concentrations, but fish and duck tissue residues remained unchanged. Attempts to leach PBB from contaminated soil with leachate quantities equivalent to 20 times the annual rainfall in Michigan resulted in a loss of less than 0.6%[30]. PBB is expected to remain in the soil for long periods of time.

PBB is toxic to mammals, birds and fish, and chronic exposure appears more toxic than an acute exposure to the same dose. The results of this experiment have shown that single high doses of PBB can be lethal and teratogenic to the rat embryo. Low doses had no observable detrimental effects on pregnancy. No human malformations, nor fetal deaths, have yet been attributed to maternal contamination with PBB.

ACKNOWLEDGEMENT

The determination of PBB was by gas chromatography performed by Environmental Research Group, Inc. Ann Arbor, MI.

References

1. Dunckel, A. E. (1975). An updating on the polybrominated biphenyl disaster in Michigan. *J. Am. Vet. Med. Assoc.*, **167**, 838
2. Robertson, L. W. and Chynoweth, D. P. (1975). Another halogenated hydrocarbon. *Environment*, **17**, 25
3. Carter, L. J. (1976). Michigan's PBB incident: Chemical mix-up leads to disaster. *Science*, **192**, 240
4. Kay, K. (1977). Polybrominated biphenyls (PBB) environmental contamination in Michigan, 1973–1976. *Environ. Res.*, **13**, 74
5. Fries, G. F. and Marrow, G. S. (1975). Excretion of polybrominated biphenyls into the milk of cows. *J. Dairy Sci.*, **58**, 947
6. Jackson, T. F. and Halbert, F. L. (1974). A toxic syndrome associated with the feeding of polybrominated biphenyl-contaminated protein concentrate to dairy cattle. *J. Am. Vet. Med. Assoc.*, **165**, 437
7. Cecil, H. C., Bitman, R. J., Lillie, R. J., Fries, G. F. and Verrett, J. (1974). Embryotoxic and teratogenic effects in unhatched fertile eggs from hens fed polychlorinated biphenyls. *Bull. Environ. Contam. Toxicol.*, **11**, 489
8. Ringer, R. K. and Polin, D. (1977). The biological effects of polybrominated biphenyls in avian species. *Fed. Proc.*, **36**, 1894
9. Babish, J. G., Gutenmann, W. H. and Stoewsand, G. S. (1975). Polybrominated biphenyls: Tissue distribution and effect on hepatic microsomal enzymes in Japanese quail. *J. Agric. Food Chem.*, **23**, 879
10. Aftosmis, J. G., Culik, R., Lee, K. P., Sherman, H. and Waritz, R. S. (1972). Toxicology of brominated biphenyls. I. Oral toxicity and embryotoxicity. *Toxicol. Appl. Pharmacol.*, **22**, 316
11. Dent, J. G., Cagen, S. Z., McCormack, K. M., Rickert, D. E. and Gibson, J. E. (1977). Liver and mammary arylhydrocarbon hydroxylase and epoxide hydratase in lactating rats fed polybrominated biphenyls. *Life Sci.*, **20**, 2075
12. Cecil, H. C., Harris, S. J. and Bitman, J. (1977). Embryotoxic effects of polybrominated biphenyls (PBB) in rats. Presented at Workshop on Scientific Aspects of Polybrominated Biphenyls PBB at Michigan State University, October 24–25
13. Corbett, T. H., Beaudoin, A. R., Cornell, R. G., Anver, M. R., Schumacher, R., Endres, J. and Szwaboska, J. (1975). Toxicity of polybrominated biphenyls (Firemaster BP-6) in rodents. *Envir. Res.*, **10**, 390
14. Allen, J. R. and Lambrecht, L. K. (1978). Responses of rhesus monkeys to polybrominated biphenyls. Presented at the 1978 Meeting of the Society of Toxicology
15. Beaudoin, A. R. (1977). Teratogenicity of polybrominated biphenyls in rats. *Envir. Res.*, **14**, 81

16. Gutenmann, W. H. and Lisk, D. J. (1975). Tissue storage and excretion in milk of polybrominated biphenyls in ruminants. *J. Agric. Food Chem.*, 23, 1005
17. Sleight, S. D. and Sanger, V. L. (1976). Pathologic features of polybrominated biphenyl toxicosis in the rat and guinea pig. *J. Am. Vet. Med. Assoc.*, **169**, 1231
18. Elcombe, C. R. and Lech, J. J. (1977). Induction of monooxygenation in rainbow trout by polybrominated biphenyls. A comparative study. Presented at the Workshop on Scientific Aspects of Polybrominated Biphenyls PBB at Michigan State University October 24–25
19. Dent, J. G., Netter, K. J. and Gibson, J. E. (1976). The induction of hepatic microsomal metabolism in rats following acute administration of a mixture of polybrominated biphenyls. *Toxicol. Appl. Pharmacol.*, **38**, 237
20. Dent, J. G., Elcombe, C. R., Netter, J. K. and Gibson, J. E. (1978). Rat hepatic microsomal cytochrome(s) P-450 induced by polybrominated biphenyls. *Drug Metabolism and Disposition: The Biological Fate of Chemicals*, **6**, 96
21. Mercer, H. D., Teske, R. H., Condon, R. J., Furr, A., Meerdink, G., Buck, W. and Fries, G. (1976). Herd health status of animals exposed to polybrominated biphenyls (PBB). *J. Toxicol. Envir. Health*, **2**, 335
22. Moorhead, P. D., Willett, L. B., Brumm, C. J. and Mercer, H. D. (1977). Pathology of experimentally induced polybrominated biphenyl toxicosis in pregnant heifers. *J. Am. Vet. Med. Assoc.*, **170**, 307
23. Cook, H., Helland, D. R., VanderWeele, B. H. and DeJong, R. J. (1978). Histotoxic effects of polybrominated biphenyls in Michigan dairy cattle. *Envir. Res.*, **15**, 82
24. Willett, L. B. and Irving, H. A. (1976). Distribution and clearance of polybrominated biphenyls in cows and calves. *J. Dairy Sci.*, **59**, 1429
25. Prewitt, L. R., Cook, R. M. and Fries, G. F. (1975). Field observations of Michigan dairy cattle contaminated with polybrominated biphenyl. *J. Dairy Sci.*, **58**, 763
26. Matthews, H. B., Kato, S., Morales, N. M. and Tuey, D. B. (1977). Distribution and excretion of 2,4,5,2′,4′,5′-hexabromobiphenyl, the major component of Firemaster BP-6. *J. Toxicol. Envir. Health*, **3**, 599
27. Detering, C. N., Prewitt, L. R., Cook, R. M. and Fries, G. F. (1975). Placental transfer of polybrominated biphenyl by holstein cows. *J. Diary Sci.*, **58**, 764
28. Tilson, H. A., Cabe, P. A. and Mitchell, L. (1978). Behavioral and neurological toxicity of polybrominated biphenyls in rats and mice. *Envir. Health Perspect.* (In press)
29. Hesse, J. L. and Powers, R. A. (1977). Polybrominated biphenyl (PBB) contamination of the Pine River, Gratiot and Midland Counties, Michigan. Presented at the Workshop on Scientific Aspects of Polybrominated Biphenyls (PBB) at Michigan State University October 24–25
30. Filonow, A. B., Jacobs, L. W. and Mortland, M. M. (1976). Fate of polybrominated biphenyls (PBBs) in soils. Retention of hexabromobiphenyl in four Michigan soils. *J. Agric. Food Chem.*, **24**, 1201

12
Acute alcohol intoxication in the pregnant rat

L. A. KENNEDY AND T. V. N. PERSAUD

INTRODUCTION

The fetal alcohol syndrome (FAS) has been described and attributed to chronic maternal alcohol abuse during pregnancy. A prospective study[2] involving 633 women revealed that infants born to heavy drinkers had twice the risk of abnormality (such as microcephaly) as those born to moderate or abstinent drinkers. Thirty-two percent of such infants had congenital defects compared to 14% in moderate, and 9% in abstinent drinkers. A mouse alcohol model[3], a beagle model[4] and a guinea pig model[5] have been reported, and Kronick[6] has shown that acute alcohol intoxication is embryolethal and teratogenic in mice. Obe and Herha[7] have suggested that alcohol is mutagenic *in vivo*. Although much is now known about the deleterious effects of chronic alcohol abuse during pregnancy, little attention has been directed to the more common problem of short-term abuse, or 'binge drinking'.

MATERIALS AND METHODS

Albino Sprague-Dawley rats of the Holtzman strain were housed in groups of three in wire-mesh cages under controlled environmental conditions (temperature, approximately 21 °C; relative humidity approximately 50%; and a 12 h light–dark cycle). They were maintained on Teklab mouse and rat diet and water, both available *ad libitum*.

All handling was done by one person. Experimental treatments and killing were performed during the first three hours of the light cycle.

Male rats were placed with virgin females (200–250 g) overnight and the following morning vaginal smears were taken. The first day of gestation was considered to be the day on which spermatozoa were found in the vaginal smear.

Each experimental treatment group consisted of six pregnant rats. For each experimental group a control group consisting of three animals was treated concurrently with equal volumes of physiological saline.

Table 12.1 Treatment schedule

Treatment group	*Number of animals*	*Treatment*
Saline control 30	3	2 ml saline
30% ethanol	6	2 ml ethanol (30% v/v) mean daily dose = 1.4 g/kg
Saline control 20	3	2 ml saline
20% ethanol	6	2 ml ethanol (20% v/v) mean daily dose = 1.1 g/kg
Saline control 10	3	2 ml saline
10% ethanol	6	2 ml ethanol (10% v/v) mean daily dose = 0.56 g/kg

The test substances were administered on days 9 through 12 of gestation. Each daily treatment consisted of the intraperitoneal injection of 2 ml of 10%, 20%, or 30% ethanol (v/v) in physiological saline (Table 12.1). Following each treatment the animals were observed for behavioural changes.

On day 20 of gestation, the animals were killed by ether administration. At autopsy the viscera were examined for signs of peritonitis, and the uterine horns were removed and opened for inspection of the contents. The total number of implantation sites were recorded as resorptions, dead or live fetuses. Maternal liver, kidney and ovary, all placentas and 75% of fetuses were placed immediately in Bouin's fixative for subsequent gross and microscopic examination. The remaining fetuses were stained with Alizarin Red S and evaluated for skeletal anomalies.

Maternal weights were recorded on day 1 of gestation, on each of the treatment days (days 9 through 12), and at autopsy (day 20).

RESULTS

Maternal toxicity

The degree of locomotor impairment was used to establish levels of intoxication which satisfactorily paralleled a binge drinking situation.

A dose-related effect was observed in the different treatment groups with respect to locomotor impairment. Those rats receiving 30% ethanol occasionally lost their righting reflex for up to 15 minutes but never lost their abdominal tonus. They displayed varying degrees of hypotonia and ataxia and generally appeared to be in a state of stupor for up to 4 h after treatment. Those rats receiving 20% ethanol displayed locomotor ataxia and hypotonia for up to 4 h after treatment. Although movements were slow and ataxic, the animals were active in contrast to the 30% group. The 10% ethanol treated group displayed no locomotor ataxia or hypotonia. They were generally more active and exploratory than the controls. The saline treated animals showed no impairment in locomotion or change in activity patterns.

Analysis of variance revealed no significant difference in weight gain between the animals of the three control groups. This observation justified the pooling of the controls into one group ($n=9$). A one-way analysis of variance revealed that maternal weight gain varied significantly over the treatment

Table 12.2 Comparison of maternal weight gain among pooled control animals and ethanol treated animals

		Mean maternal weight change (g)			
Treatment group	*Number*	*D9–12*	*D9–12/n**	*D9–20*	*D9–20/n**
Control (pooled)	9	+11.0	+1.0	+84.0	+6.0
30% ethanol	6	−3.3	−0.3	+69.0	+5.8
20% ethanol	6	+2.5	+0.2	+68.0	+6.2
10% ethanol	6	+7.0	+0.5	+77.0	+6.4
'F' ratio observed		4.477	4.082	1.873	0.740
Significance $F_{0.95}$ (3.23)=3.30		$p<0.05$	$p<0.05$	N.S.	N.S.

* n = number of live fetuses in litter
N.S. = not significant

period (day 9 to 12) among the treatment groups (Table 12.2). Multiple comparison analysis demonstrated that the animals treated with the two highest doses of ethanol gained significantly less weight than did the pooled controls (Figure 12.1). The analysis also revealed, however, that this difference had been eliminated by the end of gestation.

Microscopic changes were observed in the livers, kidneys, and ovaries of ethanol treated animals, which were related both to the blood vascular system and parenchyma. Generally, there was a dose-related effect with the highest dose of ethanol being associated with the most severe and extensive lesions. In the livers, these changes included an increase in nucleated blood elements in the blood channels, the presence of aggregates of inclusion-laden macrophages in the parenchyma (Figure 12.2) and an increase in degenerating erythrocytes and pigment fragments both in the blood channels and throughout the liver parenchyma. A marked increase in the number of mitotic figures in the liver cords adjacent to the central veins was further evidence of cellular damage and repair following acute intoxication with ethanol. There was no evidence of fatty change.

In the kidneys of ethanol treated animals, parenchymal changes were observed in both the cortex and the medulla but involved only the tubular portion of the nephron. Changes in the vascular system included marked periarteriolar cuffing, venous disruption and haemorrhage, and the presence of pigment fragments in the larger vessels of some tissues (Figure 12.3). Morphological changes in the parenchyma involved only the tubular portion of the nephron. In the cortex, degeneration of the convoluted tubules was observed but associated glomeruli were never affected. The most pronounced changes, however, were observed in the medullary tubules. Marked degeneration and sloughing off of the tubular epithelium were consistently observed in the kidneys of all ethanol treated animals, especially at the base of renal papilla. The tubules of the papilla, however, never displayed any sign of damage although debris was sometimes present in their lumina.

There was no morphological difference observed in the follicular development or supporting stroma of the ovaries of ethanol or saline treated rats. However, the corpora lutea of ethanol treated animals were distinctly con-

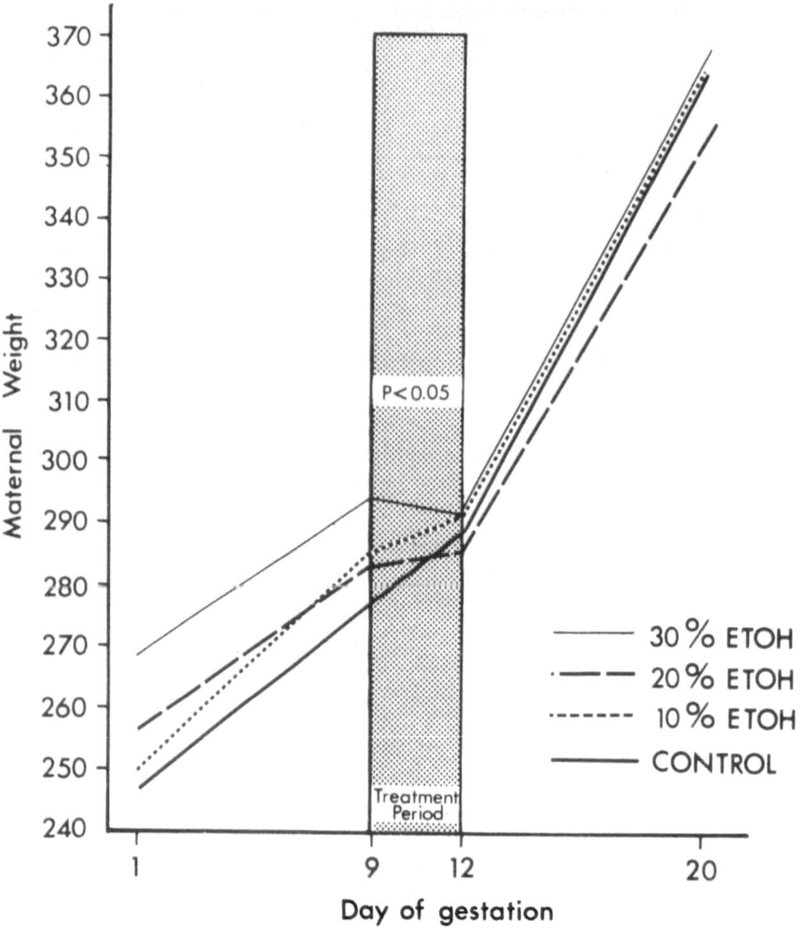

Figure 12.1 Comparison of maternal weight gain following acute gestational intoxication with alcohol

gested in the peripheral venous channels, the luteal sinusoids and the large central sinusoid. In the tissue immediately adjacent to the congested central sinusoids, there was evidence of damage to the luteal cells themselves and an increased number of macrophages (Figure 12.4).

Fetal toxicity

There was no significant variation in the weights and crown–rump lengths of fetuses of the four treatment groups as revealed by a nested analysis of variance. The placental weights of the 30% ethanol group, however, were significantly lower than were the control weights. There was no treatment-related difference in fetal mortality rates. Examination of whole fetuses under a dissecting microscope revealed no gross external malformations in the

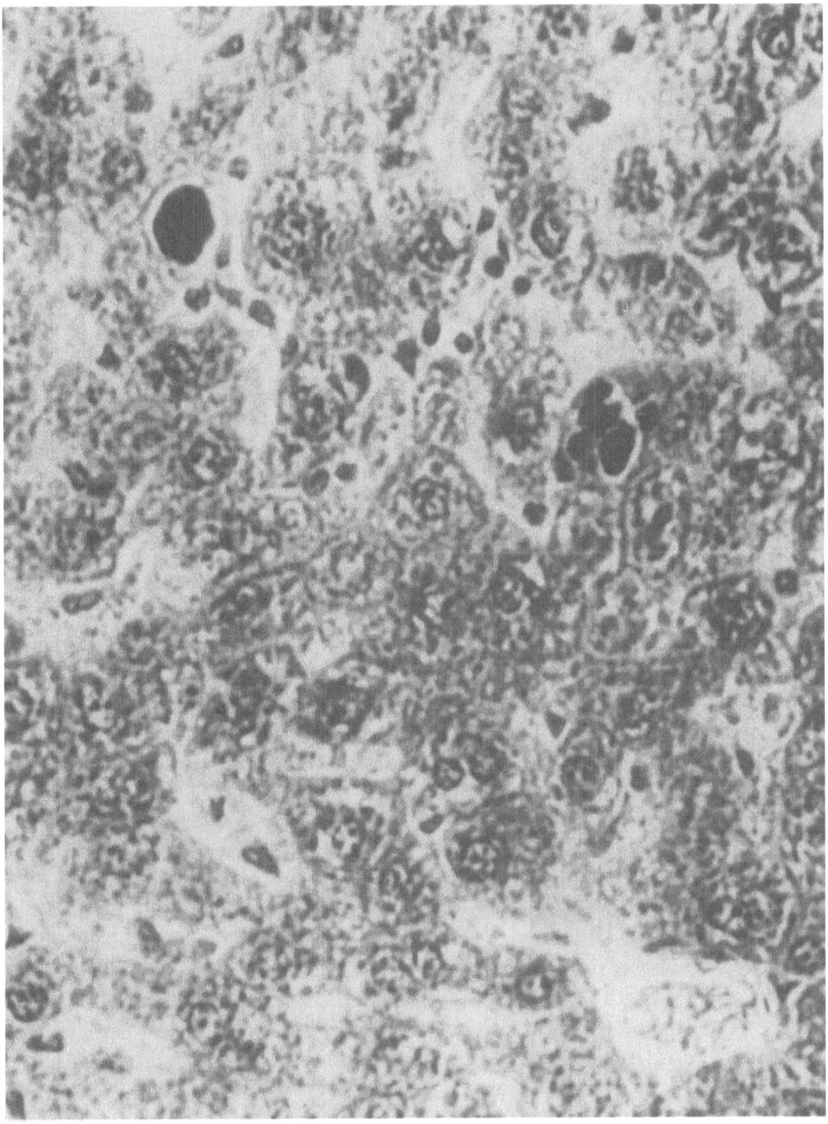

Figure 12.2 Aggregates of inclusion-laden macrophages in maternal liver following acute intoxication with 30% ethanol H&E during pregnancy (+614)

offspring exposed prenatally to ethanol and serial sectioning revealed no visceral anomalies. There were ossification deficiencies or delays in the supra-occipital and interparietal bones of the cranium (Figure 12.5), and minor deficiencies in the distal limb segments. There were no skeletal deformities detected, and it is unknown whether the ossification deficiencies would have continued postnatally.

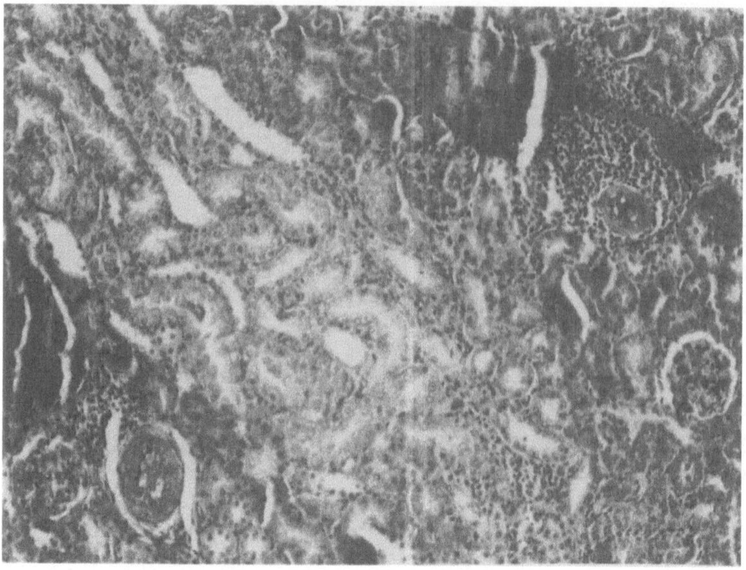

Figure 12.3 Periarteriolar cuffing and haemorrhage, parenchymal macrophages and tubular epithelial damage observed in the maternal renal cortex following acute intoxication with 10% ethanol H&E (× 120)

Based on the reports in the literature of the effects of alcohol, various fetal tissues were processed routinely with haematoxylin and eosin and examined for microscopic changes. At day 20 of gestation, fetal liver, kidney, bone and adrenal gland are immature structures with little functional or structural resemblance to the adult tissues. The liver is primarily a haematopoietic organ with indistinct cords of hepatocytes. The kidney is at an early metanephric stage with few tubules, glomeruli and definitive blood vessels. The marrow cavity of fetal limbs is just being established and is clearly not involved in haematopoiesis. Zonation between the cortex and medulla of the fetal adrenal gland is indistinct, and cortical tissue predominates. The liver, kidney and bone of offspring exposed prenatally to alcohol were indistinguishable from control tissues. Although there were subtle changes in the adrenal glands which were suggestive of fetal stress[8], further morphometric, histochemical or ultrastructural studies are required before any conclusive statement can be made.

The chorioallantoic placenta of the rat, after about day 17, is composed of a trophoblastic labyrinth with both fetal and maternal circulation, and a basal zone which contains only maternal blood. The basal zone has three cellular elements, namely, the chorionic giant cells, small basophils, and nests of glycogen cells. Overlying the basal zone is a layer of maternal decidua basalis. The fetal blood circulates in the trophoblastic septa of the labyrinth and is separated from the maternal sinusoids by cytotrophoblastic epithelium[9].

Towards the end of gestation normal degenerative changes were observed in control tissues such as cytolysis, liquefaction and hyalinization of the giant cells, necrosis of the glycogen nest cells, capsularization of the small

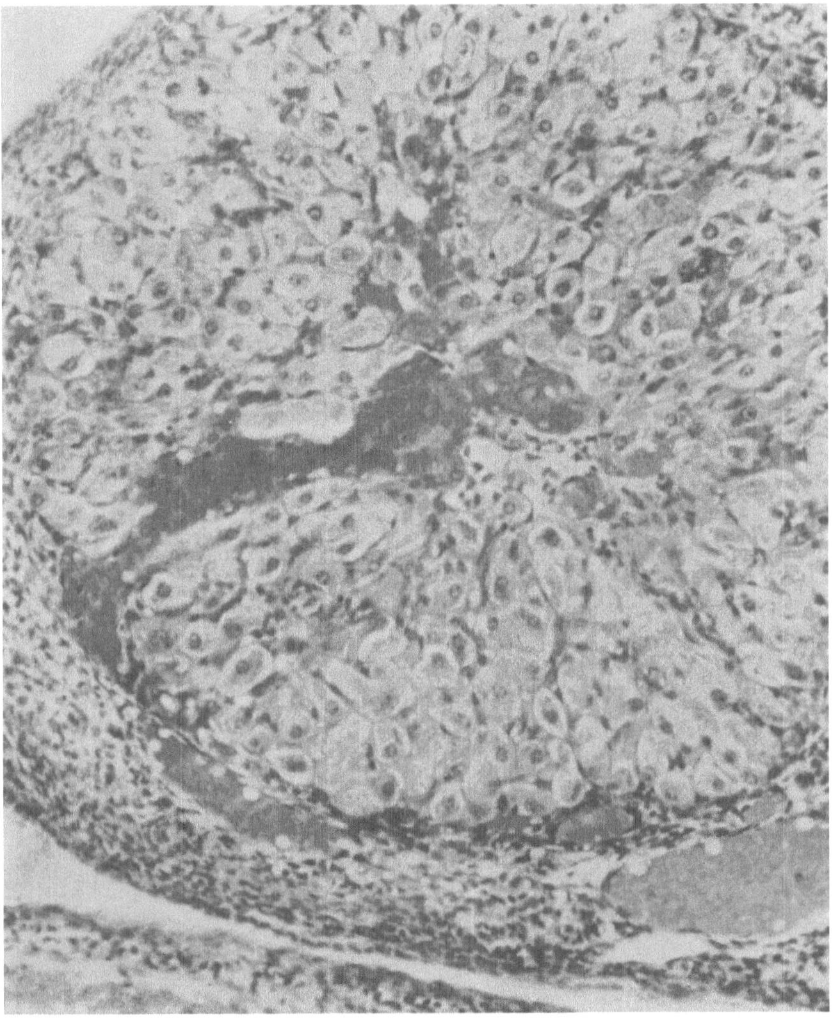

Figure 12.4 Congestion in the sinuses of the corpus luteum and minimal luteolysis around a central sinusoid following treatment with 10% ethanol H&E (× 216)

basophils with an acidophilic, PAS-positive material and fibrosis in the decidua basalis. These reflect a normal regression of placental activity.

In the placentas of ethanol-treated animals, these degenerative changes were much more extensive and pronounced than in control placentas (Figure 12.6). This was evident in tissues stained both with H & E and PAS. Severe vascular congestion was observed in the maternal labyrinthine sinusoids at the interface with the basal zone, and of the maternal sinusoids in the basal zone. A large number of nucleated blood elements (macrophages, immature monocytes, and polymorphs) were present in the basal sinusoids. Polymorphs were observed laminating the edges of the sinusoids and emigrating in large

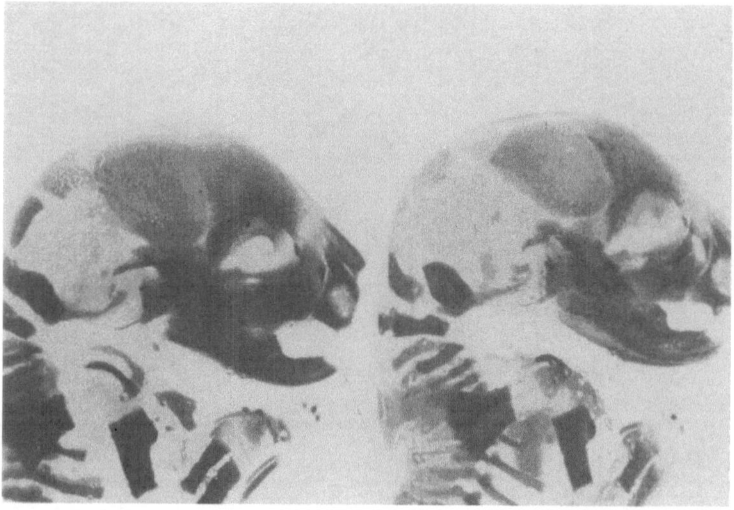

Figure 12.5 Delay or deficiency of ossification in the supraocciput and interparietal bone of fetuses exposed prenatally to ethanol. Control fetus on the left (Alizarin Red S)

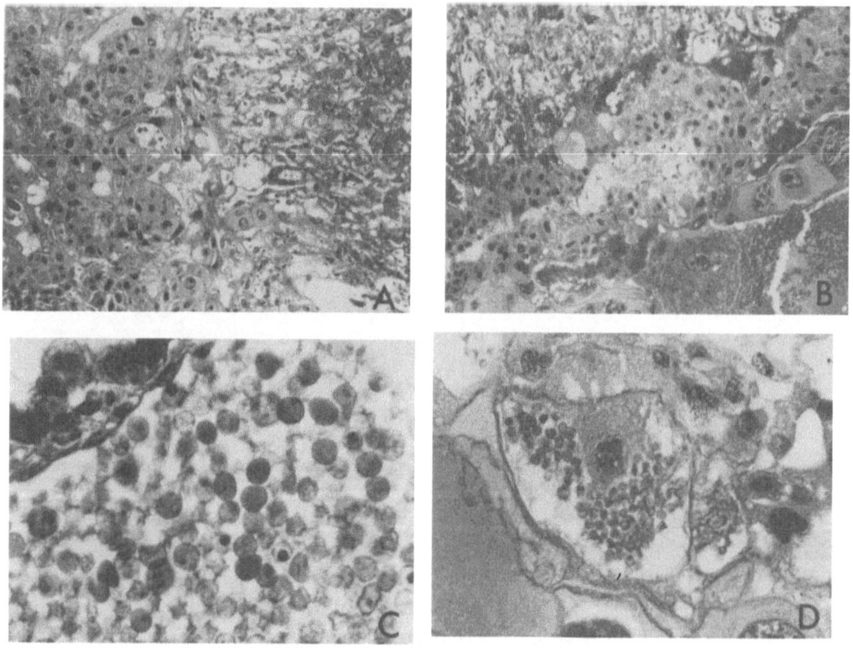

Figure 12.6 Placental changes observed following acute maternal intoxication with ethanol. A, Normal maternal and fetal circulation at the junction between the labyrinth and the basal zone. H&E (× 67); B, Severe vascular congestion in the maternal sinusoid of the basal zone and at the basal-labyrinthine junction following treatment with 10% ethanol H&E (× 67); C, Degeneration of bloodl cells in a fetal blood vessel. Dosage: 30% ethanol H&E (× 67); D, A giant cell with cytoplasm displaying extensive vacuolization and containing many blood cells. Dosage: 30% ethanol H&E (× 268)

numbers into the surrounding tissues. Cytological changes in the basal zone frequently accompanied these haematological changes. There was a distinct increase in the frequency and severity of giant cell liquefaction/hyalinization, as well as the debris contained within these cells. Large areas of amorphous eosinophilic/PAS-positive material were closely related to the cytological and haematological changes.

In the large fetal blood vessels near the fetal surface of the placenta, large numbers of blood cells appeared to be degenerating.

DISCUSSION

The administration of test substances to experimental animals in toxicological experiments presents many methodological problems. Toxicity is known to vary according to any one or combination of the following: number of doses administered, dose dilution, solvent vehicle, route of administration, degree of tolerance or habituation, animal species, as well as the age, sex and health status of the individual animal and environmental factors. Therefore, preliminary investigations were carried out to establish the following:

(a) The relative toxicity of three different doses of ethanol administered intraperitoneally to rats in different dilutions, on four consecutive days, and

(b) A desirable range of ethanol intoxication for subsequent experiments.

The rapid onset of distinctly treatment-related levels of intoxication, as evidence in locomotor impairment, indicated that the intraperitoneal route of administration allowed a rapid and efficient absorption of ethanol into the bloodstream. Furthermore, since there were no gross signs of general or severe peritonitis, it was concluded that the intraperitoneal administration of ethanol, in concentrations of 10%, 20% and 30% (v/v), would be a useful experimental model for subsequent investigations.

The mean daily doses of ethanol administered in the teratological experiments ranged from 0.56 g/kg to 1.4 g/kg (Table 12.1). The LD_{50} for single doses of ethanol, administered intraperitoneally to rats is 5 g/kg[10]. The equivalent doses for an 80 kg human being would range from 5 to 14 oz. respectively of hard liquor (40% alcohol v/v). As a general index of the level of intoxication the degree of locomotor impairment was observed. There was a distinctly dose-related effect on both the severity and duration of locomotor impairment, and the doses administered produced a state of intoxication which satisfactorily paralleled a binge drinking situation. Because ethanol was administered at the beginning of the light schedule, and since the most severe intoxication never exceeded 4 h duration, the peak activity and feeding periods of the animals were not interrupted.

Patterns of maternal weight gain were studied as an indirect indication of food intake and general well-being of the animals. Multiple comparison analyses revealed a significant reduction in weight gain over the treatment period in animals treated with 30% and 20% ethanol, compared to control animals. This difference in weight gain was temporary, however, since at the end of gestation there was no significant variation in maternal weight relative to treatment. The number of live fetuses in each litter did not affect these observations. Changes in maternal weight gain, therefore paralleled the dose-related effect of ethanol on behavioural impairment. In contrast, there was no treatment-related variation in fetal weight or crown–rump length in this investigation.

The nutritional status of the mother can influence the growth and development of her offspring[11]. Many studies in humans[2] and in laboratory animals[12] have shown, however, that the effects of chronic maternal alcohol consumption on progeny occur independently of diet and are the direct result of the ingestion of alcohol. In these and other investigations[1,3] low fetal birthweights have been reported without any mention of changes in maternal weight gain.

Kronick[6] studied the effects of single doses of ethanol during pregnancy in mice and found that maternal weight loss following treatment was temporary and that there was no significant reduction in the fetal weights at term associated with prenatal exposure to acute ethanol intoxication. This is consistent with the observations of the present investigation in the rat. Neither investigation determined, however, if there was an immediate effect of acute exposure to ethanol on fetal weight gain *in utero*. If, in fact, the normal term weights of these offspring do reflect catch-up growth, this would be inconsistent with the irreversible pre- and postnatal growth retardation observed in infants following chronic gestational exposure to ethanol.

Increased incidence of prenatal mortality and maldevelopment in mice has been reported following chronic and acute exposure to ethanol during gestation. Chernoff[3] reported a dose–response effect and strain differences in susceptibility to chronic alcohol intake. Anomalies at the lower dose levels included deficient occipital ossification and neural anomalies, and at the higher levels, cardiac and eyelid dysmorphology were observed. Following acute prenatal exposure to ethanol, Kronick[6] found that coloboma of the iris and forepaw ectrodactyly were the most frequent defects. In the present experiment both external and visceral examination failed to reveal any congenital defect related to ethanol exposure. This could be explained by species variation in susceptibility, route of administration, differences in the duration of exposure and/or the dose levels used. For example, Maling[10] reported that the LD_{50} for ethanol in the mouse is 9.5 g/kg (p.o.) or 2.0 g/kg (i.v.), whereas in the rat it is 13.7 g/kg (p.o.), 5.0 g/kg (i.p.) or 4.2 g/kg (i.v.). It appears therefore that the rat is much less susceptible to the effects of ethanol than is the mouse, but that the intraperitoneal route increases the toxicity of ethanol compared to the oral route of administration. Rates of absorption probably account for the latter difference in toxicity.

Further difficulty is encountered in interpreting these studies since different dose units and different criteria for defining intoxication were used to present the results. Chernoff[3] administered ethanol in Metrecal liquid diet before and during gestation and reported non-pregnant blood alcohol levels of 73 to 398 mg/100 ml as the definition of the level of intoxication. Blood alcohol levels of 690 mg/100 ml can be lethal in mice, whereas in rats 890 mg/100 ml, and in man 260 mg/100 ml represent the lower lethal limits[10]. It would appear that the increased mortality and morbidity in this study were associated with chronic exposure to relatively low doses of ethanol. On the other hand, Kronick[6] treated mice (intraperitoneally) on one or two days, with 0.030 ml ethanol per gram of body weight, in a 25% v/v saline solution. This is roughly equivalent to 7.0 ml/kg of 95% alcohol, or, 6.0 g/kg, which is a massive dose, approximately the LD_{50} in mice, and could certainly explain the observed fetal mortality and morbidity. No criteria of intoxication were given except that none of the animals died as a result of the treatment.

In the present study low doses (0.56 to 1.1 g/kg) in different dilutions were administered intraperitoneally on four consecutive days to pregnant rats and behavioural impairment was used to define the level of intoxication. The absence of any significant deleterious effects on fetal growth and development or of any increased prenatal mortality could be explained by a very low level and short duration of exposure to ethanol relative to the two other studies, as well as the lower species susceptibility to ethanol. It is to be noted, however, that delayed or deficient occipital ossification was reported in both this and Chernoff's study, and microcephaly is consistently reported as characteristic of the human fetal alcohol syndrome, thereby providing some measure of relatedness or overlap with the present investigation.

The effects of short-term alcohol intoxication during pregnancy were further studied microscopically. In contrast to the preliminary investigation, there was no evidence of fatty change in the maternal liver. An increase of nucleated cells in the liver sinusoids and the pavementing of the hepatic endothelium with leukocytes suggested the persistence of an inflammatory process. The increased frequency of mitotic figures in the liver parenchyma suggested that repair processes were underway and provided indirect evidence of a prior ethanol-related damage to the liver. Numerous hepatocytes displaying nuclear pyknosis were observed in the vicinity of the central veins which gave additional support to this hypothesis.

Extensive intravascular haemolysis was observed in the preliminary investigation, but not in the teratological study. However, there was evidence of an earlier state of systemic haemosiderosis, in the form of aggregates of inclusion-laden cells[13]. These cells were often so engorged that cytological characteristics were obliterated. They were most frequently observed in the livers exposed to the highest dose of ethanol and were absent in all control tissues. These cells were probably free or fixed reticuloendothelial cells which became engorged with haemosiderin pigment and cellular debris during the period of treatment with ethanol. Engorged Kupffer cells are known to detach and become free in experimental or toxic situations[14]. Alcohol not only augments iron absorption, but also carries the threat of liver injury which eventually appears to enhance iron absorption[13]. Extensive haemolysis, with the release of iron from the heme pigment, would further contribute to the systemic haemosiderosis. In the case of acute ethanol intoxication the haemolysis may result as a direct cytoxic effect of ethanol on erythrocytes, or indirectly, as a result of severe central venous stasis and congestion. It is unlikely, that apart from a possible haemolytic anaemia, the observed changes in the maternal liver would have resulted in a long-term functional impairment. This assumption was supported by the absence of any effect on maternal or fetal growth parameters at the end of gestation. It is interesting to note, however, that haemoglobin levels of 70 to 80% of normal have been found in the first generation offspring of alcoholic guinea pigs, which falls to 40 to 50% in the second gestation[5].

In contrast to maternal liver, there was no morphological evidence of a toxic effect on the fetal liver. The fetal liver was either capable of a much faster recovery, or was, unlike the maternal liver, not susceptible to the toxic effects of ethanol. Since it has been well documented that the exposure of the fetal liver to ethanol at least equals the exposure of the maternal liver[15], it is possible that metabolic immaturity may have protected the fetal liver from damage observed in the maternal liver in these experiments.

In the maternal kidney, vascular changes were suggestive of an inflammatory process and severe venous congestion. Morphological changes in the parenchyma involved only the tubular portion of the nephron. Since the tubules are generally the most active and thus the most sensitive portion of the nephron, and can take part in extrahepatic ethanol metabolism[16], the observed changes may have been the result of a direct cytotoxic effect of ethanol on the renal tubular epithelium. However, it is also possible that the altered renal haemodynamics associated with ethanol infusion[17] result in an hypoxic state in the kidney and thus indirectly damage the tubular epithelium. As in the liver, the damage in the maternal kidney was neither extensive nor severe, repair processes were underway, and there was no evidence of scarring or fibrosis which would produce long-term functional impairment. The fetal kidney displayed no pathological changes which paralleled those observed in the maternal kidney. Again, the functional and morphological immaturity of the fetal rat kidney, especially of the tubular component, may have been the factor which protected the fetal kidney from the deleterious effects of exposure to ethanol.

Since the ovaries are involved in the maintenance of pregnancy, and since there is a documented increased incidence of premature deliveries among alcoholic women[18] it was decided to examine the ovaries for morphological evidence of ethanol toxicity. Short-term ethanol intoxication had no detectable effects on the ovarian follicles or stroma of the rat. Minimal changes were noted in the corpora lutea of ethanol treated animals which, however, were not present in controls. The marginal blood vessels of the corpora lutea were frequently congested and necrotic changes were observed in the luteal cells surrounding the central sinus. The extent of the tissue changes observed would suggest little, if any, physiological luteolysis or endocrine impairment following short-term alcohol intoxication. It would be interesting to know, however, if this minimal luteolysis would be more extensive after chronic alcohol abuse, if it would also be indicative of a physiological luteolysis which could result in premature delivery, and if these regressive changes represent a direct effect of ethanol on the luteal cells or a withdrawal of endocrine support secondary to a more central effect of ethanol.

The changes observed in the placenta were difficult to interpret. Throughout gestation the placenta is maturing, then ageing and undergoing structural and functional changes which presumably reflect the changing requirements of the developing organism. At term, morphological and metabolic properties suggest a reduction in placental function[19]. In this experiment ethanol was administered at the time that the primitive labyrinth was forming. From this, it would have been expected that the labyrinth would be the most affected placental component. On the contrary, on day 20, the placentas which had been exposed to ethanol during their development, displayed advanced degenerative changes in the basal zone but no effect in the labyrinth. Although the chorionic giant cells of the basal zone do evolve from the basophils of the labyrinth[9], there was no detectable reduction in cell number or basal zone thickness which would have been expected if the formative giant cells had been damaged as a result of the treatment. The observations made in day 20 tissues, however, were not the initial response to the treatment, but rather, the result of a complex and ongoing process. Four factors should be considered when reviewing these placental changes, namely, the proximity of the chorionic giant cells of the maternal blood vessels, the function of these cells, the presence of the yolk sac placenta during the treatment period, and placental metabolic capacity.

Since ethanol readily crosses the placenta, the fact that there were morphological changes in the basal zone and not in the labyrinth cannot be explained by differential ethanol concentrations in the maternal and fetal circulation. It is conceivable, however, that placental haemodynamics were affected by ethanol intoxication, resulting in severe vascular congestion and stasis in the maternal sinusoids of the placenta. Nothing is known about the fetal vascular response to ethanol. Congestion of the maternal vessels in the placenta, however, would have resulted in a prolonged exposure of the formative giant cells to blood ethanol. On the other hand, the presence of a functional yolk sac placenta during most the treatment period may have offered some protection to the rat fetus. The yolk sac epithelium can influence the passage of physiological and foreign chemicals and thus the well-being of the developing embryo. Although the yolk sac has been detected in the human placenta as late as the fifth month, there is no evidence that the function of the human yolk sac epithelium parallels that of the rat[20].

Little has been reported about the normal function of the chorionic giant cells. Davies and Glasser[9] have postulated an endocrine function based on ultrastructural characteristics. Being fetal cells in close proximity to the maternal decidua and blood vessels, it would be logical to attribute some protective function to these cells as well, since they likely form the placental barrier. Evidence in support of this function was seen in these experiments, in the form of giant cells filled with erythrocytes and other cellular debris, and giant cells actually in the process of phagocytosis. It appears that the advanced degenerative changes in the chorionic giant cells and the basal zone which were observed in the ethanol-exposed placentas were related to their close proximity to the maternal sinusoids. Rather than reflecting a direct cytotoxic effect of ethanol, these changes may reflect an ongoing, post-treatment clean-up of the cellular debris resulting from extensive intravascular haemolysis in the maternal vessels following ethanol intoxication.

Whereas the proximity of the giant cells to the maternal blood vessels was probably a key factor in the placental changes, another factor, metabolic capacity, may also have been involved. The placenta has been shown to be capable of many drug biotransformations and to contain alcohol dehydrogenase[21]. There is also much controversy about whether unchanged ethanol or acetaldehyde is the compound exerting the toxic effects observed in alcohol abuse. Since in these studies, only those organs with proven ethanol metabolizing capacity, i.e. the adult liver, the adult kidney and the placenta, displayed detectable tissue damage, it is conceivable that this capacity renders a tissue more susceptible to the deleterious effects of alcohol intoxication. This would imply that, at least in these tissues, acetaldehyde, rather than unchanged alcohol, is the toxic substance.

In the fetus, the metabolic, functional and anatomical immaturity of the liver and kidney may well have protected these tissues from paralleling the damage observed in the maternal rat. It is reasonable to assume, however, that all the factors discussed, namely, changes in the maternal haemodynamics producing severe stasis, a direct cytotoxic effect of ethanol or acetaldehyde, metabolic capacity and functional maturity, and proximity to maternal blood may all play a role in producing the deleterious effects observed to be associated with acute ethanol intoxication, but to varying degrees in different organ systems.

In summary, it is evident from these experiments, that acute maternal ethanol intoxication in the rat during the period of organogenesis, at levels producing realistic behavioural impairment, has no long-term deleterious effect on the morphological parameters of fetal growth and development *in utero*. This study has not eliminated the possibility of postnatal functional and/or growth impairment subsequent to binge alcohol abuse. Furthermore, the results of these experiments have suggested new directions for the investigation of the pathogenesis of the fetal alcohol syndrome known to be associated with chronic maternal alcohol abuse, namely, fetal stress, and placental and ovarian dysfunction.

SOCIOMEDICAL CONSIDERATIONS

There are several differences between the state of acute alcohol intoxication induced in this investigation and that which would generally occur in a pregnant woman 'on a binge'. For example, the adult rat has several times the capacity of the adult human to metabolize ethanol, and is therefore capable of clearing ethanol from its tissue much more rapidly. Furthermore, the human fetus has some drug metabolizing capacity in the first half of gestation whereas this is not acquired in the rodent until the end of gestation[22].

Secondly, there are both structural and functional differences in the ovary and placenta between the two species. Despite these differences, alcohol is known to cross the placenta readily in many species and chronic maternal alcohol ingestion during pregnancy has produced comparable fetal alcohol syndromes in dogs, mice, guinea pigs and humans, leaving little doubt as to the teratogenicity of ethanol in all of these species.

Thirdly, humans on a binge tend to take alcohol orally and over a longer period of time. In this particular experimental situation, a single dose was administered intraperitoneally, which would have resulted in a rapid absorption and a sudden high level of blood alcohol, but a more rapid metabolic degradation of the drug. The offspring of these animals would therefore have been exposed to higher levels of alcohol for shorter periods of time, in contrast to a lower, more continuous exposure of the human fetus.

Despite the fact that maternal behaviour was severely impaired at the highest dose level and there was demonstrable tissue damage in the mother, the fetus was largely spared the characteristic morphological anomalies. Evidence of skeletal ossification deficiencies, however, suggests that the same mechanism which induces microcephaly and growth retardation following chronic prenatal exposure to ethanol in the various species is operating even during short periods of intoxication. The microscopic changes in the placenta and maternal ovary implicate placental insufficiency and/or a reduced endocrine support of pregnancy in the pathogenesis of growth retardation and the increased incidence of prematurity in pregnancies complicated by alcohol abuse. It is also important to note that the present study has not ruled out the possibility of subtle, postnatal functional and developmental deficits as consequences of acute intoxication with alcohol, especially if the fetal adrenal gland is, in fact, affected.

Most studies of this subject to date have been descriptive, and although maternal ingestion of ethanol has been implicated as the primary aetiological factor, the pathogenesis of the fetal alcohol syndrome is not understood. It is unlikely to be a simple question of whether the primary teratogen is the unchanged alcohol or one of its metabolites. Rather, each adult and fetal tissue seems capable of its own unique response. Such factors as metabolic competence and vascularity, the haemodynamic changes in pregnancy as well as in the pharmacological response to ethanol must all be considered with regard to their possible roles in the teratogenicity of ethanol. Whereas a very high maternal metabolic competency, the yolk sac placenta, and fetal metabolic and vascular immaturity may protect the developing rat from the damaging effects of ethanol during the period of organogenesis, the mid-trimester human fetus would not have the benefit of these protective mechanisms. For these reasons, the minimal teratogenic effects of acute intoxication with ethanol observed in these experiments may underestimate considerably the effects on the human fetus similarly exposed.

References

1. Jones, K. L., Smith, D. W. and Hanson, J. W. (1976). The fetal alcohol syndrome: clinical delineation. *Ann. N.Y. Acad. Sci.*, **273**, 130
2. Quellette, E. M., Rosett, H. L., Rosman, N. P. and Weiner, L. (1977). Adverse effects on offspring of maternal alcohol abuse during pregnancy. *N. Eng. J. Med.*, **297**, 528
3. Chernoff, G. F. (1977). Fetal alcohol syndrome in mice: An animal model. *Teratology*, **15**, 223
4. Ellis, R. W. and Pick, J. R. (1976). Beagle model of the fetal alcohol syndrome. *Pharmacologist*, **18**, 190
5. Papara-Nicholson, D. and Telford, I. R. (1957). Effects of alcohol on reproduction and fetal development in the guinea pig. *Anat. Rec.*, **127**, 438
6. Kronick, J. B. (1976). Teratogenic effects of ethyl alcohol administered to pregnant mice. *Am. J. Obstet. Gynecol.*, **124**, 676
7. Obe, G. and Herha, J. (1975). Chromosomal damage in chronic alcohol users. *Humangenetik*, **29**, 191
8. Thliveris, J. A. and Connell, R. S. (1973). Ultrastructure of the fetal rat adrenal gland at full-term and during prolonged gestation. *Anat. Rec.*, **175**, 607
9. Davies, J. and Glasser, S. R. (1968). Histological and fine structural observations on the placenta of the rat. I. Organization of the normal placenta. *Acta Anat.*, **69**, 542

10. Maling, H. M. (1970). Toxicity of single doses of ethyl alcohol. In: Tremolieres (ed.). *Alcohol and Derivatives*, **Vol. 2**, pp. 277–295. (Oxford: Pergamon Press)
11. Naeye, R. L. (1965). Malnutrition: probable cause of fetal growth retardation. *Arch. Pathol.*, **79**, 284
12. Tze, W. J. and Lee, M. (1975). Adverse effects of maternal alcohol consumption on pregnancy and foetal growth in rats. *Nature* (*Lond.*), **257**, 479
13. Robbins, S. L. (1974). *Pathologic Basis of Disease.* (Toronto: W. B. Saunders Company)
14. Stöhr, P., Möllendorff, W. and Goerttler, L. (1969). *Lehrbuch der Histologie*, p. 409. (Jena: Gustav Fischer Verlag)
15. Ho, B. T., Fritchie, G. E., Idänpää-Heikkilä, J. E. and McIsaac, W. M. (1972). Placental transfer and tissue distribution of ethanol-1-^{14}C. A radioautographic study in monkeys and hamsters. *Qu. J. Stud. Alc.*, **33**, 485
16. Lieber, C. S. (1976). The metabolism of alcohol. *Sci. Am.*, **234**, 25
17. Tost, H., Balint, T. and Kover, G. Y., *et al.* (1971). Effect of ethyl alcohol on kidney function. *Int. Urol. Nephrol.*, **3**, 53
18. Green, H. G. (1974). Infants of alcoholic mothers. *Am. J. Obstet. Gynecol.*, **118**, 713
19. Thliveris, J. A. (1976). Fine structure of the placental labyrinth in the rat at term and during prolonged gestation. *Virchows, Arch.* (*Zellpathol.*), 21, 169
20. Wilson, J. G. (1973). *Environment and Birth Defects.* p. 118. (New York: Academic Press)
21. Juchau, M. R. (1972). Mechanisms of drug biotransformation reactions in the placenta. *Fed. Proc.*, 31, 48
22. Pikkarainen, P. H. (1971). Aldehyde oxidizing capacity during development in human and rat liver. *Ann. Med. Exp. Biol. Fenn.*, **49**, 151

13

Experimental studies on the influence of male alcoholism on testicular function, pregnancy and progeny

R. W. KLASSEN AND T. V. N. PERSAUD

INTRODUCTION

The problem of alcohol abuse has reached alarming proportions. Despite the increasing number of reports dealing with the medical, social and economical aspects of alcoholism, the production of alcoholic spirits continues to increase by nearly 10% annually[1,2]. The relationship of alcohol consumption to disease has been well documented; however, its total involvement indicates the necessity for further investigation[3].

A possible adverse influence of parental alcoholism on the development of the offspring has been suspected for decades[4–11], but only in recent years has there been any substantial experimental evidence available in support of this viewpoint[12–18]. The 'fetal alcohol syndrome' resulting from maternal alcoholism prior to and during gestation has been described in several recent reports[15–20].

Recent studies have indicated that nearly half of all infertility cases may be attributed to the male[3]. Spermatogenesis and the male reproductive tract have been studied with respect to age[21,22], nutritional state[23,24], heat[25] and various toxic agents[26–33].

It has been shown that alcohol adversely affects spermatogenesis[34,35], possibly resulting in abnormal gametogenesis and abnormal intra-uterine development. Several investigators have directed their attention to these problems[8–11]; the experimental results obtained seem to suggest that paternal alcoholism may be teratogenic[16]. It has been reported that of all infants with congenital malformations, 1–3% cannot be accounted for by classical teratology. It is important to investigate paternal as well as maternal exposure to drugs before and during pregnancy[36]. This report is concerned with the influence of male alcoholism on testicular function, pregnancy and the development of the offspring in the rat.

MATERIALS AND METHODS

The study of alcoholism and its medical implications have been hindered by an undeniable fact – alcoholism is limited to man. Due to the natural aversion to alcohol in experimental animals, an intense search for a suitable animal analogue of human alcoholism continues[37–48]. From a review of previous reports it became evident that a simple experimental procedure for the induction of chronic alcoholism is a prerequisite for studies on the reproductive aspects of male alcoholism. A technique first described by Freund[41] was adopted on account of its ease of administration (with a minimum of stress to the animal), moderate cost and apparent success.

Twelve male (mean weight 318 g; range 292–342 g) and 25 female (mean weight 259 g; range 225–293 g) albino rats of the Sprague-Dawley strain were obtained from Bio-Labs, St. Paul, Minnesota. The male animals were individually housed in steel mesh cages in an environmentally controlled room (temperature: 22 °C $\pm$ 3 and relative humidity: 45% $\pm$ 10). The female animals were housed in pairs under the same conditions. A diurnal cycle from 08.00 to 20.00 h and nocturnal from 20.00 to 08.00 h was maintained constantly. All cages were fitted with standard drinking bottles and stainless steel spouts. Plastic cups were fastened to the inside of the cages containing male animals, directly beneath the drinking spouts, in order to receive all drops and thereby minimize errors in the measurement of fluid consumed.

After a period of one week adaptation to the laboratory conditions, each male rat was placed with two females during the hours of darkness to determine the fertility of both sexes. Successful matings were confirmed by the discovery of vaginal plugs and/or the presence of spermatozoa in the vaginal smears. After the breeding capacity of the male animals was determined, they were randomly divided into two groups of equal numbers. The experimental group was placed on a liquid diet of 'Metrecal' and ethanol. The constituents of 'Metrecal' are listed in Table 13.1. Chocolate 'Metrecal' was the flavour of

Table 13.1 Constituents of 'Metrecal'*

Protein	7.4 g
Fat	2.1g
Carbohydrates	11.6g
Choline	28 mg
Methionine	170 g
Thiamine	0.18 mg
Riboflavin	0.27 mg
Niacinamide	1.37 mg
Pyridoxine	0.18 mg
Pantothenic acid	0.91 mg
Vitamin B_{12}	0.18 μg
Calcium cyclamate	
Calories	95

* Manufactured by Mead-Johnson, Montreal, Canada

choice but on several occasions a vanilla flavour was substituted for reasons of availability. The animals were placed on a 6% ethanol diet which was increased to 10% a week later, thus increasing the proportion of ethanol-derived calories. Another group of male rats, which served as the controls, were also placed on a diet of 'Metrecal' but with an isocaloric amount of sucrose substituted for the alcohol. The diets were prepared daily according to the schedule in Table 13.2. All animals were given the liquid diets *ad libitum* and the daily consumption of each rat was recorded. The mothers received tap water and standard rat chow pellets *ad libitum*.

Table 13.2 Preparation of liquid diets

	Ethanol (95% v/v) 5.25 cal/ml		*'Metrecal'* 0.95 cal/ml		*Sucrose water* (87% v/v) 3.5 cal/ml		*Water*	
Diet	%	cal. (%)	%	cal. (%)	%	cal. (%)	%	cal/ml diet
Experimental	6.0	35	61	65	0	0	33	0.9
Control	0	0	61	65	9	35	30	0.9
Experimental	10	58	39	42	0	0	51	0.9
Control	0	0	39	42	15	58	46	0.9

After Freund[41]

After fifteen days each male rat was again placed into a cage with two females during the hours of darkness. It was not desirable to expose the male rats to the tap water and chow pellets which the females were receiving, nor did we wish the female animals to be exposed to the liquid diets. For these reasons no food, water, or 'Metrecal' diet was provided during the period the animals were together. The male rats were separated from the females and returned to their respective diets during light hours. As before, successful matings and the first day of pregnancy were determined by the presence of vaginal plugs and of spermatozoa in the vaginal smears.

All pregnancies were terminated on gestational day 20 by Caesarean section and the litter size as well as fetal mortality were assessed before fixing the fetuses and their attached placentae in Bouin's fluid. The offspring were later studied to determine their sex, fetal weight, placental weight and crown–rump length. All fetuses were examined for gross abnormalities and subsequently sectioned by the Wilson's technique[49] in order to detect visceral anomalies.

The experiment with the male rats was continued for five weeks during which period the animals were observed daily for physical and behavioural changes. On day 36 of the experiment, all animals were anaesthetized with ether, and blood samples were collected by cardiac puncture 12 h after the last exposure to alcohol. Serum glucose, serum alcohol and serum testosterone levels were determined by Ames Dextrostix reagent strips, gas chromatography and radioimmunoassay respectively[22,50].

The animals were subsequently killed by exsanguination and the following organs were removed: testes, epididymides, seminal vesicles, kidneys, liver and spleen. The tissues were fixed in Bouin's fluid and stained with Harris haematoxylin and by the Mallory's trichrome technique. For the quantitative analysis of spermatogenic yield, a method of counting nuclei of Sertoli cells, type A spermatogonia, preleptotene spermatocytes, pachytene spermatocytes and spermatids in cross-sections of seminiferous tubules at stage VII of spermatogenesis was used[51,52]. Capsular thickness and the diameter of seminiferous tubules were also measured. Student's *t*-test was used to determine the levels of significance between experimental and control values.

RESULTS

General observations

The male alcoholic rats showed signs of intoxication from the beginning of treatment, with the animals being noticeably less excitable than the control group and offering less resistance to handling. Throughout the course of the investigation the control rats showed considerable weight gain (mean 94.3 ± 6.7 g), large bright eyes, smooth shiny hair, increased activity and extensive exploratory behaviour. Animals of the experimental group, however, lost weight (mean 131.8 ± 13.3 g), became lethargic, and presented dull, ruffled hair and small pale eyes. Unlike the control group, the alcoholic animals were generally ataxic, lacking co-ordination and showed little or no exploratory behaviour.

The mean body weights, rate of diet consumption, and the amount of ethanol consumed on various days of the experiment are given in Figure 13.1. Although the alcoholic animals maintained a steady rate of diet consumption, their absolute intake of ethanol declined after a transient increase. The control rats, however, showed a general increase in the rate of diet consumption during the entire course of the experiment. No noticeable effects resulted from the substitution of vanilla for the chocolate 'Metrecal' diet.

When placed with female rats during hours of darkness, the alcoholic males showed little or no interest in the females, confining themselves to one corner of the cage. The control males, however, demonstrated much grooming and mounting behaviour. These observations were confirmed by only six successful matings among the twelve females bred to experimental males while all thirteen of the females placed with control animals were successfully mated within a period of 21 days.

During the hours of mating, when no diet was provided, the alcoholic males presented obvious withdrawal symptoms. Excessive activity, excitability, and tremor were observed with much resistance to being handled, whereas the control group showed none of these signs.

On day 35 of the experiment, one of the alcoholic rats died. Although a post-mortem examination failed to reveal the cause of death, there were signs suggestive of an alcohol withdrawal seizure. It was therefore decided to terminate the experiment at this stage.

Biochemical studies

Results of the serum glucose, alcohol and testosterone levels are given in Table 13.3. The alcoholic group of animals were noticeably hypoglycaemic compared to the controls. Serum alcohol levels were minimal (mean 0.5 mg %)

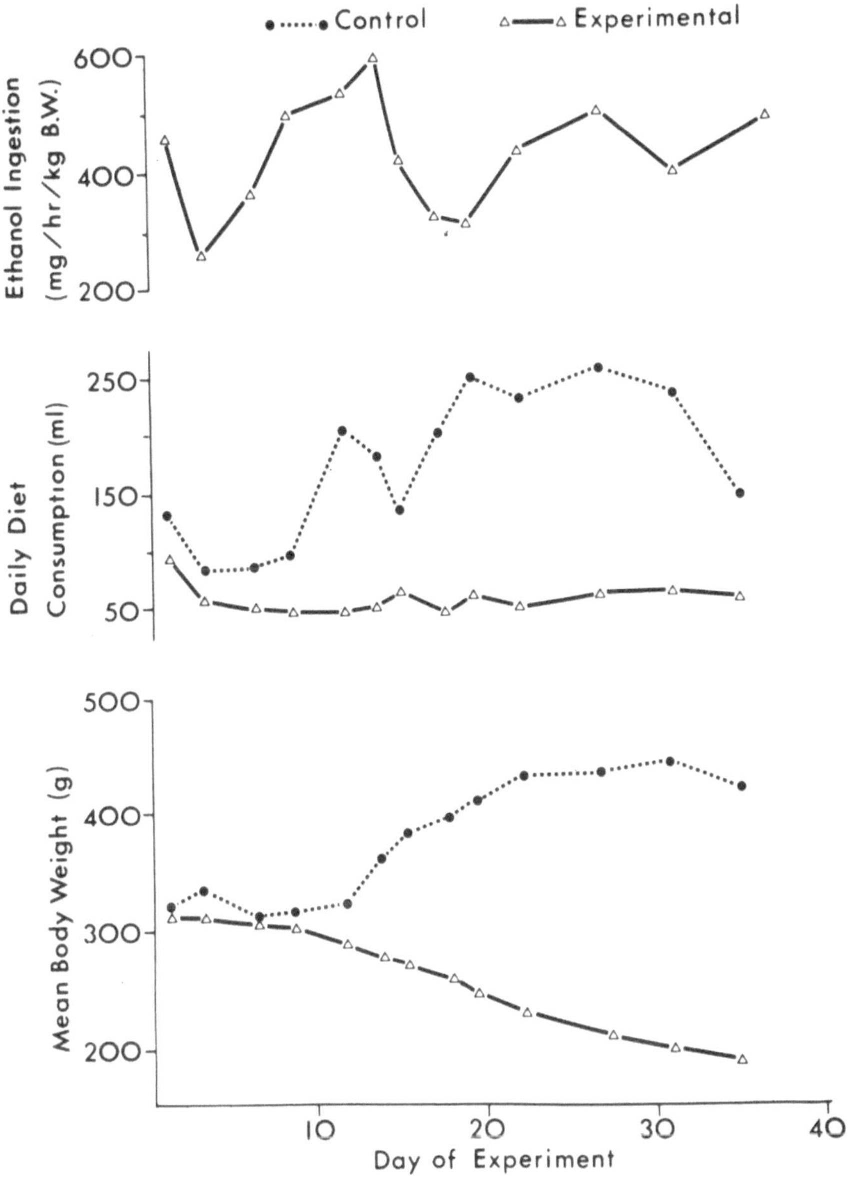

Figure 13.1 Effect of the 'Metrecal-ethanol' diet on body weight, daily consumption and ethanol ingestion during experimental period

with none being detected in two of the animals. The serum testosterone levels showed a significant difference ($p < 0.05$) between experimental and control groups.

Table 13.3 Glucose, alcohol and testosterone levels in rat serum

Groups	*Serum glucose* mg/100 ml	*Serum alcohol* mg %	*Serum testosterone* ng %
Experimental	62 ± 13.6*	5 ± 0	166.6 ± 9.2
Control	208 ± 8.3	—	214.8 ± 19.1
Significance†	$p < 0.001$	—	$p < 0.05$

* Mean ± SE
† Two-sample t-test, Wilcoxon-Mann Whitney Test and the Kolmogorov–Smirnov Two-sample Test were used to determine the levels of significance between experimental and control values

Pregnancy and progeny

Table 13.4 summarizes the results of the experimental and control matings. The number of successful matings with control males and the number of offspring per litter were significantly greater than that of females mated with alcoholic animals. A higher incidence of early abortions were detected in the experimental group than in the corresponding control. Neither the incidence of fetal mortality nor the number of abnormal offspring showed any significant differences between the two groups. The sex ratio was greater in the experimental group, but not significantly different from the control value.

The prenatal growth and development of the offspring of both the alcoholic and control male rats are shown in Table 13.5. The mean fetal weight and

Table 13.4 Influence of male alcoholism on pregnancy and fetal development

Groups	*No. females mated*	*Litter size*	*Total no. offspring recovered*	*Early resorptions*	*Fetal deaths*	*No. abnormal offspring*	*Sex ratio**
Experimental	6	7.0 ± 2.5†	21	28	0	0	47.0 ± 19.9
Control	13	12.1 ± 0.5	158	9	1	0	34.3 ± 3.5
Significance‡		$p < 0.01$		$p < 0.01$			$p < 0.3$

* $\text{Sex ratio} = \dfrac{\text{No. of males}}{\text{Males + Females}} \times 100$
† Mean ± SE
‡ Significance between experimental and control values determined by either Two-sample t-test or Chi-Square Test

Table 13.5 Influence of male alcoholism on fetal growth and development

Groups	*Fetal weight*	*Crown–rump length*	*Growth index**	*Placental index†*
Experimental	2.840 ± 0.102‡	3.01 ± 0.04	0.726 ± 0.138	17.631 ± 0.560
Control	2.288 ± 0.068	2.76 ± 0.04	0.838 ± 0.115	21.549 ± 0.533
Significance§	$p < 0.01$	$p < 0.01$	$p < 0.5$	$p < 0.01$

* $\text{Growth index} = \dfrac{\text{Mean fetal weight} \times \text{Mean crown–rump length} \times \text{Litter size}}{100}$
† $\text{Placental index} = \dfrac{\text{Placental weight}}{\text{Fetal weight}} \times 100$
‡ Mean ± SE
§ Significance between experimental and control values determined by Two-sample t-test

crown–rump length were significantly greater in the experimental group than in the controls. The mean growth index, which takes litter size into consideration, showed a smaller value for the experimental group than the control, although the difference was not statistically significant. However, the placental index, which relates fetal weight to that of the placenta, was significantly reduced in the experimental group.

Reproductive tissues

Table 13.6 shows the testes, epididymides, and seminal vesicles of the alcoholic rats to weigh significantly less than the corresponding controls. However, when corrected for body weight, only the value for testes remains statistically significant (Table 13.7). The testes of the alcoholic rats were found to be more firm than those of the corresponding controls. The capsule of the testes of alcohol-treated rats also appeared oedematous with considerable separation of tubules occurring centrally. The control animals showed large, well-defined tubules whereas those of experimental animals were atrophic and distorted (Figures 13.2A, B). This observation was confirmed by measurement of the cross-sectional diameter of these tubules (Tables 13.8). The basement membrane of seminiferous tubules in the experimental animals appeared

Table 13.6 Organ weights of control and alcohol-treated male rats

	Experimental	*Control*	*Significance of differences**
Testes	1.11 ± 0.32†	1.71 ± 0.11	$p<0.001$
G.S.I.‡	4.07 ± 1.71	14.41 ± 1.05	$p<0.001$
Epididymides	0.33 ± 0.03	0.72 ± 0.05	$p<0.001$
Seminal vesicles	0.36 ± 0.13	1.11 ± 0.29	$p<0.001$
Kidneys	0.80 ± 0.12	1.15 ± 0.24	$p<0.001$
Spleen	0.17 ± 0.07	0.56 ± 0.75	$p<0.001$
Liver	4.77 ± 1.39	12.52 ± 1.79	$p<0.001$

* Significance of the difference was determined by Two-sample *t*-test, Wilcoxon-Mann Whitney Test, and the Kolmogorov–Smirnov Two-sample Test
† Mean ± SE
‡ Gonadosomatic index = (testes weight × body weight)/100

Table 13.7 Organ weights per 100 g weight of control and alcohol-treated rats

	Experimental	*Control*	*Significance of differences**
Testes	1.12 ± 0.28†	0.82 ± 0.07	$p<0.01$
Epididymides	0.36 ± 0.04	0.35 ± 0.07	$p<0.9$
Seminal vesicles	0.21 ± 0.10	0.27 ± 0.08	$p<0.1$
Kidneys	0.89 ± 0.06	0.58 ± 0.16	$p<0.01$
Spleen	0.09 ± 0.02	0.13 ± 0.02	$p<0.02$
Liver	2.63 ± 0.45	2.99 ± 0.32	$p<0.2$

* Significance of difference was determined by the Two-sample *t*-test, Wilcoxon-Mann Whitney Test, and the Kolmogorov–Smirnov Two-sample Test
† Mean ± SE

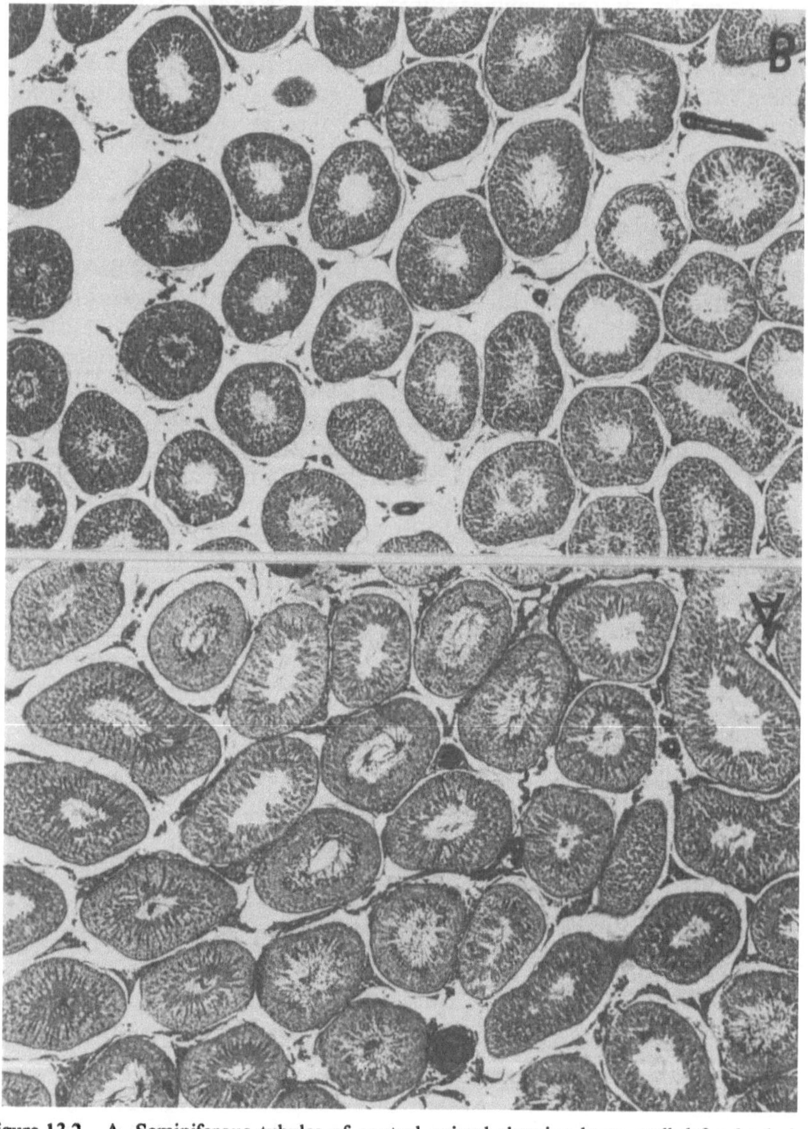

Figure 13.2 A, Seminiferous tubules of control animal showing large, well-defined tubules, thick germinal epithelium and abundant interstitial tissue (× 14.4); B, Seminiferous tubules of experimental animals with evidence of oedema, atrophic tubules, damaged germinal epithelium and reduced interstitial tissue (× 14.4)

fragmented and serrated, compared to that of the controls where it was smooth and continuous. The germinal epithelium of the alcohol-treated rats was damaged and reduced in thickness.

The results of differential nuclear counts at different stages of spermatogenesis are given in Table 13.9. In the alcohol-treated animals a significant

Table 13.8 Capsular thickness and cross-sectional diameters of seminiferous tubules in testes of experimental and control animals

	Capsular thickness	*Tubular diameters*
Alcohol-treated	57.77 ± 0.32μm*	223.31 ± 0.27 μm
Control	45.31 ± 0.29 μm	260.29 ± 0.29 μm
Significance of the difference†	$p < 0.01$	$p < 0.001$

* Mean ± SE
† Significance of the differences was determined by the Two-sample *t*-test

Table 13.9 Number of various types of germinal and Sertoli cells at stage VII of spermatogenesis

	Alcohol-treated	*Control*	*Significance of differences**
Sertoli cells	6.35 ± 0.45†	6.90 ± 0.35	$p < 0.5$
Type A spermatogonia	3.55 ± 0.27	3.37 ± 0.39	$p < 0.8$
Preleptotene spermatocytes	41.65 ± 1.05	51.60 ± 1.51	$p < 0.001$
Pachytene spermatocytes	45.10 ± 1.92	59.85 ± 2.57	$p < 0.001$
Spermatids	143.40 ± 4.68	196.35 ± 6.48	$p < 0.001$

* Significance of the differences was determined by the Two-sample *t*-test
† Mean ± SE

decrease in the number of preleptotene and pachytene spermatocytes and spermatids was observed. However, the Sertoli cells and type A spermatogonia did not differ significantly between the two groups. In addition to the presence of numerous degenerate spermatocytes with hypochromic nuclei, fewer mitotic figures were seen in the alcoholic group of animals. Many multinucleated giant cells and a general sloughing-off of the germinal epithelium were evident in the lumen of the seminiferous tubules of experimental animals. In addition to these desquamated cells, many fragmented spermatozoa were present. These changes were not found in the control animals.

The interstitial tissue of the control group was abundant with normal Leydig cells distributed evenly in cords throughout the sections. In the alcoholic animals, however, degenerate interstitial tissue was seen distributed in small clumps between the seminiferous tubules with only few Leydig cells present. No changes in the morphology or general distribution of capillaries were detected nor was there any evidence of interstitial haemorrhage.

The epididymides of the alcoholic rats showed changes similar to those of the testes. The ductules of this group were smaller and more irregular than the controls (Figure 13.3A, B). The epithelial lining of these ductules was also damaged and showed much sloughing of cells into the lumen. The lumen of the experimental organs contained very few spermatozoa (Figure 13.3B); the majority of these revealed broken heads and curled tails. No changes were

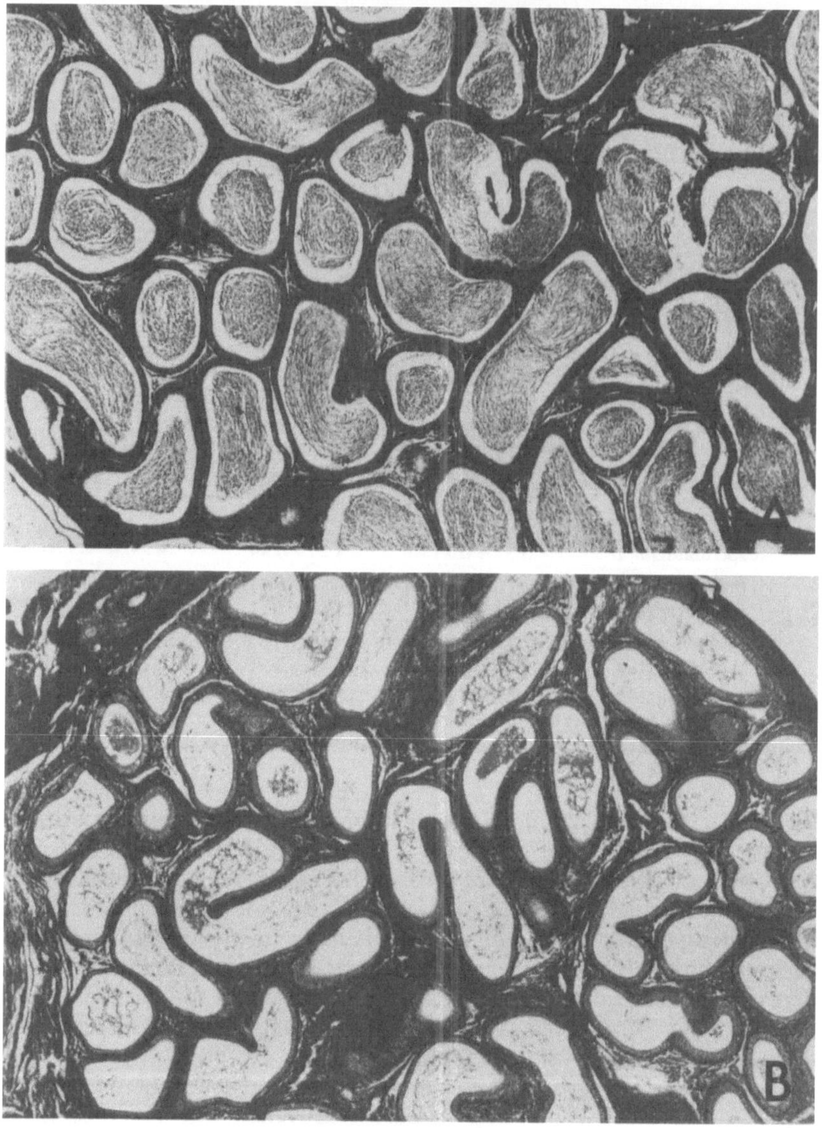

Figure 13.3 A, Epididymis of control rat showing large, well-defined ductules containing abundant spermatozoa (× 11.7); B, Epididymis of alcohol-treated animal demonstrating small, distorted ductules, containing fewer spermatozoa (× 11.7)

noted in the stereocilia lining the ductules. The interstitial connective tissue and the amount of capsular fat appeared to be increased. The control epididymides did not show such changes.

Compared to the controls, the seminal vesicles of experimental animals were atrophic. The normally extensively folded lining of these elongated sacs was reduced to small degenerate ridges extending into the small lumen. Unlike the controls which retained a healthy pseudostratified columnar epithelium the experimental group revealed a reduced, nearly cuboidal mucosa (Figure 13.4A, B).

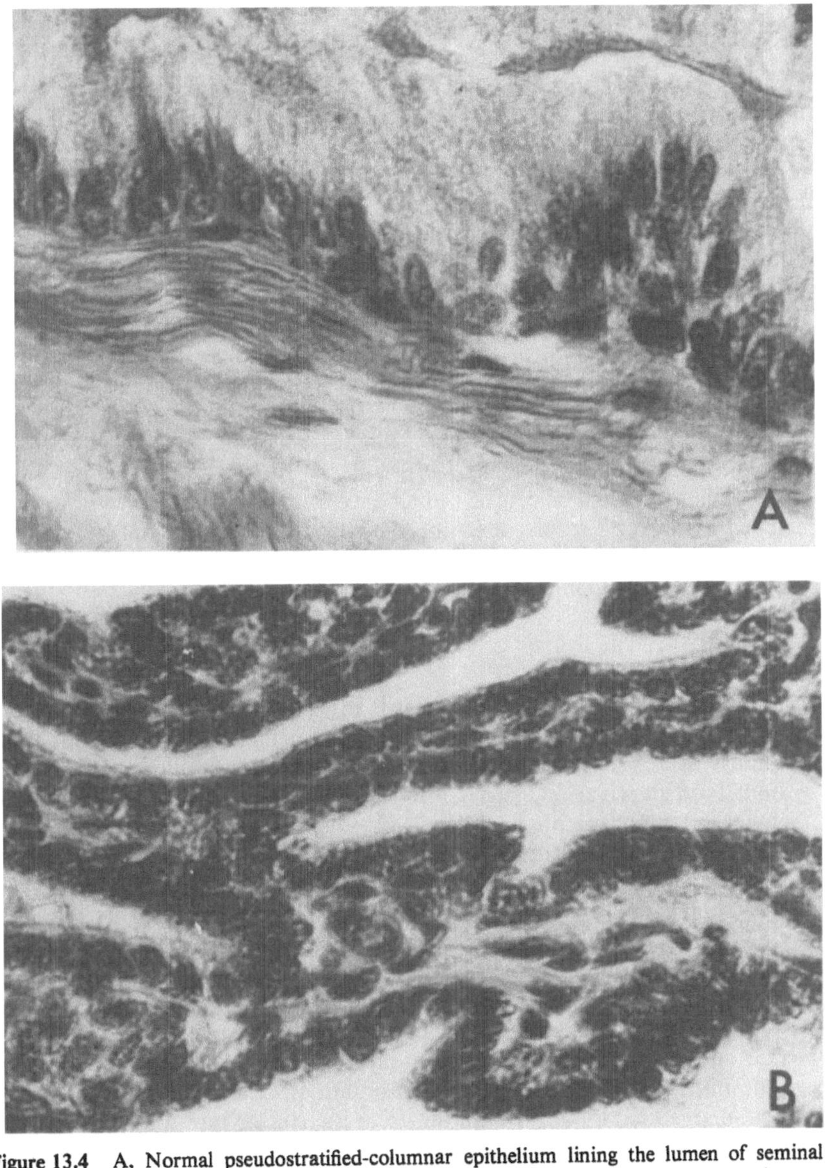

Figure 13.4 A, Normal pseudostratified-columnar epithelium lining the lumen of seminal vesicle of control animal (×216); B, Epithelium of seminal vesicle of alcoholic rat demonstrating a reduced, nearly cuboidal mucosa

Non-reproductive tissues

Table 13.6 shows the kidneys and spleen to weigh significantly less in the experimental animals than the controls, unlike the weight of the liver which was markedly increased. When corrected for body weight, however, only the values for the kidneys and spleen remained significantly different (Table 13.7). Histological examination of the kidneys showed evidence of inter- and intratubular haemorrhage in nearly all experimental animals. In several of these, large haemorrhagic areas were seen. Small degenerate glomeruli were noted in the kidneys of several of the alcoholic animals. The control group did not show these changes, but multiple large cystic spaces were found within the renal medulla in one of these animals.

Evidence of haemorrhage was noted in the zona fasciculata and reticularis of the adrenal gland in most experimental animals, but not in the controls.

Liver damage caused by alcohol consumption was evident. The mean weight of the liver in alcohol-treated animals was significantly increased compared to that of the control group. This increase in liver size was supported by the presence of fatty degeneration in the hepatocytes. Although evidence of fibrosis was minimal, the normal architecture of the hepatic lobules appeared to have been lost in some cases. These changes were not present in all animals of the alcoholic group.

The changes in the spleen of the alcohol-treated animals were few. These included a thicker capsule, more fibrous stroma, and large splenic nodules, many of these containing germinal centres. Well-defined germinal centres were not seen in the control group.

DISCUSSION

An animal model of human alcoholism is urgently needed. Cicero and Smithloff[46] proposed four criteria which a true experimental model of alcohol addiction must meet. These are: 1) The alcohol must be self-administered orally without food deprivation, intoxication to be determined by analysis of behaviour and blood alcohol levels; 2) A tolerance to the effect of chronic alcohol administration should be demonstrated; 3) A withdrawal syndrome should be evident following cessation of alcohol administration; and 4) The animal should present signs of 'psychological dependence' in its efforts to obtain alcohol. Few investigators have been able to satisfy all of these criteria.

The results of our investigation have shown that the method of inducing alcoholism in mice as described by Freund[41] may also be of value in the rat. Although a test for psychological dependence was not part of this experimental design, obvious withdrawal symptoms and daily behavioural observations indicated a chronic alcoholic situation. Oral self-administration of ethanol-derived calories was maintained at a constant rate despite the significant weight loss of the experimental animals compared to the controls. The low serum alcohol levels detected in the rats 12 h after the last administration may be due to the rapid elimination (250–300 mg/h/kg body weight) in this species[53]. We have used a semiquantitative method for the determination of serum glucose levels in the rats. Nevertheless, the findings showed significant differences which demonstrated a condition of marked hypoglycaemia, reported to be present in alcoholic animals[54].

Previous investigations relating to the reproductive aspects of alcoholism have suggested a possible teratogenic effect of this most frequently administered drug. As a result of the more recent interest directed to the problems of maternal alcoholism, a 'fetal alcohol syndrome' has been described and is now well documented. Jones *et al.*[15] reported a 17% perinatal mortality, deficiencies in prenatal growth (length, weight, head circumference), joint anomalies, as well as cardiac and ocular disorders. Similar observations were made by Tenbrinck and Buchin[17].

The adverse effects of paternal alcoholism on the offspring has received relatively little attention and should also be taken into consideration. Stockard[8] reported a high incidence of early abortions and stillbirths among the progeny of alcoholic male and female guinea pigs. Of these, more than 25% resulted from the mating of an alcoholic male to a normal female. Pearl[9] found that fewer successful fertilizations resulted from matings of alcoholic animals but that prenatal mortality was actually increased in the control group.

In the present investigation a high incidence of early fetal mortality among the offspring of alcoholic males was observed, confirming previous observations[16]. Litter size was noticeably reduced among matings of normal females with alcoholic males. Similar findings were also reported by Stockard[8] and Badr and Badr[16].

Previous investigators have detected no significant anomalies among the offspring of alcoholic males[8,9]. Although Badr and Badr[16] reported the recovery of a fetus presenting with abnormal bulgings of the head and underweight for its age, they were unable to rule out other causative factors due to the absence of a control study.

An insignificant increase in the number of male offspring has been noted in the experimental group as compared to the control in agreement with previous observations[9,10,55]. It is known that certain genetic and physiological factors may have an effect on the sex of the offspring[56]; at this time however, the possible involvement of ethanol is only the subject of speculation.

The results of the developmental studies on the offspring showed subtle differences between the experimental and control groups. Although the mean weight and crown–rump length of fetuses of the experimental group were greater than those of the control, these findings are explained by the smaller litter sizes of the experimental animals. Taking this into account, the mean growth index indicated smaller offspring from treated than from untreated males, although the differences were not significant. Similar results were reported by Pearl[9] and MacDowell and Lord[10]. The significantly smaller placental index found in the experimental group of animals remains to be explained.

Hypogonadism has been associated with chronic alcoholism in humans for more than a century. As early as 1837, Rösch[57] described testicular degeneration in alcoholics. Similar observations have been made in mice[58], and rats[34,35]. Loss of testicular weight relative to body weight is a clear indication of the toxic influence of ethanol on the testes. The small firm testicles of habitual drinkers, as described by Bertholet[59], is in agreement with the atrophic tubules and sclerotic stroma found in our histological study. The decreased cross-sectional diameter of seminiferous tubules which we have observed has also been described by other investigators[34,35] and are undoubtedly related to the degeneration of the germinal epithelium.

From the results of the differential cell counts in the germinal epithelium of the alcoholic rats, one might deduce that alcohol adversely affects cellular differentiation beyond the primary spermatocyte stage. Bouin and Garnier[60] found that degeneration of germinal epithelium occurs inversely to spermatogenesis, in their study of alcohol-treated rats – the spermatozoa are first affected, then the spermatids, followed by spermatocytes. The spermatogonia and Sertoli cells are most resistant to toxic influences as was confirmed in our study. It has been suggested, however, that these changes are non-specific and can be found in the seminiferous tubules of rats in other unrelated conditions[61]. This indicates a common mode of action for a variety of antispermatogenic agents.

The presence of fragmented spermatozoa and desquamated germ cells in the lumen of the rat seminiferous tubules reflects damage to the germinal epithelium caused by alcohol. Doepfmer and Hinckers[62] reported similar changes in human sperm morphology and described reduced motility, backward oscillations, and distended midsection. A reduced sperm count in alcohol-treated dogs was also observed by Teitelbaum and Gantt[63]. Multinucleated giant cells, as found in our study, have been described by other investigators[61,64]. Reddy and Svoboda[61] concluded that these giant cells are derived from incomplete division of the cytoplasm at the primary spermatocyte stage of spermatogenesis.

That chronic alcohol ingestion significantly decreases the levels of serum testosterone is certain[35,65]. The mechanism whereby this occurs, however, remains uncertain. Whether alcohol exerts its primary toxic effect on the Leydig cells, or the degenerate interstitium is secondary to the influence of ethanol on the hypothalamic–pituitary axis, remains to be determined. Testicular atrophy as a consequence of epinephrine-mediated vasoconstriction, secondary to alcohol induction of the adrenal medulla, may account for some of the pathological changes[66]. Loss of androgen support for spermatogenesis may also be involved[67].

Atrophy of the epididymides and seminal vesicles, seen in the present study, might have resulted from the decrease in testosterone level. Arlit and Wells[34] observed desquamation of the epithelial lining of the epididymis of alcohol-treated rats and concluded that the histological changes in the epididymis reflect the degree of testicular damage. The reason for the decreased amount of seminal fluid seen in the epididymis is uncertain. It is unlikely that fragmentation of spermatozoa in the seminiferous tubules precluded their migration through the genital tract[62]. The atrophic ridges and degenerate epithelium found in the seminal vesicles of the alcoholic animals might have resulted from a lack of androgen support and are in agreement with the findings of Van Thiel *et al.*[35]

The presence of multinucleated giant cells, fewer mitotic figures, and germ cells beyond the spermatogonal stage suggest that alcohol disturbs meiosis. Translocation-type chromosomal aberrations were detected in peripheral leukocytes of chronic alcoholics[68,69]. Badr and Badr[16] also reported induction of dominant lethal mutations in alcoholic male mice. However, the mechanism involved in alcohol induced mutagenicity is not known. Van Thiel *et al.*[70] have proposed that ethanol inhibition of retinol activation to retinal, an essential prerequisite for sperm formation[71] may account for abnormal spermatogenesis. Obe and Herha[69] pointed out that the action may be an indirect one, suggesting a 'synmutagenic' role for alcohol.

Although genetic inheritance of alcohol-induced mutations via nuclear chromatin appears most likely, the subject of cytoplasmic inheritance warrants some consideration. It has been determined that all cells contain DNA in the cytoplasm, most of which is located in the mid-piece which penetrates the ovum during fertilization. Although ultrastructural studies have shown that the mid-piece of spermatozoa eventually disintegrates, the fate of the contained DNA is not known. Since mitochondrial DNA is capable of replication, transcription of RNA and protein synthesis, the possibility of cytoplasmic mediated teratogenesis is conceivable.

CONCLUSIONS

Chronic alcoholism was induced in male albino rats by an oral self-administration technique. Obvious signs of intoxication, withdrawal symptoms and reduced reproductive performance, as well as decreased serum glucose and serum testosterone levels, were observed in these animals.

Notable differences were detected in prenatal mortality, mean litter size and sex ratio between the offspring of male alcoholic rats and the corresponding controls. In the alcohol-treated group the placental index was markedly reduced. However, the mean growth index showed no significant differences between the experimental and control groups.

Testicular atrophy and degeneration of the accessory sex organs in chronic alcoholism have been well documented, but the aetiology of this hypogonadism remains unsettled. The atrophic seminiferous tubules, degenerate germinal epithelium and fragmented spermatozoa show that ethanol is toxic to these tissues. Decreased levels of serum testosterone may account for the smaller epididymides and seminal vesicles found in the alcoholic animals.

The higher incidence of prenatal mortality in the offspring of alcoholic rats might be attributed to the toxic influence of ethanol on the germ cells of these animals. It would, therefore, appear that the mutagenic potential of alcohol is transmitted via the genome of the spermatozoa and causes abnormal development of the progeny. The precise mechanism whereby alcohol produces these changes is not known; whether the inheritance is nuclear or cytoplasmic also remains to be elucidated.

References

1. W.H.O. (1975). Alcohol: A growing danger. *W.H.O. Chronicle*, **29**, 102
2. W.H.O. (1975). Problems of non-medical drug use. *W.H.O. Chronicle*, **29**, 97
3. Seixes, F. A., Williams, K. and Eggleston, S. (1975). Medical consequences of alcoholism. *Ann. New York Acad. Sci.*, **252**, 10–396

4. Stockard, C. R. (1910). The influence of alcohol and other anaesthetics on embryonic development. *Am. J. Anat.*, **10**, 369
5. Stockard, C. R. (1912). An experimental study of racial degeneration in mammals treated with alcohol. *Arch. Int. Med.*, **10**, 369
6. Stockard, C. R. and Craig, D. M. (1912). An experimental study of the influence of alcoholism on the germ cells and the developing embryos of mammals. *Arch. f. Entwickl.*, **35**, 569
7. Nice, L. B. (1912). Comparative studies on the effects of alcohol, nicotine, tobacco smoke and caffeine on white mice. I. Effects on reproduction and growth. *J. Exp. Zool.*, **12**, 133
8. Stockard, C. R. (1913). The effect on the offspring of intoxicating the male parent and the transmission of the defects to subsequent generations. *Am. Nat.*, **47**, 641
9. Pearl, R. (1917). The experimental modification of germ cells: The effect of parental alcoholism, and certain other drug intoxications, upon the progeny. *J. Exp. Zool.*, **22**, 241
10. MacDowell, E. C. and Lord, E. M. (1927). Cited in Willis, R. A. (1962) *The Borderland of Embryology and Pathology*. (London: Butterworth)
11. Durham, F. M. and Woods, H. M. (1932). Alcohol and inheritance: An experimental study. *Med. Res. Coun, Spec. Report Series No.* **168** (London)
12. Sandor, S. and Elias, S. (1968). The influence of ethyl-alcohol on the development of the chick embryo. *Revue Roumaine d'Embryologie et de Cytologie – Serie d'Embryologie*, **5**, 51
13. Sandor, S. (1968). The influence of ethyl alcohol on the developing chick embryo II. *Revue Roumaine d'Embryologie et de Cytologie – Serie d'Embryologie*, **5**, 167
14. Sandor, S. and Amels, D. (1971). The action of ethanol on the prenatal development of albino rats. *Revue Roumaine d'Embryologie et de Cytologie – Serie d'Embryologie*, **8**, 37
15. Jones, K. L., Smith, D. W., Streissguth, A. P. and Myrianthopoulos, N. C. (1974). Outcome in offspring of chronic alcoholic women. *Lancet*, i, 1076
16. Badr, F. and Badr, R. (1975). Induction of dominant lethal mutation in male mice by ethyl alcohol. *Nature* (*Lond.*), **253** 134
17. Tenbrinck, M. S. and Buchin, S. Y. (1975). Fetal alcohol syndrome. Report of a case. *J. Am. Med. Ass.*, **232**, 1144
18. Tze, W. J. and Lee, M. (1975). Adverse effects of maternal alcohol consumption on pregnancy and foetal growth in rats. *Nature* (*Lond.*), **257**, 479
19. Clarren, S. K. and Smith, D. W. (1978). The fetal alcohol syndrome. *N. Eng. J. Med.*, **298**, 1063
20. Clarren, S. K., Alvord, E. C., Sumi, M., Streissguth, A. P. and Smith, D. W. (1978). Brain malformations related to prenatal exposure to ethanol. *J. Pediat.*, **92**, 64
21. Lutzen, L. and Ueberberg, H. (1973). A study on morphological changes in the testes of old albino rats. *Beitr. Path.*, **149**, 377
22. Stearns, E. L., MacDonnell, J. A., Kaufman, B. J., Padua, R., Lucman, T. S., Winter, J. S. D. and Faiman, C. (1974). Declining testicular function with age. *Am. J. Med.*, **57**, 761
23. Siperstein, D. (1920). The effects of acute and chronic inanition upon the development and structure of the testis in the albino rat. *Anat. Rec.*, **20**, 355
24. Mauer, S. and Mason, K. (1975). Antisterility activity of d-α-tocopheryl hydroquinone in the vitamin E-deficient male hamster and rat. *J. Nutr.*, **105**, 491
25. Bowler, K. (1972). The effect of repeated applications of heat on spermatogenesis in the rat: a histological study. *J. Reprod. Fertil.*, **28**, 325
26. Hildebrand, D., Der, R., Griffin, W. and Fahim, M. (1973). Effect of lead acetate on reproduction. *Am. J. Obstet. Gynecol.*, **115**, 1058
27. Lee, I. and Dixon, R. (1973). Effects of cadmium on spermatogenesis studied by velocity sedimentation cell separation and serial mating. *J. Pharmacol. Exp. Ther.*, **187**, 641
28. Cooper, E., Jones, A. and Jackson, H. (1974). Effects of α-chlorohydrin and related compounds on the reproductive organs and fertility of the male rat. *J. Reprod. Fertil.*, **28**, 379
29. Ahluwalia, R., Gambhir, K. and Sekhon, H. (1975). Distribution of labeled retinyl acetate and retinolic acid in rat and human testes. A site of retinyl acetate incorporation in rat testes. *J. Nutr.*, **105**, 467
30. Cicero, T., Bell, R., Wiest, W., Allison, J., Polakoski, K. and Robins, E. (1975). Function of the male sex organs in heroin and methadone users. *N. Eng. J. Med.*, **292**, 882

31. Kalla, N. and Bansal, M. (1975). Effect of carbon tetrachloride on gonadal physiology in male rats. *Acta Anat.*, **91**, 380
32. Nagy, F. and Edmonds, R. (1975). Cellular proliferation and renewal in the various zones of the hamster epididymis after colchicine administration. *Fertil. Steril.*, **26**, 460
33. Urry, R. and Dougherty, K. (1975). Inhibition of rat spermatogenesis and seminiferous tubule growth after short-term and long-term administration of a monoamine oxidase inhibitor. *Fertil. Steril.*, **26**, 232
34. Arlit, A. H. and Wells, H. G. (1917). The effect of alcohol on the reproductive tissues. *J. Exp. Med.*, **26**, 769
35. Van Thiel, D., Cavaler, J., Lester, R. and Goodman, M. (1975). Alcohol-induced testicular atrophy. *Gastroenterology*, **69**, 326
36. Lutwak-Mann, C. (1964). Observations on progeny of thalidomide treated male rabbits. *Br. Med. J.*, **1**, 1090
37. Clay, M. L. (1964). Conditions affecting voluntary alcohol consumption in rats. *Qu. J. Stud. Alcohol*, **25**, 36
38. Myers, R. D. (1966). Voluntary alcohol consumption in animals: Peripheral and intracerebral factors. *Psychosom. Med.*, **28**, 484
39. Essig, C. F. and Lam, R. C. (1968). Convulsions and hallucinatory behaviour following alcohol withdrawal in the dog. *Arch. Neurol.*, **18**, 626
40. Deneau, G., Yanagita, T. and Seevers, M. R. (1969). Self administration of psychoactive substances by the monkey. *Psychopharmacologie*, **16**, 30
41. Freund, G. (1969). Alcohol withdrawal syndrome in mice. *Arch. Neurol.*, **21**, 315
42. Goldstein, D. B. and Pal, N. (1971). Alcohol dependence produced in mice by inhalation of ethanol: Grading the withdrawal reaction. *Science*, **172**, 288
43. Ellis, F. W. and Pick, J. R. (1970). Experimentally induced ethanol dependence in Rhesus monkeys. *J. Pharmacol. Exp. Ther.*, **175**, 88
44. Woods, J. H., Ikomi, F. and Winger, G. D. (1971). The reinforcing property of ethanol. In: M. K. Rooch, W. M. McIssac and P. J. Creaven (eds.). *Biological Aspects of Alcohol*, pp. 371–388. (Texas: Austin University Texas Press)
45. Falk, J. L., Samson, H. and Winger, G. (1972). Behavioural maintenance of high concentrations of blood ethanol and physical dependence in the rat. *Science*, **177**, 811
46. Cicero, T. J. and Smithloff, B. R. (1973). Alcohol and self administration in rats: Attempts to elicit excessive intake and dependence. In: M. M. Gross (ed.). *Advances in Experimental Medicine and Biology*, **Vol. 35**: *Alcohol Intoxication and Withdrawal: Experimental Studies*. (New York and London: Plenum Press)
47. Erickson, C. K., Koch, K. I., Mehta, C. S. and McGinity, J. W. (1978). Sustained release of alcohol: Subcutaneous silastic implants in mice. *Science*, **199**, 1457
48. Craig, J. R., Munsat, T. L. and Chuang, M. (1977). Programmed feeding as a model of chronic alcoholism in the rat. *Ann. Neurol.*, **2**, 311
49. Wilson, J. G. (1973). *Environment and Birth Defects*, pp. 227-232. (New York and London: Academic Press)
50. Reyes, F. I., Boroditsky, R. S., Winter, J. S. D. and Faiman, C. (1974). Studies on human sexual development II: Fetal and maternal serum gonadotropin and sex steroid concentrations. *J. Clin. Endocrinol. Metab.*, **38**, 612
51. Leblond, C. and Clermont, Y. (1952). Definition of the stages of the cycle of the seminiferous epithelium in the rat. *Ann. N.Y. Acad. Sci.*, **55**, 548
52. Croft, B. and Bartke, A. (1976). Quantitative study of spermatogenesis in vasectomized mice. *Int. J. Fertil.*, **21**, 61
53. Vitale, J. J., Di Giorgio, J., McGrath, H., Nay, J. and Hegsted, D. M. (1953). Alcohol oxidation in relation to alcohol dosage and the effect of fasting. *J. Biol. Chem.*, **204**, 257
54. Lieber, C. S. (1975). Interference of ethanol in hepatic cellular metabolism. In: F. A. Seixes, K. Williams and S. Eggleston (eds.). Medical Consequences of Alcoholism. *Ann. N.Y. Acad. Sci.*, **252**, 24
55. Stockard, C. R. and Papanicolaou, G. (1916). A further analysis of the hereditary transmission of degeneracy and deformities by the descendants of alcoholized mammals. *Am. Nat.*, **50**, Part I, 65–88; Part II, 144–177
56. Rorvik, D. M. and Shettles, L. B. (1970). *Your Baby's Sex: Now You Can Choose*. (Toronto: Dodd Mead and Co.)

57. Rösch (1837). Cited by Arlit, A. H. and Wells, H. G. (1917). The effect of alcohol on the reproductive tissues. *J. Exp. Med.*, **26**, 769
58. Mirone, L. (1952). The effect of ethyl alcohol on growth, fecundity and voluntary consumption of alcohol by mice. *Qu. J. Stud. Alc.*, **13**, 365
59. Bertholet, E. (1919). Cited in Arlit, A. H. and Wells, H. G. (1917). The effect of alcohol on the reproductive tissues, *J. Exp. Med.*, **26**, 769
60. Bouin and Garnier (1900). Cited in Monterosso, B. (1912). L'azione del diguinoe dell' apitelio del tubo seminifero del topo. *Archiv. de Biol.*, **27**, 35
61. Reddy, K. and Svoboda, D. (1967). Alterations in rat testes due to an antispermatogenic agent. *Arch. Pathol.*, **84**, 376
62. Doepfmer, R. and Hinckers, H. (1965). Zur Frage der Keimschädigung im akuten Rausch. *Z. Haut-u. Geschlechtskr.*, **39**, 94
63. Teitelbaum, H. and Gantt, W. (1958). The effect of alcohol on sexual reflexes and sperm count in the dog. *Qu. J. Stud. Alcohol*, **13**, 394
64. Schopper, K. (1911). *Frankfurter Z. Path.*, **8**. Cited in Arlit, A. H. and Wells, H. G. (1917). The effect of alcohol on the reproductive tissues. *J. Exp. Med.*, **26**, 769
65. Badr, F. and Bartke, A. (1974). Effect of ethyl alcohol on plasma testosterone level in mice. *Steroids*, **23**, 921
66. Levin, J., Lloyd, C., Lobotsky, J. and Friedrich, E. (1967). The effect of epinephrine on testosterone production. *Acta Endocrinol.*, **55**, 184
67. Weddington, S., Hansson, V., Ritzen, E., Hagenas, L., French, F. and Nayfeh, S. (1975). Sertoli cell secretory function after hypophysectomy. *Nature* (*Lond.*), **254**, 145
68. Bregman, A. (1971). Cytogenetic effects of ethanol in human leukocyte cultures. *EMS Newsletter*, **4**, 35
69. Obe, G. and Herha, J. (1975). Chromosomal damage in chronic alcohol users. *Humangenetik*, **29**, 191
70. Van Thiel, D., Lester, R. and Sherins, R. (1974). Hypogonadism in alcoholic liver disease: evidence for a double effect. *Gastroenterology*, **67**, 1188
71. Clausen, S. (1938). The pharmacology and therapeutics of vitamin A. *J. Am. Med. Ass.*, **111**, 144

14
Teratogenicity of inhalation anaesthetic agents

W. D. B. POPE

INTRODUCTION

Over the past fifteen years, the possibility of potential teratogenic properties of inhalation anaesthetics has raised serious doubts about the safety of exposing pregnant women to these agents. The use of halocarbons and various ethers to induce anaesthesia has been a part of the practice of medicine for nearly 150 years, and apart from direct physiological side effects, the agents were considered to be innocuous. However, in 1937, Goldschmidt[1] showed that starved dogs exposed to chloroform developed liver necrosis after one exposure. This was the same agent which had gained wide acceptance nearly sixty years earlier when Queen Victoria took it during the births of her last two children, Prince Leopold and Princess Beatrice. The discovery of this side effect indicated that anaesthetics may have various toxic effects beyond direct depression of the respiratory, cardiovascular or central nervous systems. This chapter deals with the teratogenic potential of inhalation anaesthetics.

It is necessary, however, to determine the questions raised, that is, are these substances teratogenic, and if so, what differences occur between people chronically exposed to low doses and those given single exposures to full anaesthetizing concentrations? From the teratological point of view the major emphasis has been placed on the effects of chronic exposure.

EARLY STUDIES

During the 1950s and early 1960s, investigators in a number of fields showed that the potent inhalation anaesthetics could seriously affect many systems. Some of the earliest work was done with flowers, when Claude Bernard showed that high concentrations of diethyl ether inhibited plant growth and cell division[2]. Many papers have shown that high concentrations of anaesthetic agents exhibit the same depressant effects as colchicine, an antimitotic agent that halts mitosis in metaphase[3]. For example, Nunn and his coworkers[4–6] have shown that halothane reduces cell division in the ciliate protozoon *Tetrahymena pyriformis*[4], in the fava bean[5], and in Chinese hamster lung fibroblasts[6], when these are exposed to surgical levels of anaesthetic agents. Equally important, all these effects are reversible. However, although it was shown that exposure of cells to anaesthetics could produce changes in the structure of the cell, at most clinical concentrations human cells in culture are affected only minimally by the halogenated anaesthetics halothane, chloroform and methoxyflurane.

Furthermore, Cohen[7] showed that mitochondrial function as measured by oxygen uptake could be markedly diminished by exposure to clinical concentrations of anaesthetics, and that this dose-related decrease in both mitochondrial electron transport and mitochondrial respiratory control was also totally reversible. This is seen with all four of the major volatile anaesthetics – diethyl ether, halothane, methoxyflurane and enflurane. With halothane, the reduction begins at clinical concentrations [approximately 1% (v/v)] and continues to a maximum 75% decrease at 3%. Below 2.0 to 2.5 MAC*, this depression is reversible with all the agents. Chronic exposure to even the seemingly innocuous anaesthetic nitrous oxide could cause a severe and significant depression of cellular production by bone marrow, both in experimental animals[8,9] and in human beings[10]. To ascertain any involvement of RNA and DNA in this depression, the incorporation of tritiated thymidine in bone marrow cells was measured in animals chronically exposed to various agents[11]. It was found that although the proliferation and maturation of bone marrow cells was decreased with chronic exposure, this phenomenon did not involve interference with DNA or RNA synthesis *per se*.

It was also shown that the inhalation anaesthetics, which had previously been thought to be excreted unchanged from the lungs, were in fact metabolized, in some cases to a significant degree[12] and that some of these metabolic byproducts could in fact be toxic in their own right. For example, Mazze showed that a severe dose-related nephrotoxicity (manifest as a high-output type of renal failure) appeared in both rodents[13] and humans[14] when the flouride ion concentration in blood reached a certain level following the biodegradation of inhaled methoxyflurane. It was also found that differing pathways of metabolism could create dangerous toxic byproducts. For example, fluroxene[15], a flammable volatile anaesthetic which seemed innocuous in humans, was uniformly fatal at high inhaled levels in animals. A different breakdown pathway produced a substance causing massive hepatic necrosis and this was enhanced when substances were given that induced the production of liver enzymes. At the same time the report of the National Halothane Study[16] suggested massive hepatic necrosis as a possible sequella to (repeated) exposure to halothane or cyclopropane. All of these studies raised the spectre of severe systemic toxicity following exposure to inhalation anaesthetics.

* MAC – The minimum alveolar concentration of an agent that will prevent response to a surgical stimulus in 50% of subjects. This is a measurement used to compare the potencies of anaesthetics

ANIMAL STUDIES

Naturally, this raised certain fears about the safety of these agents for pregnant women. A number of studies were done using animal models. The earliest of these employed chick embryos. Diethyl ether[17], cyclopropane[18], methoxyflurane[19], halothane[20] and nitrous oxide were all found to increase chick embryo mortality and significantly reduce the chick neural tube mitotic index at greater than anaesthetic concentrations[21]. When lower, more clinical doses were used, the results were variable and the anomalies that did occur were primarily minor and associated with skeletal development. Various postulates for a mechanism for this included intracellular hypoxia, decreased cell division and depression of cellular metabolism.

With these conflicting results, researchers turned to the pregnant rodent and this has remained the major animal model for the last decade. In 1956 Lassen[10] had demonstrated a decrease in human bone marrow cell proliferation after chronic exposure to nitrous oxide. Moreover, Parbrook[22] had shown that rats exposed to 48 h of nitrous oxide (60%) in the first week after mating had a significant decrease in the number of live pups. In 1967, Fink[23] published a paper showing that exposure of pregnant Sprague-Dawley rats to 0, 2, 4 or 6 days of nitrous oxide (50%) starting on day 8 of gestation resulted in a dose-related increase in the number of both resorptions and survivors with skeletal malformations. The skeletal defects were similar to those seen when pregnant mice were subjected to hypoxia.

In 1968, Basford and Fink[24] exposed pregnant female Sprague-Dawley rats to an anaesthetizing concentration of halothane 0.8% (v/v) or 8000 parts per million (ppm) for a single 12 h exposure period during days 8–12 of gestation. The animals were killed at term and it was found that a higher level of bony abnormalities was present in those fetuses exposed on days 8, 9.5 and 10. However, these statistics are slightly misleading as the numbers of abnormal exposed animals were only statistically higher than their own day controls. No difference was noted when compared to other controls. Furthermore, no increase in resorptions could be found. In fact, there was an indication that halothane had a protective effect on implantation. Although many queries could be made concerning this study, it did raise doubts about the safety of inhalational anaesthetic exposure to pregnant women.

In 1973, several related studies appeared. Bruce[25] exposed three sets of mice to air and halothane 16 ppm (0.0016% v/v) both males and females, for 7 h per day, 5 days a week for 6 weeks. The animals were then mated and daytime exposure continued. All were sacrificed at day 18 of gestation. The animals were examined for number of implantations and resorptions; histological studies were carried out on the liver, spleen and testis; fetuses were fixed in absolute alcohol and subsequently stained with alizarin Red for skeletal examination. No evidence of any deleterious effects was found, including no difference in the total number of implantations.

However, Corbett[26] in 1973, found slightly differing results with nitrous oxide (N_2O). Routinely, concentrations in the inhalational zone of the anaesthetist were around 1000 ppm (0.1%) with a range of 300 to 10000 ppm (0.03%–1.0%). Pregnant rats were exposed to 0, 100, 1000 or 15000 ppm of nitrous oxide for 8 or 24 h a day on various days of gestation (i.e. days 8–13, 12–19, 10–13, 14–19 or 10–19). Some of those exposed had a greater fetal death rate and a decrease in the number of implantations, both at 8 and at 24 h exposure periods. At that time this raised real fears about the safety of the operating room environment for pregnant women. Also Katz and Clayton[27] in a paper presented at the annual meeting of the American Society of Anesthesiologists in 1973, suggested that the perinatal (up to 14 days after delivery) death rate was higher in rats exposed to low concentrations of halothane (10 ppm) during mating, gestation and lactation. Despite a very high control perinatal death rate of 51%, this study still emphasized a possibly dangerous situation. Bussard *et al.*[28] exposed pregnant hamsters to 60% nitrous oxide and 0.6% (6000 ppm) halothane for a single 3 h exposure on either day 9, 10 or 11 of pregnancy. This was intended to simulate a full general anaesthetic with the most frequently used combination of agents. Those exposed on day 11 had a higher resorption rate and a decreased fetal weight and length.

EPIDEMIOLOGY

These studies were given further emphasis by a report[29] in 1974 from an Ad Hoc Committee on the effect of trace anaesthetics on the health of operating room personnel entitled 'Occupational Disease among Operating Room Personnel', and was sponsored by the American Society of Anesthesiologists. This extensive study was prompted by a number of reports from the USSR[30], Denmark[31], the United Kingdom[32], and the USA[33], that women who worked in an operating room environment had more spontaneous abortions, a greater number of children with congenital abnormalities and a higher incidence of cancer than did women who worked in a non-operating room hospital situation. The 1974 study canvassed 20467 women who worked in the operating room, comprising female members of the American Society of Anesthesiologists, the American Association of Nurse Anesthetists and the Association of Operating Room Nurses/Technicians. Control questionnaires were sent to members of the American Academy of Pediatrics and the American Nurses Association, and responses here numbered 7199. Although some of the statistical interpretations are questionable, this study indicated that women who worked in the operating room have a 1.3–2.0 times increase in the number of spontaneous abortions and a twofold increase in the number of congenital abnormalities among liveborn babies.

RECENT ANIMAL STUDIES

Pope *et al.*[34] exposed pregnant Sprague-Dawley rats to various increasing concentrations of halothane on days 8–12 of gestation. These concentrations ranged from 50 to 3200 ppm and were given for 8 h per day. Particular care was taken to avoid hypoxia, hypercarbia, hyperthermia or too high a relative humidity. Two large perspex chambers were built into which the normal breeding cages could be placed to avoid unnecessary stress on the animals. Careful 15 min monitoring of the anaesthetic concentration was done with gas chromatography. Control rats were exposed in duplicate chambers at each concentration. No higher concentration was attempted since at 0.32% halothane (3200 ppm) the animals were slightly drowsy and failed to feed properly. No significant differences from control were seen in maternal weight, fetal number or fetal weight. Equally important, there was no specific increase or decrease in these variables with increasing concentrations, indicative of a lack of dose dependency, and the range of values for the resorption rates was the same in the exposed as in the control animals. Moreover, histological examination and careful skeletal staining[35] demonstrated only occasional skeletal anomalies such as supplementary lumbar ribs, defective or delayed ossification centres in the sternebrae and missing ossification centres in the lower parts of the limbs. These were seen in all groups, both control and exposure, again with no relationship between frequency and increasing dosage.

A further, more elaborate study from our laboratory[36] looked at the effects of a number of agents again on an 8 h per day regime, this time throughout gestation. Halothane at 0.16% (1600 ppm), nitrous oxide at 1%, 10% and 50% and methoxyflurane at 0.01, 0.04 and 0.08% were all investigated. No significant increase in fetal loss was noted (Table 14.1). The major finding was a decrease in the average fetal weights following exposure to the higher subanaesthetic concentrations of all the agents, accompanied by a corresponding decrease in crown–rump length and slight developmental retardation as shown by variable decreases in the number of ossification centres (Table 14.2). No gross skeletal anomalies were present and no consistent or dose-dependent change in the sex ratio of the fetuses was seen at any time.

A pilot study with the newer volatile inhalation anaesthetic enflurane[37] has indicated similar results at a high concentration (3200 ppm), and exposures to low concentrations (10.7 and 63.7 ppm) by Strout and colleagues[38] in which the animals were allowed to litter showed no changes in litter size. A very slight decrease in fetal weight could be seen at both concentrations.

Finally, a very complete experiment by Wharton and colleagues[39] used Swiss/ICR mice at both subanaesthetic and full anaesthetic concentrations. Halothane exposures of 0.024, 0.1, 0.4, 1.2 and 4.0 MAC hours per day were employed. At the two lowest levels studied, no adverse effects on reproduction were observed. At 0.4 MAC hours there was a decreased weight gain in both the mothers and the fetuses. At the higher concentrations, both pregnancy rate and implantation rate (and thus the number of live fetuses) were decreased, but without an increased fetal loss or a fall in postnatal survival.

DISCUSSION

All of these studies leave us in something of a quandary. The earliest reports indicated a real hazard if pregnant animals were exposed to anaesthetics at full anaesthetic or even very low atmospheric concentrations of the agents. The more recent work seems to suggest that this danger is not as great as had previously been feared. Naturally it is difficult to extrapolate animal models to the human situation, and this is particularly true in the case of inhalation

Table 14.1 Litter sizes and percentage fetal losses

Agent	Concentration (v/v %)	Anaesthetic group: Pregnant rats	Anaesthetic group: Live/litter (SE)	Anaesthetic group: % fetal loss (SE)	Control group: Pregnant rats	Control group: Live/litter (SE)	Control group: % fetal loss (SE)
Halothane	0.16	7	14.0 (1.5)	1.0 (1.8)	6	14.2 (0.8)	0
Halothane	0.32	1	14.0	0	7	14.1 (0.8)	0
Halothane	0.32*	9	14.0 (1.5)	5.3 (3.7)	10	14.7 (0.8)	6.4 (3.3)
Nitrous oxide	1.0	7	12.7 (1.4)	8.2 (3.3)	8	14.3 (0.7)	0
Nitrous oxide	10.0	7	14.3 (0.6)	8.3 (3.7)	8	11.3 (1.3)	1.1 (0.8)
Nitrous oxide	50.0	10	12.0 (0.9)	10.4 (2.2)	10	13.7 (0.6)	8.1 (2.5)
Halothane + nitrous oxide	0.16 10.0	9	13.6 (0.8)	3.2 (1.7)	10	13.4 (0.7)	5.6 (1.9)
Methoxyflurane	0.01	10	15.6 (0.5)	1.3 (1.4)	11	13.0 (1.0)	2.7 (1.0)
Methoxyflurane	0.04	7	14.3 (1.1)	4.8 (2.8)	9	14.2 (0.6)	5.9 (2.5)
Methoxyflurane	0.08	5	13.2 (0.6)	5.7 (1.4)	11	12.5 (0.5)	2.8 (1.1)
'Stress group'		4	1.5 (1.5)	91.3 (8.8)			

* Exposure period: days 1–8 of gestation

Pope *et al.*[36]

Table 14.2 Fetal and placental weights

Agent	*Concentration* (v/v %)	*Anaesthetic group* Total number of live fetuses	*Anaesthetic group* Fetal weight g (SE)	*Anaesthetic group* Placental weight g (SE)	*Control group* Total number of live fetuses	*Control group* Fetal weight g (SE)	*Control group* Placental weight g (SE)
Halothane	0.16	98	4.88 (0.07)*	0.57 (0.01)	85	5.52 (0.05)	0.57 (0.01)
Halothane	0.32	14	4.26 (0.08)*	0.54 (0.02)	99	5.26 (0.07)	0.53 (0.01)
Halothane	0.32†	126	5.31 (0.05)	0.57 (0.01)	147	5.29 (0.06)	0.60 (0.01)
Nitrous oxide	1	89	5.31 (0.07)	0.51 (0.01)*	114	5.45 (0.04)	0.59 (0.01)
Nitrous oxide	10	100	4.22 (0.05)*	0.42 (0.01)‡	90	5.02 (0.05)	0.45 (0.01)
Nitrous oxide	50	120	4.35 (0.07)*	0.43 (0.01)‡	137	5.51 (0.04)	0.47 (0.01)
Halothane + nitrous oxide	10 0.16	122	4.66 (0.07)*	0.56 (0.01)	134	5.77 (0.05)	0.51 (0.01)
Methoxyflurane	0.01	156	5.18 (0.04)*	0.47 (0.01)	143	5.43 (0.05)	0.45 (0.01)
Methoxyflurane	0.04	100	4.69 (0.08)*	0.48 (0.01)	128	5.60 (0.05)	0.49 (0.01)
Methoxyflurane	0.08	66	4.47 (0.06)*	0.43 (0.01)	138	4.90 (0.03)	0.43 (0.01)
'Stress group'		6	3.25 (0.19)	0.31 (0.02)			

* $p < 0.001$
† Exposure period: days 1–8 of gestation
‡ $p < 0.01$

Pope *et al.*[36]

anaesthetics where numerous other variables are involved that are practically impossible to duplicate in an experimental situation. It is significant that in one paper[36], animals that were severely stressed during marking showed an enormous resorption rate (see Table 14.1) and equally significant that as the dose increased in concentration, the one consistent abnormality appeared to be a decrease in fetal weight. Yet indications are that these smaller fetuses quickly catch up to the controls and develop normally[38].

When paired feedings were used, it was seen that the decreased fetal weight still occurred in the exposed animals only. This would indicate that some factor other than direct food deprivation must be responsible for the decreases in weight. Also the exposed animals consumed more food for each gram of weight gain than did the controls.

There is at the moment insufficient data to postulate a definitive mechanism for those changes, but some changes in uteroplacental nutrition or perhaps blood flow could well explain the observed effects. How these effects are moderated by anaesthesia, and whether they are due to direct effects of the agent, breakdown products or the systemic side effects of the agents has yet to be adequately determined. The effect of a single high dose would also appear to be less dangerous than might have been supposed, as repeated exposures seem to be necessary to produce significant changes.

References

1. Goldschmidt, S., Ravdin, I. S. and Lucke, B. (1937). Anaesthesia and liver damage. I. The protective action of oxygen against the necrotising effect of certain anaesthetics on the liver. *J. Pharmacol. Exp. Ther.*, **59**, 1
2. Bernard, C. (1878). *Le Cons sur les Phenomènes de la Ive Communs aux Animaux et aux Végétaux.* Chapter VII, p. 259. (Paris: Ballière)
3. Anderson, N. B. (1966). The effect of CNS depressants on mitosis. *Acta Anaesth. Scand.*, **10** (Suppl. 22), 1
4. Nunn, J. F., Dixon, K. L. and Moore, J. R. (1968). Effects of halothane on *Tetrahymena pyriformis. Br. J. Anaesth.*, **40**, 145
5. Nunn, J. F., Dixon, K. L. and Lovis, J. D. (1969). Effects of halothane on mitosis. *Anesthesiology*, **30**, 348
6. Sturrock, J. E. and Nunn, J. F. (1976). Synergism between halothane and nitrous oxide in the production of nuclear abnormalities in the dividing fibroblast. *Anesthesiology*, **44**, 461
7. Cohen, P. J. (1973). Effect of anesthetics on mitochondrial function. *Anesthesiology*, **39**, 153
8. Green, C. D. and Eastwood, D. N. (1963). Effects of nitrous oxide inhalation on hemopoiesis in rats. *Anesthesiology*, **24**, 341
9. Bruce, D. L. and Koepke, J. A. (1966). Changes in granulopoiesis in the rat associated with prolonged halothane anesthesia. *Anesthesiology*, **27**, 811
10. Lassen, H. C. A., Henriksen, E., Neukirch, F. and Kristensen, H. S. (1956). Treatment of tetanus; severe bone marrow depression after prolonged nitrous oxide anaesthesia. *Lancet*, i, 527
11. Bruce, D. L., Koepke, J. A. and Traurig, H. H. (1968). Studies of DNA synthesis in the halothane treated rat. In: B. R. Fink (ed.). *Toxicity of Anesthetics*, pp. 123–129. (Baltimore: Williams and Wilkins)
12. Van Dycke, R. A. and Chensweth, M. D. (1965). Metabolism of volatile anesthetics. *Anesthesiology*, **26**, 348
13. Mazze, R. I., Cousins, M. J. and Kosek, J. C. (1972). Dose-related methoxyflurane nephrotoxicity in rats: A biochemical and pathologic correlation. *Anesthesiology*, **36**, 571
14. Cousins, M. J., Mazze, R. I., Kosek, J. C., Hitt, B. and Love, F. V. (1974). The etiology of methoxyflurane nephrotoxicity. *J. Pharmacol. Exp. Ther.*, **190**, 530

15. Cascorbi, H. F. and Sergh-Amaranath, A. V. (1972). Fluroxene toxicity in mice. *Anesthesiology*, **37**, 480
16. Summary of the National Halothane Study. (1966). *J. Am. Med. Ass.*, **197**, 775
17. Smith, B. E., Gaub, M. L., Usubiago, L. and Lehrer, S. (1968). Teratogenic effects of diethyl ether. In: B. R. Fink (ed.). *Toxicity of Anesthetics*, 1st Ed. (Baltimore: Williams and Wilkins)
18. Anderson, N. B. (1966). The effect of CNS depressants on mitosis. *Acta Anaesth. Scand.*, **22** (Suppl.), 1
19. Smith, B. E. and Lehrer, S. B. (1967). Teratogenic effects of anaesthesia in the mammal. (Unpublished)
20. Smith, B. E., Gaub, M. L. and Moya, F. (1965). Investigations into the teratogenic effects of anaesthetic agents. The fluorinated agents. *Anesthesiology*, **26**, 260
21. Smith, B. E., Gaub, M. L. and Moya, F. (1965). Teratogenic effects of anaesthetic agents: nitrous oxide. *Anaesth. Analg.*, **44**, 726
22. Parbrook, G. D. (1965). Studies of nitrous oxide toxicity in experimental animals. *Br. J. Anaesth.*, **39**, 119
23. Fink, B. R., Shepard, T. H. and Blandau, R. J. (1967). Teratogenic activity of nitrous oxide. *Nature (Lond.)*, **214**, 146
24. Basford, A. B. and Fink, B. R. (1968). The teratogenicity of halothane in the rat. *Anesthesiology*, **29**, 1167
25. Bruce, D. L. (1973). Murine fertility unaffected by traces of halothane. *Anesthesiology*, **38**, 473
26. Corbett, T. H., Cornell, R. G., Endres, J. L. and Millard, R. I. (1974). Effects of low concentrations of nitrous oxide on rat pregnancy. *Anesthesiology*, **39**, 299
27. Katz, J. and Clayton, W. (1973). Fetal mortality in rats chronically exposed to low concentrations of halothane. Abstract of Scientific papers. American Society of Anesthesiologists Annual Meeting, p. 57
28. Bussard, D. A., Stoelting, R. K., Peterson, C. and Ishaq, M. (1974). Fetal changes in hamsters anesthetized with nitrous oxide and halothane. *Anesthesiology*, **41**, 275
29. Cohen, E. N., Brown, B. W. Jr., Bruce, D. L., Cascorbi, H. F., Corbett, T. H., Jones, T. W. and Witcher, C. E. (1974). Occupational disease among operating room personnel: A national study. *Anesthesiology*, **41**, 321
30. Vaisman, A. I. (1967). Working conditions in surgery and their effect on the health of anesthesiologists. *Eksp. Khir. Anesteziol.*, **3**, 44
31. Askrog, V. and Harvald, B. (1970). Teratogen effekt of inhalations-anestetika. *Saertyk Nord. Med.*, **83**, 490
32. Knill-Jones, R. P., Moir, D. O., Rodrigues, L. V. and Spence, A. A. (1972). Anaesthetic practice and pregnancy: A controlled survey of women anaesthetists in the United Kingdom. *Lancet*, **i**, 1326
33. Corbett, T. H., Cornell, R. G., Lieding, K. and Endres, J. L. (1973). Incidence of cancer among Michigan nurse anesthetists. *Anesthesiology*, **38**, 260
34. Pope, W. D. B., Halsey, M. J., Lansdown, A. B. G. and Bateman, P. E. (1975). Lack of teratogenic dangers with halothane. *Acta Anaesthesiol. Belg.*, **26**, 169
35. Lansdown, A. B. G., Pope, W. D. B., Halsey, M. J. and Bateman, P. E. (1976). Analysis of fetal development in rats following maternal exposure to subanesthetic concentrations of halothane. *Teratology*, **13**, 299
36. Pope, W. D. B., Halsey, M. J., Lansdown, A. B. G., Simmonds, A. and Bateman, P. E. (1978). Fetotoxicity in rats following chronic exposure to halothane, nitrous oxide or methoxyflurane. *Anesthesiology*, **48**, 11
37. Pope, W. D. B. and Persaud, T. V. N. (1978). Foetal growth retardation in the rat following chronic exposure to the inhalation anaesthetic enflurane. *Experientia*, (In press)
38. Strout, C. D., Nahrwold, M. L., Taylor, M. D. and Zagon, I. S. (1977). Effects of subanesthetic concentrations of enflurane on rat pregnancy and early development. Abstracts of Scientific Papers, American Society of Anesthesiologists Annual Meeting, p. 555
39. Wharton, R. S., Mazze, R. I., Baden, J. M., Hitt, R. A. and Dooley, J. R. (1978). Fertility, reproduction and postnatal survival in mice chronically exposed to halothane. *Anesthesiology*, **48**, 167

15
Effects of prenatal exposure to ultrasound

M. R. SIKOV AND B. P. HILDEBRAND

INTRODUCTION

It is generally accepted that the prenatal animal is susceptible to many forms of insult, including ionizing radiation, microwaves, organic and inorganic environmental pollutants, and deficiencies or excesses of specific nutrients. The most dramatic effects caused by these agents include death of the conceptus or the production of gross or histological malformations. Of equal importance, however, are effects which are manifested as changes in postnatal physiological or behavioural fitness and survival.

Many of these effects appear at dose levels below those required to produce significant alterations in the adult. The response of the prenatal or neonatal organism thus becomes a sensitive means, as well as a necessary parameter, for evaluating the potential hazard of physical agents. In view of increasing use of ultrasound for clinical examination of the human conceptus, it is important that the effects of such exposure be studied in detail.

The nature of ultrasound

Ultrasound is produced by a vibrating element and propagated as pressure waves in a fluid or solid medium. Ultrasound does not, however, differ from ordinary audible sound except that it is of higher frequency – in the 0.5–10 MHz range, as used clinically – and since attenuation increases with frequency, it is poorly propagated in a gaseous medium. In general, the medium through which ultrasound passes vibrates at the same frequency as the transducer and the instantaneous local pressures and densities in the medium also change at the same frequency. Ultrasound thus differs from most physical agents of biological concern, which are electromagnetic in nature.

When the vibratory motion of the emitting element is maintained without interruption, it is said to be operating in the continuous mode or to be emitting continuous wave (CW) ultrasound. In this case, the rate of power or energy delivered to the system is constant over time.

Alternatively, the transducer may be operated in the pulsed or burst mode. In this case, the device operates in a series of short bursts, each consisting of a few cycles, separated by an 'off' period. The length of this burst is referred to as 'burst duration' and is usually in the order of μsec and the 'burst repetition frequency' gives the number of bursts per second, typically 10^3. The reciprocal of the repetition frequency is referred to as the 'burst repetition period' and product of burst duration and burst repetition frequency is the 'duty factor', the decimal fraction of the time that sound is being emitted. In this mode, power varies with time and in an oversimplified view may be considered to be the 'peak power' during the burst and zero between bursts. An important measure, time-average power (TA), is the product of the peak power times the duty factor.

Power density or intensity is the measure of power per unit cross-sectional area of the beam and is of particular importance in the interpretation of experiments on biological effects. A crude approximation of the spatial average (SA) intensity (Ia) is obtained by dividing power by the area of the beam, but it is more complicated in practice. Accordingly, these details will be considered in conjunction with the calibration of the apparatus used in our experiments (see below).

Clinical uses of ultrasound

The use of ultrasound in clinical medicine has increased markedly in the past few years and further increases may be anticipated. These uses presently include a variety of diagnostic procedures and therapeutic (diathermy) applications[1–3].

A multitude of diagnostic procedures using ultrasound are presently available; some of these are widely used while the availability of others is restricted to a few major centres[1,3]. Perhaps the most common categories of use are those associated with obstetric practice[4,5]. A number of these applications are based on the reflection of pulsed ultrasound and include placental localization, determination of fetal placement, and estimation of the size of fetal organs. These instruments typically have a SATA intensity of about 3–10 mW/cm^2. Continuous-wave ultrasound is used in the form of instruments which employ the Doppler shift phenomenon to monitor fetal function or distress or to determine the viability of early pregnancy from the presence of a fetal heart beat and are associated with SATA intensities in the order of 20 mW/cm^2.

Higher intensities of CW ultrasound (0.5–3 W/cm^2) are commonly used for diathermy. No deleterious effects have been reported when therapeutic ultrasound is used according to accepted techniques, but damage has been found to result from injudicious or incorrect use of these procedures[6]. A survey of medical diathermy devices (about one-half of which were ultrasonic) in one county in Florida found that 13 000 patients per month were treated of a population of about 500 000 people in the county. This survey found several abuses including use of unacceptable procedures, non-uniformity of controls for regulating intensity, and operation by untrained personnel[7].

Historical perspective

In early studies, continuous wave ultrasound was found to produce malformations in *Drosophila*; the relative sensitivity of the different developmental stages approximated that found with X-rays[8,9]. It was also shown that insonation of amphibian embryos could produce dissociation of the germ layers[10], and more recent studies have demonstrated a malformative action[11,12]. These considerations resulted in concern about the potential embryotoxic effects of ultrasound, as used clinically, and led to a number of laboratory studies to assess the effects of ultrasound during pregnancy.

Most of the earlier laboratory studies employed exposure to continuous wave ultrasound, usually from a clinical Doppler instrument operating at a frequency between 2.25 and 2.5 MHz. These investigations include those of Takeuchi *et al.*[13] who exposed rats to 150 mW/cm^2 for 20 min on the third and fourth days of gestation and found essentially no mortality among either the control or exposed fetuses and no gross abnormalities. McClain *et al.*[14] exposed rats to 10 mW/cm^2 for 30 or 120 min on 8–10 or 11–13 days of gestation (d.g.). The fetuses were examined after 20 days of gestation; no consistent changes in the incidence of mortality or abnormality were detected. Mannor *et al.*[15] studied the effects of 5- and 60-minute exposures at intensities ranging from 164 to 1050 mW/cm^2 given at a single unstated (but variable) time between 8 and 16 days, or as five daily exposures between 8 and 20 d.g. Some of the pregnant mice died following exposure to the highest intensity; the authors attribute these deaths to heating. The fetuses from the surviving females at this intensity, as well as all of those at lower intensities (164–490 mW/cm^2), appeared to be grossly and microscopically normal. Shoji *et al.*[16] used 5 h exposures of mice on 8 d.g. to presumably CW ultrasound at 40 mW/cm^2. They reported a statistically significant increase in prenatal mortality relative to the controls, although not in comparison to a sham-exposed group. There was a tendency towards an elevated incidence of fetal abnormality in the exposed fetuses, although it did not reach statistical significance. Murai *et al.*[17,18] used a similar unit to expose immobilized rats to 20 mW/cm^2 for 5 h at 9 d.g. Although there were indications of deficits in the postnatal development of the exposed offspring and in their emotional and cognitive behaviour, it is difficult to distinguish between the effects of ultrasound and immobilization.

Warwick, Woodward, and their colleagues[19,20] were the first to use pulsed ultrasound for prenatal exposure. They examined the effects of ultrasound on pregnant mice over a wide range of frequencies, intensities, duty-cycles, and stages of gestation and concluded that there were no consistent effects on the incidence of mortality or abnormality. It appears from their data, however, that those groups which were exposed to the highest time-average intensity were those in which an effect was observed.

Akamatsu[21] used a focused 2.25 MHz pulsed beam of about 30 mW/cm^2 to insonate preimplantation rat and mouse embryos *in vitro*. A 12 h exposure did not affect subsequent *in vitro* development or the development of those transferred to recipient dams.

A number of papers have quoted the earlier report of Smyth[22] as a primary citation. It is of some interest to note that they seem to include common elements of misquotation which give the impression that prenatal exposure to ultrasound was found to be without teratological effects and did not affect postnatal behaviour or reproduction effects in rats and mice. Actually, as reported by Smyth[22], the prenatal mice were not exposed to ultrasound. The testes of the male parent were exposed (10 mW/cm^2, 2.25 MHz) daily during a 5-day premating period and for the 10 days of a subsequent mating period. The right ovary only of the mothers was insonated during the same period and for 10 minutes a day for the first 17 days of gestation. No congenital anomalies were noted on postnatal examination of either the offspring or their (F_2) offspring in subsequent reproduction studies. The effect of ultrasound on behaviour of adult rats was studied in a separate experiment with negative findings.

Clinical evaluations have also been performed on children whose mothers had received ultrasonic examinations during pregnancy. There was no evidence of an increased rate of abortion or abnormality, although as indicated by Scheidt and Lundin[23], the information was limited. On the basis of a feasibility study by Thompson and Olson[24], they have recently undertaken a more intensive follow-up study of children insonated *in utero*.

Ongoing laboratory studies

Most of the earlier studies described in the foregoing section are not valid in the classical toxicological sense. Usually some multiple or multiples of clinical operating-time was selected, exposures made, and the failure to produce a detectable effect taken as indicating the safety of the current clinical procedure. In general, attempts were not made to obtain relationships between intensity and effect or to define a progression of effect with change in experimental conditions. Such data are necessary for a rational assessment of the hazard associated with prenatal exposure to ultrasound.

The more recent studies in our laboratory and others are designed to identify deficits which are detectable in the prenatal or postnatal period and to obtain quantitative data which will allow derivation of exposure–response relationships and perhaps estimation of thresholds. These studies include a wide range of well-defined intensities, which extend into the range at which definite embryotoxicity is observed. In general, they involve a system in which the anaesthetized and shaved mouse is immobilized and its torso and limbs immersed in a water-filled tank for transcutaneous exposure.

The ongoing studies of which we are aware include those at the Bureau of Radiological Health, U.S. Food and Drug Administration, which use broad beams of continuous wave ultrasound. A flat transducer is mounted at one end of the tank and the animal holding assembly is mounted in the far-field, with an absorbing material behind it. The beam characteristics and exposure times are controlled and monitored by a minicomputer[25,26]. Although the results have not been reported in detail, to date these studies have shown a depression of fetal weight, neonatal mortality, and altered organ weights in the postnatal period[27–29].

A collaborative study between Indiana University, the Indianapolis Center for Advanced Research, and the University of Illinois includes the 8 d.g. mouse embryo among the biological systems being investigated. A moving beam of focused, pulsed ultrasound under computer control is used so as to cover a selected matrix over the abdomen. A gradation of exposure is obtained by changing beam area, peak intensity and burst time[30]. These studies have demonstrated prenatal death and anomaly production in relation to beam size, burst duration, and intensity[30]. A study at Stanford Research Insitute also employs a similar exposure system, which can operate in either pulsed or CW modes and provide focused or unfocused beams[31]. A preliminary report indicates that prenatal exposure results in increased prenatal mortality and depressed fetal weight[32].

The present report will concentrate on the ongoing studies in our laboratory with rats. Because of (or despite) the differences in technique and species, it is becoming evident that these experiments and those in other laboratories are providing complementary results.

Our experiments employ three well-defined conditions of frequency and pulse-repetition representative of clinical situations: 2–3 and 0.7–0.9 MHz CW, and 2–3 MHz pulsed ultrasound to simulate Doppler, therapeutic and echographic conditions, respectively. Nine d.g. has been used for one series of experiments to evaluate ordinary teratological measures, since it corresponds to a stage of maximum sensitivity to production of malformations in rats. Exposure at 15 d.g., which has been found to be sensitive to production of postnatal deficits by X-irradiation, is being used in experiments to evaluate postnatal growth and development. In both series, the transducer was coupled to the exteriorized uterus to allow accurate definition of the energy incident on the fetus.

In each case, pilot studies were performed to establish the intensity at which distinct embryotoxic effects could be noted. These were followed by experiments to quantitate the relationship between exposure intensity, changes in ultrasonic conditions, and the effects produced. The results of these studies, together with those from experiments to explore embryological and physical mechanisms will form a matrix which should allow partial evaluation of the interactions between physical variables (intensity, pulse regimen and frequency) and the related biological variables, including stage of gestation.

METHODS

Accurate, reproducible, and meaningful experiments to evaluate teratogenicity or embryotoxicity require proper selection, definition, and statement of experimental conditions. These include (1) knowledge and statement of the embryological stages at exposure, (2) a detailed characterization of the agent administered and its dose levels, and (3) the times and methods used to evaluate the selected end-points. It will be evident to the teratologist that some of the earlier studies reported in the literature (see above) fail to meet criteria (1) and (3) although the requirements for criterion (2) may be less obvious.

It has been long accepted that a statement of exposure conditions and a measurement of tissue dose were a requisite in studies with ionizing radiations and appropriate methodologies have been available for their determination.

Commonly accepted physical procedures for biological studies with microwaves are presently emerging. With ultrasound, as with other forms of energy which have been studied less intensively, e.g. electric and magnetic fields, there is no accepted unit of 'dose'. It therefore becomes imperative that exposure conditions be stated in adequate detail to allow replication in other laboratories. It should be noted the inherent variability between similar transducers and the general lack of calibration of commercial instruments lead to indefinite exposure conditions in the absence of calibration by the investigator.

The information required includes a statement of whether CW or pulsed ultrasound was used, its frequency, and in the latter case, burst duration and repetition rate. It is important to know the intensity and where it was measured, pulse shape and breadth, beam uniformity, and whether it was focused or flat. As will be considered in greater detail below, these factors and the method by which they are determined can have a marked impact on the interpretation of results.

Exposure system

A transducer assembly was developed for teratological evaluations of the effects of exposures of individual rat embryos. Initially, this used an air-backed PZT-5A ceramic bowl although more recently a flat transducer and an acoustic lens have been used. For experiments at 3.2 MHz, a transducer with a 10 cm radius of curvature and a diameter of 5 cm was used, giving an F/2 ultrasonic beam. A water-filled plastic cone was attached to the bowl and the end sealed with a thin plastic membrane (Figures 15.1 and 15.4). The diameter of the cone was 2 mm at the focal plane and 4 mm at the exit port, with the exit port situated 4 mm beyond the focal plane to provide a beam of ultrasound of 4 mm diameter incident on the uterus. In addition, the 2 mm neck at the focal plane blocked any side-lobes which might interfere with interpretation of the biological data. For experiments with CW ultrasound the transducer was driven by a 100 watt RF Power amplifier, with input from a Hewlett-Packard signal generator. The dimensions of the cone were changed, as appropriate, for use with ultrasound of other frequencies. For pulsed or burst mode exposure, the transducer was driven with a Velonex pulse generator or by gating the Hewlett-Packard generator and driving with the RF Power amplifier.

The applicator for the initial studies on the postnatal sequelae consisted of a flat airbacked PZT-5A ceramic element driven at 710 KHz CW by a 500 W ENI amplifier and the same signal generator. A plastic cylinder, capped with a thin membrane, was attached to the transducer to provide coupling (Figure 15.1). The length of the cylinder was chosen by measuring the beam profile at various distances from the transducer and accepting that value where it was most uniform (far-field). This turned out to be about 13 cm for the particular transducer used. The cylinder was filled with degassed water under gravity feed so that the pressure in the cylinder could be varied by raising or lowering a reservoir above or below the level of the transducer. This assured that intimate contact between membrane and uterus was maintained over the whole aperture by a slight positive pressure.

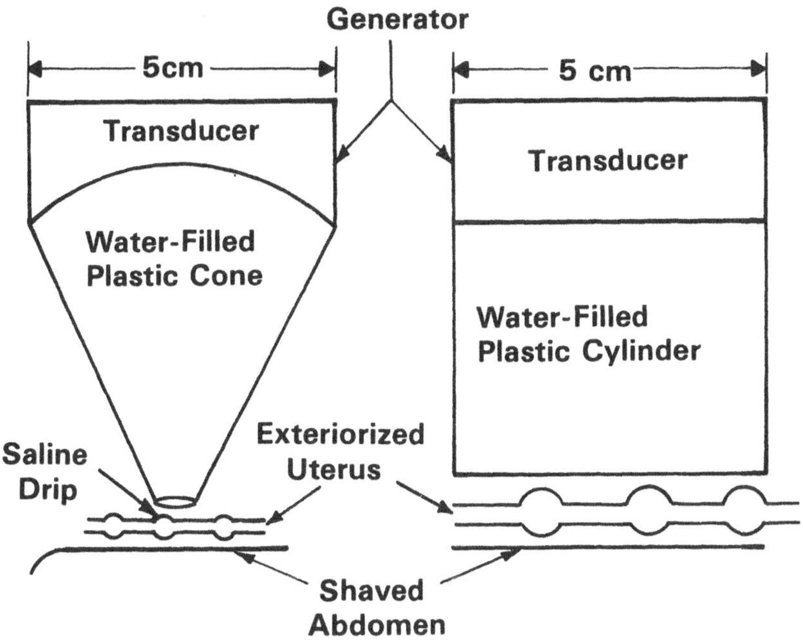

Figure 15.1 Diagrammatic representation of exposure system: (left) focused transducer used for exposure of individual embryos; (right) unfocused transducer used for exposure of the entire uterine horn

The same unit is being used for exposures at higher frequencies, e.g. 2.5 MHz. Rather than the cumbersome marked extension of the water column which would be required to allow use of the far-field (1 m), we measured the beam shape in planes closer to the transducer (near-field) until one with a flat profile was found.

Animal procedures

Young adult female rats (Hilltop Farms, Wistar derivation) were housed in groups of four in climatically controlled quarters, providing 12 h of artificial light per day. They were caged overnight with males of the same strain and the presence of sperm in the vaginal smear the following morning was taken as evidence of mating; 9:00 a.m. of that day was denoted as day 0 of gestation.

All exposures were performed between 7:00 a.m. and 12:00 noon on the designated day. In each case the pregnant rat was anaesthetized by intraperitoneal injection of pentobarbital sodium (Nembutal) and the uterine horns exteriorized through a midline incision.

The implantation site or uterine horn was placed so that it lay on the shaved maternal skin to provide a nonreflecting exit path. A saline drip was used to ensure acoustical coupling and provide cooling. The ultrasound exposure was administered at the designated intensity and time. Following insonation, the uterine horn or horns were repositioned in the abdominal cavity, the abdominal muscles sutured with gut, and the skin closed with stainless steel wound clips.

Studies of prenatal effects

The embryotoxic level was identified, as indicated above, to define the range for more detailed study. A number of exposure intensities ranging downward from the embryotoxic level were then selected, and were used for 5 or 15 min exposures to a minimum of 20 embryos[33,34]. To allow meaningful comparisons, two sham-exposed groups of embryos, which were handled identically with the exception that the transducer was not energized during the 5- or 15-minute period, were also included. Toward the end of the study, it appeared desirable to have data at intermediate levels and additional embryos were exposed at these levels.

Most of our experiments have been performed at 9 d.g. More limited experiments are in progress in which the embryos are insonated at 10 or 12 d.g. using a restricted number of intensities or at 15 d.g. using 10 W/cm^2. Two implantation sites from each exteriorized horn were exposed or sham-exposed according to a previously established random assignment scheme.

The selection of embryos was not random, however; criteria of apparent viability and accessibility were imposed and adjacent embryos were not included. The tip of the transducer was positioned against a selected implantation site and an adjustable stainless steel ring positioned to maintain the embryo in mid-field, care being taken to avoid trauma or compression of the fetoplacental unit (Figure 15.1).

The rats were exsanguinated under ether anaesthesia after 20 days of gestation. The exposed and sham-exposed fetuses were examined for viability and external malformations, sexed, and weighed, and their crown–rump length measured. The residual unexposed population, which is not a valid control because of the selection procedures used, was also examined. Each treated and sham-exposed fetus from the CW experiments was fixed for subsequent teratological examination according to its earlier random assignment. Approximately one-half were placed in Bouin's fixative and free-hand razor blade sections made and examined following the procedure described by Wilson[35]. The others were fixed in 95% ethanol, cleared in KOH and the skeletons stained with alizarin red for examination. With the pulsed exposures, detailed skeletal study was omitted because the results in the preliminary experiments indicated that only visceral malformations were produced.

Studies of postnatal effects

For studies of postnatal effects the rats were exposed at 15 d.g., a stage which has been found to be sensitive to other physical insults[36]. Procedures were designed to minimize exposure of the spinal cord of the dam to permit uncompromised rearing of the offspring[37]. To accomplish this, one uterine horn was excised through a midventral incision and the other horn was exposed. The back of the rat was placed in a water bath with absorbers in the exit path to avoid reflection. A flat PZT-5A transducer was coupled to the uterus via a water-filled plastic cylinder which was closed by a thin plastic membrane (Figure 15.1). A 5 min exposure time was used, covering an intensity range of 0.01 to 1 W/cm^2, together with sham-exposed controls. The individual rats, which were assigned randomly to their exposure group, were allowed to undergo parturition and rear their litters. The offspring were weighed and examined at birth and at several subsequent times. A number of measures of physical development, including eye and ear opening, testes descent, and vaginal opening, were also evaluated at appropriate intervals.

Neurological status and neuromuscular development in the neonatal period were evaluated using several measures[36]. These include the reflexive grasping of a fine wire laid against the palmar surface of the forepaw and the retraction of the paw when electrically stimulated. At intervals spontaneous behaviours were determined by observing the rats while they were in an open lucite box measuring 30 × 30 cm, ruled into nine squares. The occurrence of spontaneous movement, such as departure from the centre square, grooming, and attempts and success at escape from the box and the elicited startle response using a brief flash of light, were noted. Head lift was defined as positive if the rat held the head above the surface for five consecutive seconds. The upright response was scored in terms of ability to raise their abdomens off the bottom of the box for five seconds during the initial 60 second test period.

The normal neonatal rat will show a reasonably coordinated ability to turn over and right itself if placed on its back (surface righting); whether or not successful righting was accomplished in 1 minute and the time required was recorded. A related phenomenon, the righting reflex, was measured in terms of the ability to land feet-down after being dropped from a height of 12 inches in the dorsum-down position. A positive response was defined as two or three successful responses out of three trials.

Subsequent to removal from the test box, the animals were examined for the placing and hopping reactions, for evidence of ataxia or spasticity of the limbs, and for gait defects. Performance on the inclined plane (a measure of negative geotropism) was determined by placing the animals at the base of a wire screen, 10 inches in length, set at a 40° angle to the horizontal. The time in seconds to reach the top was recorded, allowing a maximum of 1 minute on the test. Overall strength was evaluated in terms of the ability of the rat to hold to a horizontal rod by its forefeet.

The animals were tested for their susceptibility to audiogenic seizures at 55 days of age by exposing them to approximately 120 db of mixed frequency sound. The fraction of the animals showing a clonic–tonic seizure during a 1 minute test period is recorded and used as a measure of the sensitivity of the CNS to this stimulus.

Calibrations and dosimetry

The concept of acoustic power was presented briefly above. This concept and its implications for mechanisms by which biological effects are produced may be better appreciated by considering the various methodologies employed for its direct measurement.

Absorbed ultrasonic energy is converted to thermal energy so that power may be measured by calorimetry. The ultrasound is beamed into a thermally-insulated vessel containing a highly absorbing liquid so that all energy is converted into heat. The rate of temperature rise is measured and power calculated from the equivalence of 1 watt and 0.239 calories per second.

Since ultrasound involves propagation of pressure, a beam will exert a force on an object on which it impinges. This force may be measured in terms of the apparent change in the weight of an absorber attached to a balance and power calculated by multiplying by a constant, i.e. $W = 14.8 \times$ force (gram-weight). This force may also be measured in terms of the deflection of a ball suspended by a fine filament or a vane radiometer.

Power measurements

Although conceptually simple, the experimental apparatus for accurate measurements may become complex and cumbersome. In the initial phases of our research, total power was determined from a radiation force measurement using an absorber made up of camel's hair embedded in RTV rubber attached to an analytical balance. We have compared calibrations obtained with this absorber against those using a stainless steel reflector and have found very close agreement. This procedure required that the calibration curve be run quickly so that the temperature of the absorber remained relatively constant in order to obtain reproducible results.

More recently, we have developed a simpler method which may be of value to workers beginning studies in this area. This uses a Scientec electronic weighing system operating in the pull mode and interfaced with an HP97 desk calculator (Figure 15.2). The transducer is directed towards a cone-shaped target with an included angle of 115°, which causes reflected energy to be absorbed in the side chamber of the tank. The reflector was fabricated by sandwiching a sheet of paper between two layers of thin brass sheet to provide a nearly perfect reflection. With this angle, 117 mg force represents 1 watt of acoustic power. The electronic balance has a fast settling time so that a calibration curve can be run quickly and easily. As voltage is applied to the transducer, the weighing system is activated and the ultrasonic force on the balance is recorded by the calculator. If desired, the calculator can be programmed to correct for the beam area to obtain intensity directly. It can also be programmed to take multiple readings and average them for increased accuracy.

We have found that the calibration curves from both systems are in good agreement with one another and that they fit the theoretical square law curve remarkably well. Thus, for a check, we can measure the intensity at a single drive voltage level and determine that the transducer is operating correctly.

Concepts of intensity

Published reports on the biological effects of ultrasound have not employed a common definition of intensity. As indicated above, intensity in watts/cm^2 can be derived from total power and the cross-sectional area of the beam or aperture. Average intensity (SA) over this aperture is found by $I_a = p/a$ where p = measured power and a = area of aperture. This value would also be the peak intensity (spatial, SP) if the beam were cylindrical. It should be noted that such cylindrical beams are not always uniform and that effect of non-uniformity is more pronounced in the near field.

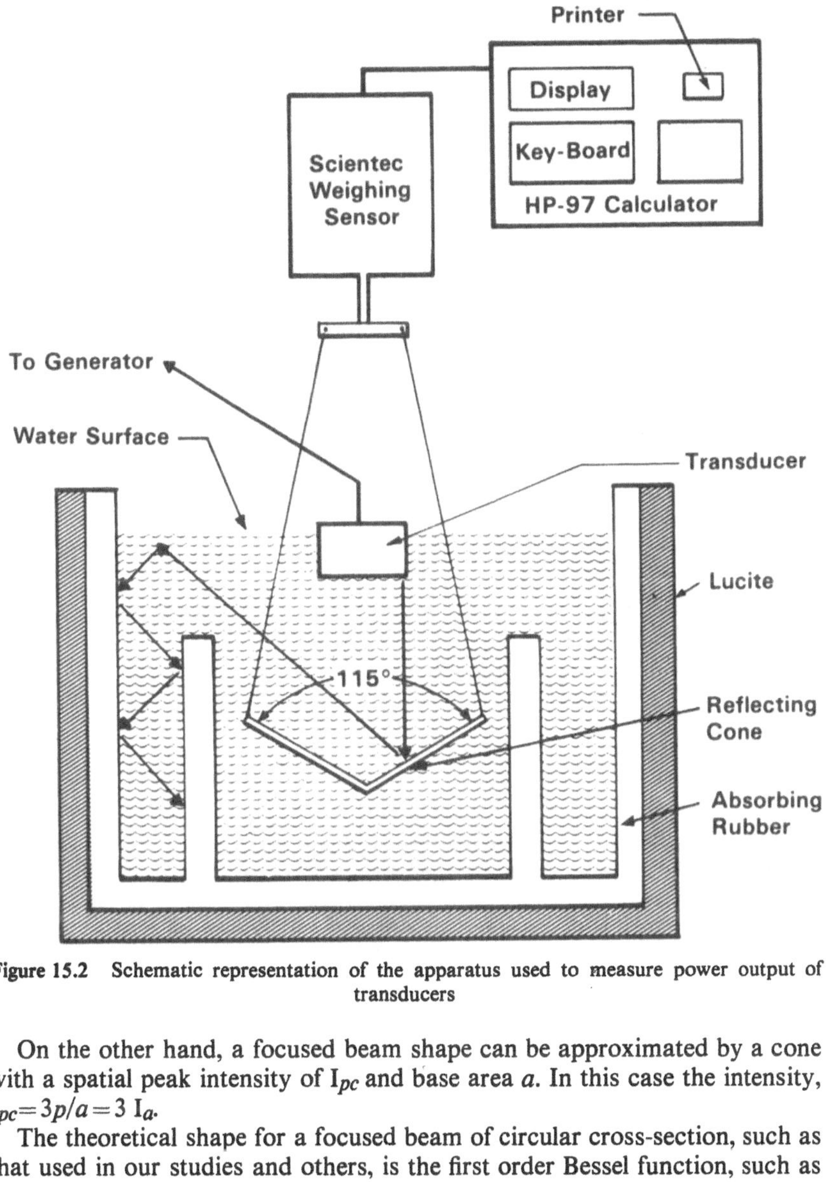

Figure 15.2 Schematic representation of the apparatus used to measure power output of transducers

On the other hand, a focused beam shape can be approximated by a cone with a spatial peak intensity of I_{pc} and base area a. In this case the intensity, $I_{pc}=3p/a=3\ I_a$.

The theoretical shape for a focused beam of circular cross-section, such as that used in our studies and others, is the first order Bessel function, such as that shown in Figure 15.3. When the beam is of the Bessel function shape, with area under the first zero equal to a, the peak intensity of $I_{pb}=4.382\ p/a=4.382\ I_a$.

Still other workers report results in terms of the average intensity in an area defined by the half-power points of the beam. In this case the average intensities for the conic and Bessel function beam shapes are $I_{ac}=0.66\ I_{pc}$ and $I_{ab}=0.73\ I_{pb}$, respectively.

It is obvious from these formulae that the means of expressing intensity

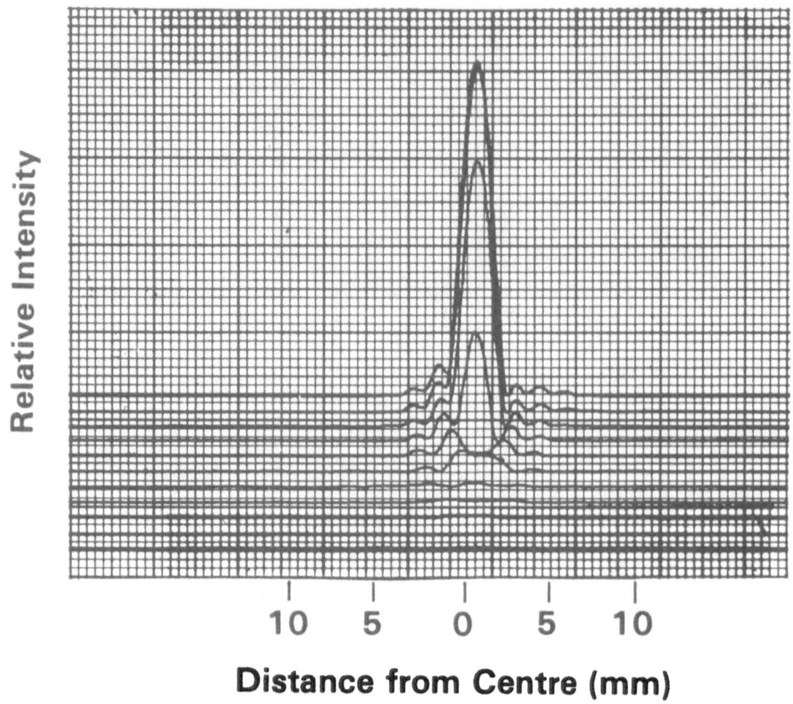

Figure 15.3 Isometric scan of the beam pattern of the focused transducer assembly used for exposure at 3.2 MHz

will affect the interpretation of the experimental results. Use of the cylindrical beam approximation will yield the lowest apparent intensity, I_a, and hence a low damage threshold. Reporting spatial peak intensity, I_{pb}, yields the highest value of intensity and hence a high damage threshold. A good compromise is probably average intensity within the half-power region of the beam, I_{ab}.

If the burst mode is used, further complication arises due to the intermittent operation of the transducer. In this case the power varies with time, necessitating further definition of power levels. The radiation force measurement is by nature a time average. Thus, if no mathematical compensation is made, a measure based on radiation force is both a spatial and a temporal average (SATA). This is not a good measure, since the power during the burst can be very high but will be averaged over the, usually long, off time. Thus, another measure, spatial average temporal peak (SATP) is sometimes used. This value represents the spatial average during the burst, and may be obtained by dividing the SATA by the duty factor. Two other intensity values sometimes used are spatial peak, temporal average (SPTA) and spatial peak temporal peak (SPTP).

By way of illustration, suppose we have a focused beam of Bessel function shape, a burst length of 10 μs and a pulse repetition frequency of 100 Hz. Then the duty factor, D, is 0.001. Assuming that $SATA = I_a = 3$ mW/cm^2 we have the following values:

$SATA = I_a = 3\ mW/cm^2$
$SATP = I_a/D = 3000\ mW/cm^2 = 3\ W/cm^2$
$SPTA = I_{pb} = 4.382\ Ia = 13.146\ mW/cm^2$
$SPTP = I_{pb}/D = 13.146\ mW/cm^2/D = 13.146\ mW/cm^2 = 13.46\ W/cm^2$

Beam profiles and intensities

The beam profile used in our experiments was obtained by scanning through the plane at the tip of the plastic cone with a 0.25 mm diameter hydrophone. Two types of scans were performed: rapid qualitative visualizations of the beam pattern for purposes of adjustment of the cone, and more detailed quantitative measurements for calibration. In the first instance, the scanner is a two-dimensional mechanical device that can make isometric scans such as those shown in Figure 15.3. For accurate calibration we use a large lath-bed manual scanner. The signal processing is much more precise so that true signal amplitude and shape can be plotted. From these plots, it appears that the beam pattern used in our studies is reasonably close to the Bessel function shape. The width of the 3.2 MHz beam was found to be 4.1 mm at the first minimum.

The changes in refractive index associated with the pressure variations produced as a beam of ultrasound passes through a liquid medium makes it

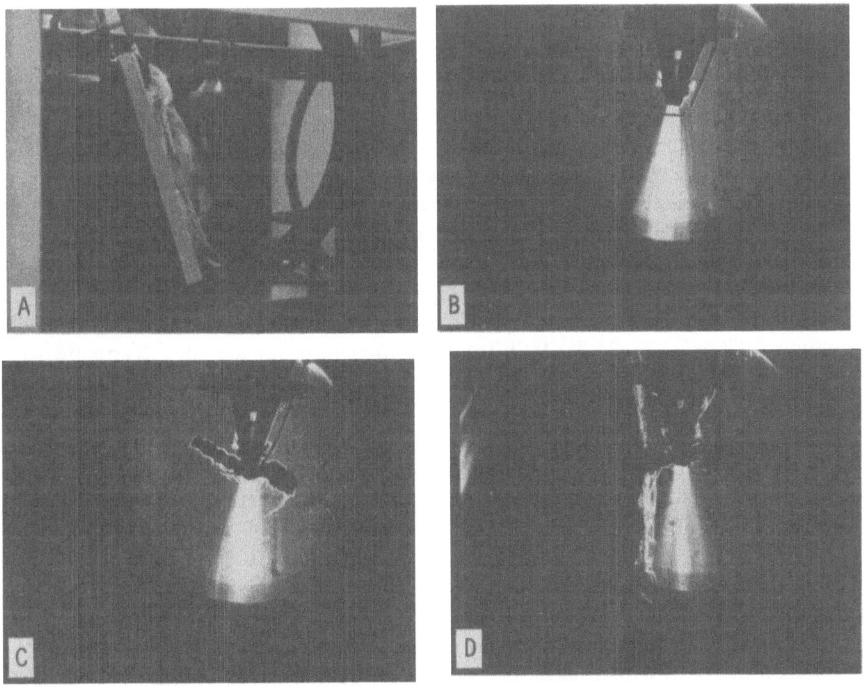

Figure 15.4 Visualization of focused ultrasound interactions with the uterus. A, Schlieren apparatus, with transducer assembly in contact with exteriorized rat uterus; B, Undisturbed beam; C, Beam interactions in passage through excised uterus; D, Beam interactions in passage through exteriorized uterus

amenable to visualization by schlieren or spark photography. A pulsed laser unit[38] designed for use with pressure variations occurring at MHz frequencies was available in our Engineering Physics Laboratory. This unit has been used to observe the interactions of the beam as it passes through the uterus (Figure 15.4). It is notable that both excised uterus and the exteriorized uterus *in vivo* produce relatively little distortion or attenuation of the beam and are not associated with gross reflection.

Most workers, however, have reported average intensity, or merely stated an undefined intensity value, which may be assumed to be the average. Accordingly, we also have expressed our results with CW in terms of the average intensity, I_a or SATA. The SATP value has been used for pulsed ultrasound. The relationships indicated above will allow re-expression of our data as desired.

RESULTS AND ONGOING STUDIES

The results of some of our studies have been published previously[33,34,36]. The illustrative material from these reports will not be repeated here except for comparison with the results of ongoing studies.

Studies of prenatal effects

Exposure at 9 d.g.

3.2 MHz CW

The results with the 3.2 MHz continuous wave exposure have been published[33,34]. In summary, there was greater mortality with the 15 min than with the 5 min exposures at higher intensities (Figure 15.5). The LD_{50} values, with 95% fiducial intervals, were determined by the methods described by Finney[39], using an iterative computer programme for calculation. The resulting LD_{50} values, 18.4 and 16.3 W/cm^2 for 5 and 15 min exposures, respectively, with 95% fiducial limits of 13.7–26.8 and 10.5–25.5, did not differ significantly. A combined analysis was then performed, providing an overall value of 17.6 (13.9–24.3). It initially seemed appropriate to extrapolate the exposure–response curves to intersect the abscissa to estimate an 'apparent threshold' for use in safety analyses. As will be discussed below, we are now somewhat uncertain of the proper statistical procedures for such extrapolation and the actual significance of the extrapolated value. Nevertheless, it is clear that essentially any extrapolation method used will result in a positive intercept on the X-axis.

The unexposed fetuses do not represent a true control population because of the necessity of utilizing selection criteria for the exposed and sham-exposed fetuses. Nevertheless, the incidence of mortality, 7.9%, and the presence of an occasional dwarfed fetus in this population, were similar to that routinely seen in this laboratory. The mean lengths and weights of the surviving exposed fetuses were not discernibly different from those in the sham-exposed or unexposed groups. Major malformations were not noted among the sham or unexposed fetuses or among those insonated at intensities below 10.5 W/cm^2. At this level and above, there were five fetuses with major

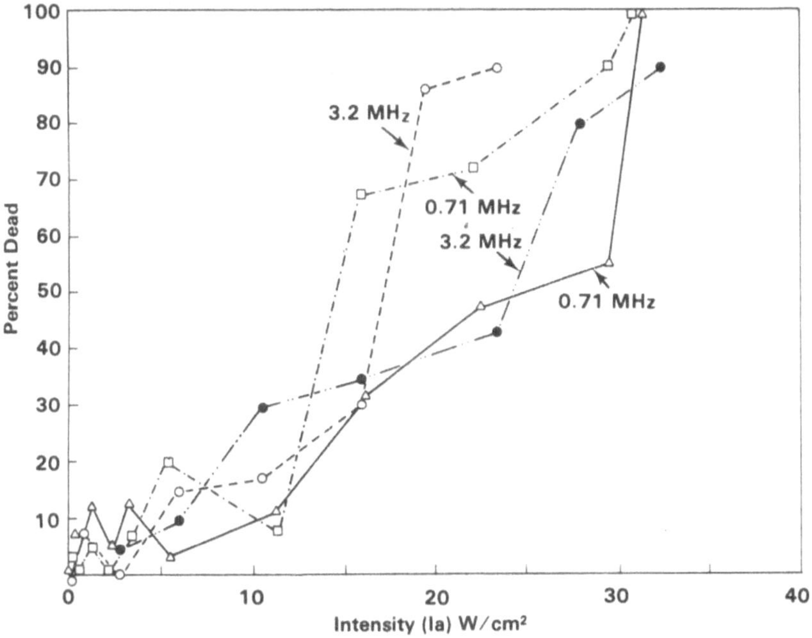

Figure 15.5 Prenatal mortality after exposure of 9 d.g. rat embryos to ultrasound. △—△ 0.71 MHz, 5 min; □–·–□ 0.71 MHz, 15 min; ●–··–● 3.2 MHz, 5 min; ○–·–○ 3.2 MHz, 15 min

malformations among a total of 89 survivors (5.7%). There were insufficient survivors in the teratogenic range to establish definite relationships between exposure intensity and incidence. With the exception of one fetus with unilateral degeneration of the eye, the abnormal fetuses had multiple malformations including cleft lip and palate, exencephaly, anophthalmia, multiple anomalies of the face, unilateral absence of the kidney, and septal defect of the heart.

0.7 MHz CW

The relationship between intensity and prenatal mortality observed after exposure to 0.71 MHz CW ultrasound was similar to that at 3.2 MHz, as has been reported previously[34]. There was again a greater effect with 15 min than with 5 min exposures. The variability of these earlier data was too great to be able to demonstrate a statistically significant difference between the two exposure times with probit analysis although analysis of variance of the linear regressions indicates a difference in slope. The response curves extrapolated to zero or to a negative intercept on the abscissa. Additional embryos have since been exposed to better define the shape of the response curve and obtain additional data at low intensities. The overall mortality values (Figure 15.5) appear to be similar to those obtained with exposure at 3.2 MHz.

Although the teratological examination of the fetuses from the more recent experiments are not yet completed, it is clear from the initial experiments that there was a greater incidence of major malformations and that they were observed at lower intensities at 0.71 MHz CW than at 3.2 MHz although there was no apparent difference in type. There was a suggestion of an increased incidence at an intensity of 2.9 W/cm^2 and a statistically significant increase at 6 W/cm^2 and above.

2.5 MHz pulsed

The studies with 2.5 MHz pulsed ultrasound concentrated on exposure to 10 μs pulses with a 1000 Hz repetition rate. The results were compared with 2 and a 16 μs pulses at selected intensities; no differences were found and the results were pooled. As we have reported[34] the highest average power (SATA), approximately 6 W/cm^2, that produced no increase in mortality was at the level where the same frequency of continuous wave ultrasound slightly elevated the incidence of prenatal death.

The major effect seen with pulsed ultrasound was the presence of cardiac defects[34]. Many of these abnormalities took either of two atypical forms: thinned ventricular walls with enlargement of the lumen or thickened ventricular walls with reduced lumen size. Histological examination of these hearts demonstrated that the fibres were disarrayed in a number of areas, and there were focal areas of necrosis. The incidence of these defects showed a better correlation with peak intensity (SATP) than with average intensity (SATA). Compact hearts were not observed in the sham-exposed animals, but there was a trend toward increased incidence of this defect with increasing peak intensity up to a plateau at about 100 W/cm^2 (SATP) and above. The unexposed animals had about a 4% incidence of dilated hearts; the incidence in the sham-exposed was about 20%, suggesting that the manipulation *per se* may have predisposed the fetuses to the production of this lesion. Nevertheless, there was a further increase in the incidence of dilated hearts with increasing intensity. The incidence of fetuses with other cardiac abnormalities such as septal defects, was also increased at higher intensities although this did not show an intensity dependence.

Studies were undertaken to study the pathogenesis of these cardiac lesions, using a serial sacrifice approach. These experiments, which employed a more refined pulse-generation system, failed to replicate the earlier results. The only difference which we have been able to identify is a cleaner pulse shape (see below) and we are presently investigating this factor. Accordingly, the results described above may thus be considered as nonreplicated.

Exposure at 10–15 d.g.

It is well known from extensive studies with X-rays that the periods of peak sensitivity for mortality and for the production of specific malformative events change with the stage of gestation. Although our experiments are still in progress, it is clear that the prenatal mortality curves for 5 min exposures to CW ultrasound in the 0.7 MHz range at 10 or 12 d.g. do not differ markedly from the curve obtained with exposure at 9 d.g. The only intensity studied at 15 d.g., 10 W/cm^2, has resulted in a lower incidence of prenatal mortality.

The incidence of malformed fetuses has a tendency to be lower with exposure later in gestation. On the basis of the limited series presently available, it is clear that there is a shift in the types of malformation detected (Figure 15.6). Each vertical column in the figure relates to a specific fetus and a filled-

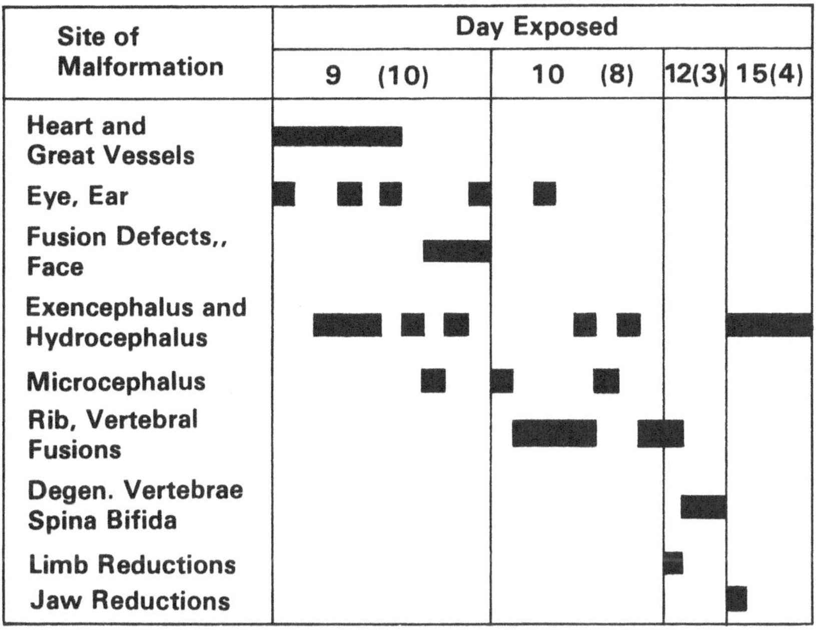

Figure 15.6 Types of malformations observed after exposure of the rat to 0.71 MHz CW ultrasound at midrange intensities

in block on two or more lines indicates multiple anomalies. It is seen that malformations of the heart and great vessels and fusion defects of the face were limited to exposure at 9 d.g. and eye defects occurred primarily at this stage. Brain defects were produced at all stages of gestation but tended to involve the forebrain more at the earlier times. Defects such as fusions of the ribs or vertebrae, spina bifida, and limb reductions were produced only by exposure at 10 or 12 d.g.

The types of malformation produced by ultrasound thus appear to follow the same pattern of change with gestational age as do those produced by X-irradiation. This does not infer anything relative to mechanism but does suggest that the critical periods are generally independent of mechanism.

Studies of postnatal effects

Earlier there appeared to be a slight increase in the incidence of prenatal mortality, which includes stillbirths, at exposures of 0.04 W/cm^2 and above[36]. More recent data suggest that this increase is only significant at intensities approaching 1 W/cm^2. Much of the mortality is attributable to high mortality in a few litters. There was no evidence of postnatal mortality or reduced growth rate. This contrasts with the postnatal mortality in the offspring of mice in which the entire abdominal area was exposed to similar ultrasound beams, without surgical intervention[28]. The differing results may be attributable to species differences. Alternatively, the difference may reflect the fact that a smaller area of the abdomen was exposed and the spine was avoided in the present studies. The latter would suggest that some of the effects obtained with the whole abdomen exposure may be derived from effects on the dam.

Deficits in several of the measures of neuromuscular development were evident by the day following birth. About 50% of control rats show a normal (positive) grasp response at 1 day of age and essentially all responded by 6 days. The fraction responding at 1 and 6 days of age was decreased in all insonated groups although the magnitude of this decrement was independent of exposure intensity. This was only a delay in development, since by 13 days of age all animals showed a positive response.

The exposed animals showed a slight increase in the time required to accomplish surface righting, which was not statistically significant. The righting reflex has not developed completely in the controls by 6 days of age although most showed this reflex by 13 days of age. There was a general and consistent decrease in the response of all exposed groups at 6 days of age. The rats of the lower intensity groups responded normally at 13 days although there was still a decrement in the 0.54 and 1.0 W/cm^2 groups. All groups showed a normal response at subsequent evaluations, again suggesting retardation of development. At 9 days, 50% of the controls showed a sustained head lift whereas less than 25% of the animals in each experimental group showed this response. Likewise, about 70% of the controls showed the upright response at 13 days. A smaller percentage of all of the animals of the experimental groups showed a similar response at this time. At subsequent times of observation, the fractions of the experimental and control animals showing either of these responses were similar. No delays were observed in the development of other measures of posture or movement.

When tested at 13 days of age, over 70% of the controls could hold to the horizontal bar for the 20-second test period. In the experimental groups, a much smaller fraction was able to hold on for this period. This test could not be repeated at subsequent times since the animals rapidly learned that they could safely drop to the padded surface below and would not maintain themselves at this distance. None of the other measures of development showed an appreciable decrement in the exposed animals.

MECHANISMS AND INTERACTIONS

A number of mechanisms have been proposed for the biological changes which have been found resulting from exposure to ultrasound. Although some of these have been explored in greater detail more recently, the general concepts presented in the excellent review by Nyborg[40] remain unchanged. He was able to differentiate between effects for which the mechanism was clear and those for which the mechanism had (and still has) not been identified. In general, these mechanisms may be classified as thermal, cavitational, and 'other', including incompletely defined phenomena such as microstreaming and shear stresses. It also seems that some interactions which take place in *in vitro* systems may not pertain in a living mammal, except under unusual conditions. As is often the case, large grey areas remain in terms of uncertainties as to the commonality of *in vivo* and *in vitro* mechanisms and the applicability and significance of *in utero* effects to the intact animal. Understanding and quantification of these mechanisms will provide information of fundamental biophysical importance. Moreover, only on the basis of knowledge of the conditions under which these interactions pertain can we establish whether or not there are threshold levels for specific effects, and in turn, it is on the basis of the existence of such thresholds that the ultimate determination of safety and/or risk-benefit assessments must be made. Accordingly, it may be profitable to consider these mechanisms and interactions, and insofar as possible, indicate their implication in embryotoxicity.

Threshold

It is not yet clear whether or not there is a threshold below which it is impossible that a given biological effect can be produced by ultrasound. This results from theoretical statistical considerations as well as uncertainties about the precise shape of the curves at low intensities and the mechanisms involved. If one is willing to accept the potential for errors, it is sometimes possible to obtain a crude estimation of the 'apparent threshold' by extrapolation of the curves relating percent response to intensity. Assuming a sigmoid relationship and using the probit calculation, we earlier[33] extrapolated the lethality results obtained with 3.2 MHz focused ultrasound at 9 d.g. to intersect the 0% mortality incidence observed in the sham-exposed embryos. Although this is not a statistically valid procedure, the intercept (3 W/cm^2, fiducial limits −0.5 and 6.0 W/cm^2) serves as a basis against which to compare the intensities associated with clinical techniques. Alternatively, although there is no biological justification for doing so, one may fit a straight line to these data by linear regression procedures which more correctly (in the mathematical sense) can be extrapolated to zero incidence – at about 1.5 W/cm^2, depending on the techniques used for fitting. Uncertainties remain since the shape of the curve is not precisely defined at low intensities. However, if a number of such approaches all extrapolate to similar large positive intercepts of the abscissa, one may use the 'apparent threshold' with some degree of confidence.

On the other hand, when the X-intercept approximates zero, as tentatively appears to be the case with similar exposures at 0.71 MHz, interpretation becomes difficult. It is here that quantitative information about mechanisms becomes most important. If it can be demonstrated that embryotoxicity (or other biological effects) is produced through specific mechanisms and if it can be further demonstrated on physical bases that there is a threshold for these mechanisms, one can more realiably accept a biological threshold.

Thermal effects

When ultrasonic energy is absorbed in tissue (or other media) it is converted to thermal energy and gives rise to temperature increases in the insonated volume. It is clear that many of the reported biological effects of ultrasound are attributable to temperature increases in the affected volume or in the whole organism. This effect is especially pertinent to CW exposure since the average energy is greater than with pulsed exposure. For a given intensity, local temperature rise becomes greater as the cross-sectional area of the beam increases since total energy absorption is greater and the opportunity for heat dissipation is less. Prolonged exposures usually produce a greater effect than do short exposures; this is due to similar physical considerations as well as biological relationships involving the time course of evolution of thermal damage.

The thermal effects of ultrasound are thus clearly related to those involved in the substantial literature on the biological effects of hyperthermia. Lele[41] has recently presented a number of analogies between ultrasound-induced teratological effects and those produced by systemic or local hyperthermia. In addition, he presented data on intrauterine temperatures in mice exposed to transcutaneous abdominal CW ultrasound for about 35 minutes. It is clear from these data that some of the deleterious effects which have been found in experiments with mice under similar exposure conditions are attributable to thermal mechanisms.

We have been measuring temperature rises under the conditions of our experiments, using implanted thermocouples and a digital thermometer. Because of the Nembutal anaesthesia, the exteriorization of the uterus, and the saline drip at room temperature, the baseline uterine temperatures were substantially below normal body temperature (Figure 15.7) although abdominal temperatures were less affected. The figure shows typical temperature curves for a normal exteriorized uterus and for a uterus in which the uterine and ovarian vessels had been ligated during exposure to focused 5.5 W/cm^2 ultra-

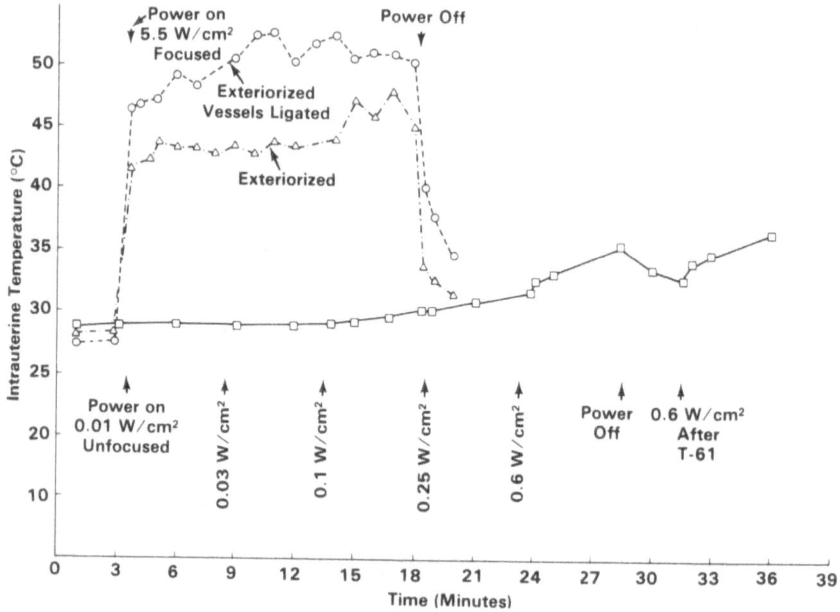

Figure 15.7 Uterine temperature curves during exposure to focused ultrasound at 5.5 W/cm^2, with and without vascular ligation (upper curves) or to unfocused ultrasound at increasing intensities (lower curve)

sound, such as used for exposures at 9 d.g. There was a marked difference between the two, attributable to heat dissipation by the circulation. It should be emphasized that these are typical curves; there are marked animal to animal variations in the actual temperatures attained and in the effect of ligation. Nevertheless, it is clear that exposure at this and higher intensities might be expected to produce thermal damage, and that the likelihood of this would increase with longer exposure times.

A typical temperature curve for the 'worst-case' situation with the broad unfocused transducer used with exposures at 15 d.g. for postnatal evaluations is also shown in Figure 15.7. The power was increased progressively at 5 min intervals to cover the range of experimental intensities. Under these conditions of depressed basal temperature, the uterine and fetal temperatures barely reached normal levels during exposure, indicating that the observed effects were probably nonthermal in nature. After a 2 min period during which the unit was turned off to allow cooling, the rat was injected with a rapidly-acting euthanasia agent (T-21) and the transducer was again energized. The temperature rise paralleled that in the living animals, suggesting that cooling via the circulatory system was minimal under these conditions. On the other hand, thermocouples placed in the abdominal cavity and along the spine often reached temperatures in excess of 45 °C and occasionally above 50 °C.

Cavitation

The propagation of ultrasound involves densification and rarefaction of the medium through which it passes. Dissolved gas in the medium may form into microscopic bubbles which will oscillate at the ultrasonic frequency or continue to increase in size and ultimately collapse. Although there are theoretical distinctions between the two, which have been denoted stable and transient cavitation, respectively – the practical distinctions are not as clear.

In general, most studies of cavitation have been undertaken with *in vitro* systems at low frequencies. The oscillating bubbles associated with stable cavitation have been shown to produce vibration of the cell surface resulting in microstreaming and movement of intracellular particles. The threshold intensities are dependent on frequency, viscosity, and the pressure in the system. Higher intensities are required to produce transient cavitation under any given set of physical conditions. The biological consequences of the collapse of the bubble, with the attendant intense hydrodynamic shearing forces and production of free radicals, are usually greater than those associated with stable cavitation.

There are reasons to consider cavitation to be an unlikely event in organized tissues, except at very high intensities. This area is under active investigation but it is not yet clear whether this is an important mechanism of effect for *in vivo* mammalian systems. Lele[42] has recently reported the presence of half-harmonic sonic emission, indicative of stable cavitation, in liver and brain isonated with 2.7 MHz ultrasound in 100 ms bursts, at intensities above 100 mW/cm^2. This did not appear to be associated with biological changes. Transient cavitation, detected by wide-band emission, was only observed above peak intensities of 1500 W/cm^2. Although biological damage was observed, it was not possible to differentiate them from the thermal damage produced at this intensity.

Mechanical

The propagation of ultrasound involves marked mechanical energy changes in the medium, such as submicroscopic displacements associated with large values of velocity and acceleration. Until recently, research on these phenomena was limited to *in vitro* systems in which the shear stresses resulting from acoustic streaming were found to produce changes in membrane permeability, disruption of intracellular organelles, and cellular lysis.

More recently, however, a probe vibrating at low ultrasonic frequencies was shown to produce local thrombogenic and vascular effects when applied directly to an exposed vessel in the mouse[43]. Under specific conditions in which standing-waves were produced, stasis in chorionic vessels of the chick embryo has been demonstrated during exposure[44].

Thermal and cavitational effects appear to be particularly unlikely with exposure to pulsed ultrasound. The high instantaneous and low average intensities with pulsed or burst exposure suggests that any observed biological effects may be attributed to mechanical phenomena. It may be further speculated that pulses of irregular shape would be more disruptive than those of regular shape. We have accepted this as a working hypothesis for further experiments relative to the cardiac changes produced by exposure to pulsed ultrasound at 9 d.g.

EXTRAPOLATIONS TO MAN

Extrapolations of results from animal experiments to man are always associated with a number of recognized uncertainties relative to commonality of mechanisms and to scaling factors. These uncertainties also pertain in the case of prenatal exposure to ultrasound although the mechanisms of action would be expected to be constant across mammalian species and metabolic differences minimal. On the other hand, one must consider whether or not the entire conceptus is exposed and the temporal relationship of exposure relative to the specific clinical application; these relationships are not yet clear.

Calculation of the scaling factors is usually straightforward. The ultrasound intensity at the level of the embryo is ordinarily measured in recent experiments, or may be readily calculated. The attenuation of the beam by absorption may be calculated from the relationship $(I/I_o)^{1/2} = e^{-\alpha x}$ where the absorption coefficient α is taken as 0.63 and 0.0018 for muscle and water, respectively, at 3 MHz[45] and x is path length in cm. For illustration of calculation for the clinical situation, one may assume: a 10 cm distance (5 cm muscle, 5 cm water) from the skin surface to the embryo, a full urinary bladder, and the absence of intestinal gas. An attenuation factor of 1.8×10^{-3} results for a plane wave or 0.13 for a 2.5 cm transducer focused at 10 cm. The output of clinical CW Doppler devices has been measured[45,46] as being about 20 mW/cm^2 (SATA) (17,18). If one uses the above values and the intensity of 20 mW/cm^2 incident on the pregnant abdomen, directed toward the embryo, values of 0.04 or 0.3 mW/cm^2 at the embryo, for plane and focused beams, respectively, would result.

Such calculated or measured values may be used as a basis against which to compare the intensities which are found to produce, or to be without effect in laboratory studies. Ongoing clinical evaluations should provide the denominator for calculation of the risk/benefit ratio; it is hoped that ongoing laboratory studies will supply the numerator.

ACKNOWLEDGEMENTS

This research was supported by National Institutes of Health Research Grant Number 1 R01 GM 20661 from the National Institute of General Medical Sciences.

We are grateful to Mrs Jean D. Stearns, Mr E. V. Allen, and Mr L. L. Kopf for technical assistance, to Dr Pamela G. Doctor and Ray L. Buschbom for statistical advice, and to Dr Marvin C. Ziskin for helpful discussions.

References

1. White, D. N. (ed.). (1977). *Recent Advances in Ultrasound in Biomedicine*, Vol. 1. (*Ultrasound in Biomedicine Series*, Vol. 3.) 256 p. (Forest Grove, OR: Research Studies Press)
2. Reid, J. M. and Sikov, M. R. (eds.). (1972). *Proc. Workshop on the Interaction of Ultrasound and Biological Tissues*. DHEW Publ. (FDA) 73-8008. (Washington, DC: US GPO)
3. Brown, R. E. (1972). Uses of ultrasound in diagnosis. In: J. M. Reid and M. R. Sikov (eds.). *Interaction of Ultrasound and Biological Tissues*, pp. 221–223. DHEW Publ. (FDA) 73-8008. (Washington, DC: US GPO)
4. Thompson, H. E. (1972). The future role of ultrasound in obstetrical care. In: J. M. Reid and M. R. Sikov (eds.). *Interaction of Ultrasound and Biological Tissues*, pp. 225–226. DHEW Publ. (FDA) 73-8008. (Washington, DC: US GPO)
5. Garrett, W. J. and Robinson, D. E. (1970). *Ultrasound in Clinical Obstetrics*. 116 p. (Springfield, Il: Charles C. Thomas)
6. Lehmann, J. F. and Guy, A. W. (1973). Ultrasound therapy. In: J. M. Reid and M. R. Sikov (eds.). *Interaction of Ultrasound and Biological Tissues*, pp. 141–152. DHEW Publ. (FDA) 73-8008. (Washington, DC: US GPO)
7. Remark, D. G. (1971). Survey of Diathermy Equipment Use in Pinellas County, Florida. Bur. Rad. Health Rept. BRH/NERHL 71-1.
8. Fritz-Niggli, H. and Böni, A. (1950). Biological experiments on *Drosophila melanogaster* with supersonic vibrations. *Science*, 112, 120
9. Counce, S. F. and Selman, G. G. (1955). The effects of ultrasonic treatment on embryonic development of *Drosophila melanogaster*. *J. Embryol. Exp. Morphol.*, 3, 121
10. Bell, E. (1959). A new approach to some problems in experimental embryology through the use of ultrasound. In: H. Quastler and W. J. Morowitz (eds.). *Proc. 1st National Biophysics Conference*, pp. 674–682. (New Haven, CT: Yale U. Press)
11. Selman, G. G. and Juran, A. (1964). An electric microscope study of the endoplasmic reticulum in newt notochord cells after disturbance with ultrasonic treatment and subsequent regeneration. *J. Cell Biol.*, 20, 175
12. Pourhadi, R., Bonhomme, C. and Turchini, J. P. (1968). Action teratogene des ultrasons sur des oeufs de batrachiens. *Arch. Anat. Microsc. Morphol. Exp.*, 57, 255
13. Takeuchi, H., Nakazawa, T., Kumakiri, K. and Kusand, R. (1970). Experimental studies on ultrasonic Doppler method in obstetrics. *Acta Obstet. Gynaecol. Jap.*, 17, 11 (Quoted by McClain *et al.*, see Reference 14)
14. McClain, R. M., Hoar, R. M. and Saltzman, M. B. (1972). Teratologic study of rats exposed to ultrasound. *Am. J. Obstet. Gynecol.*, 114, 39

15. Mannor, S. M., Serr, D. M., Tamari, I., Meshorer, A. and Frei, E. H. (1972). The safety of ultrasound in fetal monitoring. *Am. J. Obstet. Gynecol.*, **113**, 653
16. Shoji, R., Momma, E., Shimizu, T. and Matsuda, S. (1971). An experimental study on the effect of low-intensity ultrasound on developing mouse embryos. *J. Fac. Sci. Hokkaido Univ., Ser. VI, Zool.*, **18**, 51
17. Murai, N., Hoshi, K. and Nakamura, T. (1975). Effects of diagnostic ultrasound irradiated during fetal stage on development of orienting behavior and reflex ontogeny in rats. *Tohoku J. Exp. Med.*, **116**, 17
18. Murai, N., Hoshi, K., Kang, C. and Suzuki, M. (1975). Effects of diagnostic ultrasound irradiated during fetal stage on emotional and cognitive behavior in rats. *Tohoku J. Exp. Med.*, **117**, 225
19. Warwick, R., Pond, J. B., Woodward, B. and Connolly, C. C. (1970). Hazards of diagnostic ultrasonography – A study with mice. *IEEE Trans. Sonics Ultrasonics*, **Su-17**, 158
20. Woodward, B., Pond, J. B. and Warwick, R. (1970). How safe is diagnostic sonar? *Br. J. Radiol.*, **43**, 719
21. Akamatsu, N. (1977). Ultrasound irradiation effects on pre-implanted mouse and rat embryos. In: D. White and R. E. Brown (eds.). *Ultrasound in Medicine*, **Vol. 3b**, p. 1999. (New York: Plenum Press)
22. Symth, M. G., Jr. (1966). Animal toxicity studies with ultrasound at diagnostic power levels. In: C. C. Grossman *et al.* (eds.), *Diagnostic Ultrasound*, pp. 296–299. (New York: Plenum Press)
23. Scheidt, P. C. and Lundin, F. E. (1977). Investigations for effects of intrauterine ultrasound in humans. In: *Symposium on Biological Effects and Characterizations of Ultrasound Sources*, DHEW Publication (FDA) 78-8048. (Washington, DC: US GPO)
24. Thompson, H. E. and Olson, S. L. (1977). A feasibility study of possible adverse effects of ultrasound on the fetus. In: *Symposium on Biological Effects and Characterizations of Ultrasound Sources*, p. 31. DHEW Publication (FDA) 78-8048. (Washington, DC: US GPO)
25. O'Brien, W. D., Christman, C. L. and Yarrow, S. (1974). Ultrasonic biological effect exposure system. In: *Ultrasonic Symposium Proceedings*, p. 57. (IEEE Cat. No. 74 CHO 89601SU). (New York: IEEE)
26. Christman, C. L. (1977). An improved ultrasonic biological effect exposure system. In: *Symposium on Biological Effects and Characterizations of Ultrasound Sources*, p. 125. DHEW Publication (FDA) 78-8048. (Washington, DC: US GPO)
27. O'Brien, W. D., Jr. (1976). Ultrasonically induced fetal weight reduction in mice. In: D. White and R. Barnes (eds.). *Ultrasound in Medicine*, **Vol. 2**, pp. 531–532. (New York, NY: Plenum Press)
28. Curto, K. A. (1976). Early postpartum mortality following ultrasound radiation. In: D. White and R. Barnes (eds.). *Ultrasound in Medicine*, **Vol. 2**, pp. 535–536. (New York, NY: Plenum Press)
29. Stratmeyer, M. E., Simmons, L. R., Pinkavitch, F. Z., Jessup, G. L. and O'Brien, W. D., Jr. (1977). Growth and development of mice exposed in utero to ultrasound. In: D. G. Hazzard and M. L. Litz (eds.). *Symposium on Biological Effects and Characterization of Ultrasound Sources*, pp. 140–145. DHEW Publ. (FDA) 78-8048. (Washington, DC: US GPO)
30. Fry, F. J., Erdmann, W. A., Johnson, L. K. and Baird, A. I. (1978). Ultrasonic toxicity study. *Ultrasound Med. Biol.*, **3**, 351
31. Holzemer, J. F., Taenzer, J. C., Havlice, J. F., Ramsey, S. D. and Green, P. S. (1977). A facility for the investigation of the bioeffects of ultrasound. In: D. White and R. E. Brown (eds.). *Ultrasound in Medicine*, **Vol. 3B**, p. 1995. (New York: Plenum Press)
32. Stolzenberg, S. J., Torbit, C. A., Edmonds, P. D., Taenzer, J. C., Nell, D. P., Madan, S. M., Marks, D. O. and Pratt, D. E. (1978). Effects of continuous wave ultrasound on the mouse at different stages of gestation. *J. Acoust. Soc. Am.*, **63**, Suppl. 1, S27
33. Sikov, M. R. and Hildebrand, B. P. (1975). Effects of ultrasound on the prenatal development of the rat. I. 3.2 MHz continuous wave at nine days of gestation. *J. Clin. Ultrasound*, **4**, 357
34. Sikov, M. R. and Hildebrand, B. P. (1977). Embryotoxicity of ultrasound exposure at nine

days of gestation in the rat. In: D. White and R. E. Brown (eds.), *Ultrasound in Medicine*, **Vol. 3B**, p. 2009. (New York: Plenum Press)

35. Wilson, J. G. (1965). Embryological considerations in teratology. In: J. G. Wilson and J. Warkany (eds.), pp. 251–261. *Teratology*. (Chicago, IL: University of Chicago Press)
36. Sikov, M. R., Hildebrand, B. P. and Stearns, J. D. (1977). Postnatal sequelae of ultrasound exposure at fifteen days of gestation in the rat. In: D. White and R. E. Brown (eds.), *Ultrasound in Medicine*, Vol. 3B, p. 2017. (New York: Plenum Press)
37. Ritter, E. J., Scott, W. J. and Wilson, J. G. (1971). Teratogenesis and inhibition of DNA synthesis induced in rat embryos by cytosine arabinoside. *Teratology*, **4**, 7
38. Newman, Dennis R. (1973). Observations of cylindrical waves reflected from a plane interface. *J. Acoust. Soc. Am.*, 53, 1174
39. Finney, D. J. (1964). *Statistical Method in Biological Assay*, Chapter 18. (New York, NY: Hafner Press)
40. Nyborg, W. L. (1972). Summary: Effects of ultrasound on cells. In: J. M. Reid and M. R. Sikov (eds.). *Interaction of Ultrasound and Biological Tissues*, pp. 47–55. DHEW Publ. (FDA) 73-8008. (Washington, DC: US GPO)
41. Lele, P. P. (1975). Ultrasonic teratology in mouse and man. In: *Second European Congress on Ultrasonics in Medicine*) Excerpta Medica International Congress Series no. 363) p. 22. (Amsterdam: Excerpta Medica)
42. Lele, P. P. (1977). Thresholds and mechanisms of ultrasonic damage to 'organized' animal tissues. In: D. G. Hazzard and M. L. Litz (eds.). *Symposium on Biological Effects and Characterization of Ultrasound Sources*, pp. 224–239. DHEW Publ. (FDA) 78-8048. (Washington, DC: US GPO)
43. Williams, A. R. (1977). Intravascular mural thrombi produced by acoustic microstreaming. *Ultrasound Med. Biol.*, 3, 191
44. Dyson, M., Pond, J. and Woodward, B. (1972). The induction of red cell stasis in embryos by ultrasound. In: J. M. Reid and M. R. Sikov (eds.). *Interaction of Ultrasound and Biological Tissues*, pp. 139–140. DHEW Publ. (FDA) 73-8008. (Washington, DC: US GPO)
45. Hill, C. R. (1969). Acoustic intensity measurements on ultrasonic diagnostics devices. In: J. Bock and K. Ossoinig (eds.). *Proc. 1st World Congress on Ultrasonic Diagnostics in Medicine and SIDUO III*, Vol. 2, pp. 21–27. (Vienna Med. Acad.)
46. Rooney, J. A. (1973). Determination of acoustic power outputs in the microwatt-milliwatt range. *Ultrasound Med. Biol.*, 1, 13

16
Clinical and experimental aspects of prenatal virus infections

A. B. G. LANSDOWN

INTRODUCTION

In his address to the University of Chicago in 1942, the late Dr Ernest Goodpasture emphasized that cells of the avian embryo or mammalian fetus were significantly more susceptible to microbiological infections than those of the adult[1]. He showed that vaccinia and herpes simplex viruses lead to widely disseminated infections in the embryo or fetus but caused only mild lesions in the adult. Although Goodpasture was unable to explain this greater susceptibility of embryonic tissue to some infections, he suggested that the young and relatively undifferentiated cells formed more propitious media for viral replication, probably on account of their higher metabolic rate and absence of effective immunological systems.

Prenatal infection of the human fetus was recognized by 1940 and abortion attributable to severe maternal infections had been described[2]. However, it was not until Gregg and other workers in Australia showed a correlation between rubella in early pregnancy and the birth of infants exhibiting a consistent range of gross congenital defects, that viruses became recognized as causes for fetal deformity[3,4]. In subsequent years a wide range of viruses has been identified as causes of abnormal prenatal growth in man and other species of mammals.

Estimates vary as to the magnitude of the fetal risk resulting from congenital virus infections, and there is some controversy as to how many viruses are detrimental in human fetal development. A number of general reviews have appeared on the subject[5–11] and the role of viruses in causing specific patterns of fetal growth impairment involving the central nervous system[12–14], the cardiovascular system[15–17], the respiratory tract[18] or intra-uterine growth retardation[19] has been discussed. In many cases, reviewers have tended to concentrate on the epidemiology of congenital abnormalities with a presumed viral aetiology allowing less attention to be given to possible mechanisms of abnormal growth as provided by experimental studies using suitable animal models and *in vitro* techniques.

In the present review which refers mainly to human viruses, epidemiological evidence for the implication of particular viruses as causes of fetal risk is discussed in relation to relevant experimental studies conducted in animals to substantiate this risk and to elucidate possible pathogenic mechanisms. Evidence is also reviewed which supports the view that fetal systems are more prone to viral replication, alteration in cell function and damage[1,20].

VIRAL INFECTIONS AS TERATOGENIC AGENTS IN MAN

The viral infections regarded as potential causes for impaired prenatal growth in man include rubella, cytomegalovirus and herpesvirus simplex which have been associated with structural deformities[21,22]. Other infections such as influenza, Coxsackievirus A and B, mumps, measles (rubeola), and smallpox are known to lead to fetal death, stillbirth and reduced postnatal viability and may cause structural abnormalities but the available information is equivocal or insufficient. It is widely held however, that any infection which evokes a profound reduction in the state of health of a pregnant mother or injures susceptible fetal tissues is likely to be teratogenic.

For the purposes of the present review, the term 'teratogenic virus' is used in its broadest sense to define those agents which cause abnormalities in the growth or survival of a fetus following maternal infection during the period from conception until parturition. Congenital abnormalities include growth defects which arise during this period and which are manifest either as fetal loss or as structural or metabolic anomalies present at birth and which lead to lowered postnatal viability and growth potential.

Dudgeon[5] listed four criteria for identifying viral teratogens, namely:—

1. Clinical – consistency of clinical pattern
2. Epidemiological – similar pattern of fetal changes with same agent irrespective of geographical area
3. Persistence of infectious agent
4. Persistence of antibody after decline of maternal antibody – presence of antibody in the IgM fraction

These criteria were derived largely from his extensive experience from the study of rubella, but they are equally relevant in respect of other infections.

Although some infections may be teratogenic in susceptible pregnancies, reliable evidence for mild or clinically inapparent infections is hard to find[23]. Several viruses, such as rubella, mumps, measles and chicken pox are essentially illnesses of childhood and in adults are infrequently accompanied by clinical illness. Whitehead[24] in his appraisal of viruses in heart disease referred to the difficulty in demonstrating a clear relationship between cardiomyopathies and some inapparent or 'silent' infections.

Isolation and characterization of a virus from the tissues of a deformed or abnormally small baby provides unambiguous evidence for viral teratogenesis, but may not necessarily indicate when that infection was acquired. Severe illness in mothers as occurred in the widespread influenza epidemics of 1918 and 1957 followed by increased numbers of abortions is also good evidence for viral action[2,25]. Infections such as rubella, Coxsackievirus B, chicken pox and infectious hepatitis have been studied in large scale epidemics and are well documented, but in the case of sporadic infections such as mumps, the relationship between infection and congenital deformity has proved more difficult to establish. It relies on the recognition of a maternal infection in pregnancy and identification of prenatal growth defects in the infant.

In mumps and Coxsackievirus B infections which are often subclinical in affected patients, infection has been diagnosed by serum antibody titrations. A fourfold or greater rise in serum neutralizing antibodies is evidence of a recent infection[26], but in infections which are particularly ubiquitous such that most people possess antibodies to them, information taken from this source alone can be misleading[23].

Difficulties arise in establishing a viral cause for teratogenesis in situations where an infant born after an apparently uncomplicated pregnancy appears normal at birth but develops metabolic disorders or becomes mentally retarded later. This problem is particularly relevant in the case of cytomegalovirus infections.

With certain viruses which exhibit extreme antigenic variability, such as influenza, teratogenicity may be evident in some epidemics but not others. For example, central nervous system deformities were significantly increased in the course of the 1957 influenza epidemic in Europe[27,28] but not in later outbreaks[29,30]. This variation may be attributed to differences in the virus serotype, genetic differences in populations exposed or to drugs taken to combat the disease. In this latter situation, the aetiology of the congenital defects is likely to be complex, since it may not be clear as to whether they are due to the action of virus, drugs or a combined effect. The problem was illustrated in the study of complications arising out of a recent influenza epidemic in Finland[31]. This survey showed that 52% of infected pregnant women had taken drugs in the first trimester. A low rate of teratogenicity revealed in such a study may reflect a mitigation of disease through drug therapy.

RUBELLA

Clinical and epidemiological aspects

Rubella is an acute infectious disease which is now a well recognized cause of congenital deformity in man following maternal infection in the first trimester of pregnancy. N. MacAlister Gregg in 1941 was the first to observe a causal relationship between rubella and congenital deformities[3] and since then numerous epidemiological surveys have been reported and many views given as to the possible mechanisms by which these defects arise.

Rubella is essentially a childhood illness manifest by fever, headache, anorexia and respiratory distress[32]. Primary infection does occur in adults but this is frequently symptomless or accompanied by a mild erythematous rash. A primary infection in early pregnancy usually leads to placental infection and in approximately 30% of cases, fetal infection results[21]. This infection may persist until birth or even into postnatal life[33]. In the United Kingdom, rubella infection complicates 0.25–0.3 per 1000 live births and the complications seen range from stillbirth, abortion and congenital deformity to no obvious effect at all[22]. Occasionally, children appear normal at birth following maternal rubella during pregnancy but develop deafness and mental retardation later in life[34–36]. This delayed onset of neurological disorders may occur in the absence of virus in the tissues[37].

The principal complications associated with maternal rubella at susceptible stages of pregnancy include congenital heart disease, cataract, deafness and aberrant skeletal development[3,4,34]. Panencephalitis leading to postnatal neurological deformities and autism is a further complication[37]. Various estimates have been given as to the magnitude of the fetal risk in congenital rubella but according to Higgins[38], rubella accounts for no more than 1–2% of all congenital heart diseases. The risk of fetal deformity increases from 10–15% following a primary maternal infection in the first month of pregnancy to 14–25% in the second month and then 7–17% in the third month[39]. But because rubella is so frequently inapparent in pregnancy, these estimates may be low and the actual risk as high as 50%[5,21]. In a study of 139 children with perceptive or sensineural damage rubella antibodies were found in 61%[40].

Children born following maternal rubella are frequently of low birthweight and exhibit a characteristic pattern of abnormalities, the so-called 'rubella syndrome'. Important features of this are cataract, skeletal abnormalities, microphthalmia, mental retardation and the cardiac defects of patent ductus arteriosus, hypertrophy of ventricular muscle, septal defects and focal myocardial haemorrhages[3,4,21,41]. Other rubella-induced changes not recognized by earlier workers including a focal to diffuse myocarditis, necrosis and scarring of cardiac valves, haematological disorders, endocarditis and hepatosplenomegaly were identified later and became included in what is now known as the 'expanded rubella syndrome'[42–47]. Although low birthweight is a frequent complication in congenital rubella, it is more frequent when other growth defects are present, e.g. cataract or cardiac deformities[48]. Children with congenital rubella deafness may be of normal birthweight however[35,40].

Follow-up studies of low birthweight children in rubella show that catch-up growth does not occur postnatally, suggesting that virus persists in the cells and continues to inhibit mitotic activity[49]. Children with the expanded rubella syndrome are often acutely sick and less viable than normal; virus persists in infected tissues and is excreted in the urine even though circulating antibodies may be present[50].

Pathological studies in children with congenital rubella demonstrate a clear viral predilection for tissues of the eye lens, inner ear, enamel organs and vascular endothelium[51,52]. However, the actual mechanism by which malformations develop is not clearly understood. Several hypotheses have been postulated including, local tissue damage accompanied by inflammatory changes, mitotic inhibition through elaboration of a growth inhibitor, and chromosomal damage[53–55]. Each of these mechanisms would result from a direct viral action in susceptible tissues[56].

There is abundant evidence to show that rubella causes inflammatory and degenerative changes in target tissues. For example, in defective long-bone development, preosseous cartilage was hypertrophied and metaphyseal tissues disorganized[57]. Osteoclasts and osteoblasts were less numerous than usual with infected cells containing cytoplasmic inclusion bodies. Similar observations were made in a study of the influence of rubella on fetal long-bone development *in vitro*[58]. Increased ^{35}S-uptake in fetal bone explants indicated an increase in mucopolysaccharide synthesis which previously has been shown to lead to abnormal patterns of ossification[69,70].

Rubella exhibits a clear predilection for fetal vascular tissues, particularly the renal arteries, and this leads to a diffuse intimal proliferation and stenosis[52]. Virus-induced selective cell death may underlie a failure in the closure of the intraventricular septum or the obliteration of the ductus arteriosus link between the aorta and pulmonary artery[15]. This might also result from an inhibition in the synthesis of elastic tissue in the vascular laminae. Direct viral damage is a probable cause of defects in the eye (e.g. aphakia – absence of lens) and deafness through cochleosaccular changes[61,62] since virus has been isolated from these tissues at autopsy[63,64].

To examine virus–cell interactions further, explanted human fetal eye rudiments have been exposed to rubella in culture[65]. Eye rudiments cultured at the open vesicle stage (from fetuses of 4–5 weeks) showed a consistent pattern of histological changes of vacuolar degeneration in lens fibres, similar to that seen in congenitally affected fetuses. Lenticular changes were not present in eye rudiments cultured with virus from the closed vesicle stage (6 week fetuses), except where the vesicle was artificially ruptured. Thus at this stage the lens capsule forms a protective barrier around developing lens fibres.

Despite its ability to damage certain types of cell, rubella is non-cytolytic in others. That is it does not destroy them but may alter their genotype or capacity to differentiate and undergo mitosis[63]. Monkey cells cultured in the presence of infected material from a baby having congenital rubella exhibited a reduced mitotic rate and survival[66]. In the *in vivo* situation, it was noted that children with congenital rubella had relatively large hearts for their body weights but even so the organs were small for gestational age[54]. The hearts contained a lower number of cells than normal. Inhibited cell division may be attributable to the production of a 'growth inhibitor' substance in infected cells[67].

Rubella may evoke genetical changes in susceptible cells[22] but there appears to be some scepticism as to the relevance of this in the production of congenital abnormalities. Increased numbers of chromosomal breaks have been reported in cultured leukocytes and it may be that damage of this type leads to mitotic inhibition through an inability to replicate DNA[55].

Rubella-induced placental damage

In addition to causing specific patterns of fetal damage, rubella is known to evoke degenerative changes in the placenta[68]. Changes such as nuclear shrinkage, cell fragmentation and emboli formation in chorionic vessels with frank endothelial necrosis would, as have been described, lead to placental insufficiency and result in fetal death or retarded growth. A similar effect would result from villous degeneration or lack of functional villi. Desquamation of virus-containing necrotic material from chorionic vessels or from fragmented emboli is considered by some to be an important route for fetal infection[69]. Cell debris of probable chorionic origin has been identified in the fetal circulation.

Rubella vaccination in pregnancy

Vaccination is adopted as a preventive measure against congenital rubella in many countries but it is normally only given to women of childbearing age where normal contraceptive measures are taken[70]. Although the risk of congenital damage following inadvertent vaccination in pregnancy is low, fetal infection may occur[71–73]. In one prospective study of 19 women vaccinated in early pregnancy, nine offspring aborted whilst 10 proceeded to term and normal children were born[74]. One aborted fetus exhibited viral antigen and also had histological evidence of cataract and lenticular damage. A larger survey of 145 patients scheduled to have therapeutic abortions, showed that virus was present in only nine apparently normal conceptuses[73].

Animal models for the study of congenital rubella infections

In human pregnancy congenital heart disease, optic and hearing defects are associated with congenital rubella but these changes have not been regularly produced in experimental studies in animals. Fetal responses to rubella in animals have been variable and occasionally equivocal. Infection has led to antibody formation in some strains of monkeys, rats and ferrets but this has not been consistent.

The rhesus monkey is sensitive to rubella and shows a comparable antibody response and excretion pattern to that seen in man, but although fetal infection occurred development was normal and consistent with gestational age[75,76]. In another strain of monkeys injected at 3–4 weeks' gestation, transplacental infection occurred but nine of these fetuses aborted and showed lenticular degeneration[77]. Virus-like particles were identified in the lens capsule by electron-microscopy. Infected animals exhibited petechial haemorrhages on the face, thorax and trunk, and hepatomegaly. Abnormalities in the ear, skeleton and chorion of the type seen in human fetal tissues were also recognized. Abortion without gross abnormality was seen in three baboon fetuses exposed to rubella at 23–98 days gestation[78].

Injection of rats with either of two strains of rubella did not result in fetal infection[79], but an infection on the fifth or sixth day led to intra-uterine growth retardation[80] and malformations of the lens, heart and long bones.

Pregnant hamsters were not susceptible to rubella infection and fetal deformities were not detected[81].

ENTEROVIRUSES

Problems associated with enterovirus infections in human pregnancy

Enteroviruses are RNA-containing agents which include members of the ECHO-, polio- and Coxsackievirus groups which are believed to be teratogenic in man. They vary greatly in their ability to impair fetal development.

Available evidence indicates that ECHO-viruses are least injurious and may be without effect either in sporadic or epidemic scale infections[82–85]. Poliovirus infections in pregnancy can lead to abortion or premature delivery with a higher rate of neonatal mortality[86–91]. Infants surviving an *in utero* poliovirus infection are often lethargic at birth and exhibit a marked flaccidity in their muscles[9]. Small size at birth and a subsequent retardation in mental development may be an additional complication of poliovirus infections[39,92]. Coxsackieviruses are ubiquitous organisms and tend to occur in epidemic proportions. Like the polioviruses, they present risks of abortion and possibly congenital deformity. Maternal Coxsackievirus infections late in pregnancy may rarely lead to congenital disease which is not evident until later in life[90,93–95]. Since Coxsackievirus infections are frequently inapparent in pregnancy, their role in causing fetal distress is hard to substantiate[23,97]. Prospective studies on Coxsackievirus B infections in pregnancy in the United States, indicate that illnesses are inapparent in at least 50% of patients but that there is a statistically significant correlation between B3 and B4 infections and congenital heart disease[97]. In 50 stillborn infants, five exhibited myocarditis with Coxsackievirus B present in the tissues. Similar observations were made earlier in three infants dying neonatally with encephalohepatomyocarditis where virus was isolated from the liver and spinal cord[99,100]. Coxsackievirus B infection in early pregnancy may also be a cause of endocardial fibroelastosis[101], but this lesion has been more especially linked with congenital mumps infection and is discussed further in that section.

More recent epidemiological studies on the complications associated with congenital Coxsackievirus infections suggest a relationship between types B3 and B4 and cardiac defects similar to those identified with rubella, namely patent ductus arteriosus, septal and valvular defects[26,84]. The estimated risk is 1.3% for Coxsackievirus B3, but this figure could be higher allowing for undiagnosed maternal infections. Other malformations associated with Coxsackievirus infections include urogenital anomalies (B2 and B4) and deformities in the digestive tract[84]. Coxsackievirus B5 and B6 appear to be without detrimental effects in the human fetus even when the pregnant mother becomes acutely ill[102]. Severe hepatitis in a mother infected with B5 was associated with the birth of an apparently normal baby who was seen to be mentally retarded at the age of 10 months[103].

Experimental studies in animals of Coxsackievirus infection in pregnancy

Coxsackieviruses were originally isolated and characterized in suckling mice and this species has been used extensively in experimental studies on their pathogenicity and teratogenicity[104,105]. Clear differences have been demonstrated in the capacities of viruses in groups A and B to impair fetal development. Coxsackieviruses A8, A9, A13 and A18 injected into pregnant mice failed to influence fetal development adversely and virus was not isolated from fetal tissues[106–108], but with types A6 and A7 fetal infection did occur and fetal wastage was reported[109,110]. Offspring from A7 infected dams became sick 2–18 days after birth and developed a fatal paralysis. Virus was recovered from necrotic muscle fibres which showed a marked proliferation in the sarcolemmal region.

Fetal infection results from some Coxsackievirus B infections in pregnant mice but these viruses differ in their propensity to evoke fetal growth impairment[110–115]. Mice infected at mid-pregnancy with B3 or B4 showed a marked increase in fetal wastage with growth retardation of those fetuses surviving until near-term[115]. A less severe fetal effect was noted when dams were injected with B1 or B5 viruses and no significant effect occurred with B2 or B6 infections. In small fetuses, the state of development of skin and thymus was less advanced than in normal fetuses of the same age, and it was related to the body weight rather than gestational age[116,117]. Plasma protein profiles were also consistent with fetal body weight and were 'immature' for gestational age[118]. Although Coxsackieviruses B are known to have a clear predilection for cardiac tissues in neonatal mice (and human babies), there is no evidence that B3 at least, causes myocarditis in prenatal infections even though virus has been identified in fetal myocardia by specific immunofluorescence[119]. Abnormal heart development does occur (Figure 16.1) but this is probably related to the generally retarded state of fetal development rather than to any direct viral effect.

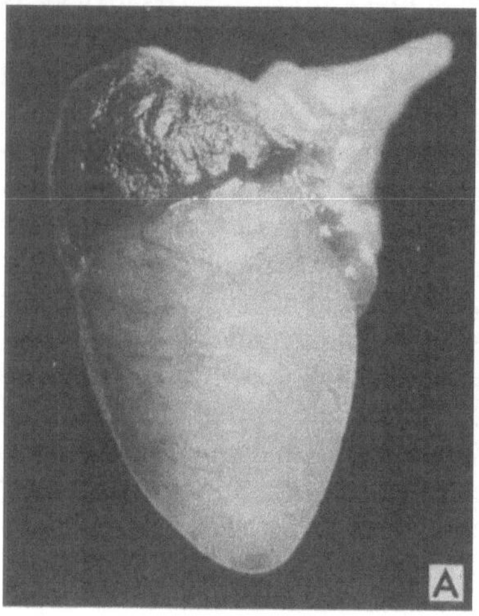

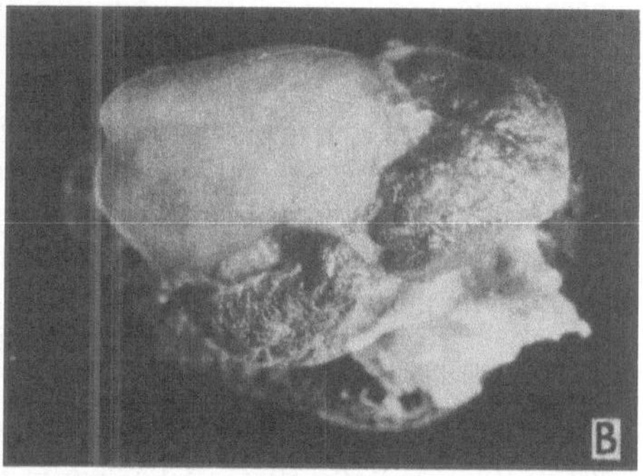

Figure 16.1 A, Heart from a normal one-day old mouse (× 10); B, Heart from a one-day old mouse born to a mother infected on day 12 of pregnancy with Coxsackievirus B3 showing hypertrophy of the right ventricle and a bifid apical condition (× 10)

Although fetal infection occurs in mice infected with Coxsackieviruses B, wastage and retarded fetal growth is largely due to an indirect effect, that is, through virus induced maternal illness[120]. A severe maternal pancreatitis was regularly seen within 2–3 days after infection and this lead to a pancreatic exocrine insufficiency and an inability of the pregnant mothers to digest adequate dietary protein for full maintainance of pregnancy. This virus-induced metabolic defect resulted in the mobilization and depletion of maternal reserves of labile protein and hence a fetal protein deprivation[121,122]. As a consequence fetal growth was retarded and in some instances fetal death occurred. In subsequent studies it was shown that only those B-viruses that caused a frank maternal pancreatitis in mice resulted in fetal growth impairment. The level of that growth impairment was closely related to the severity of the pancreatic damage[115]. When pregnant mice infected with Coxsackievirus B3 were fed a diet containing hydrolysed protein to overcome the effects of the pancreatitis, fetal development was not significantly different from normal and fetal plasma protein profiles were characteristic for the stage in pregnancy (Figure 16.2)[122, 123].

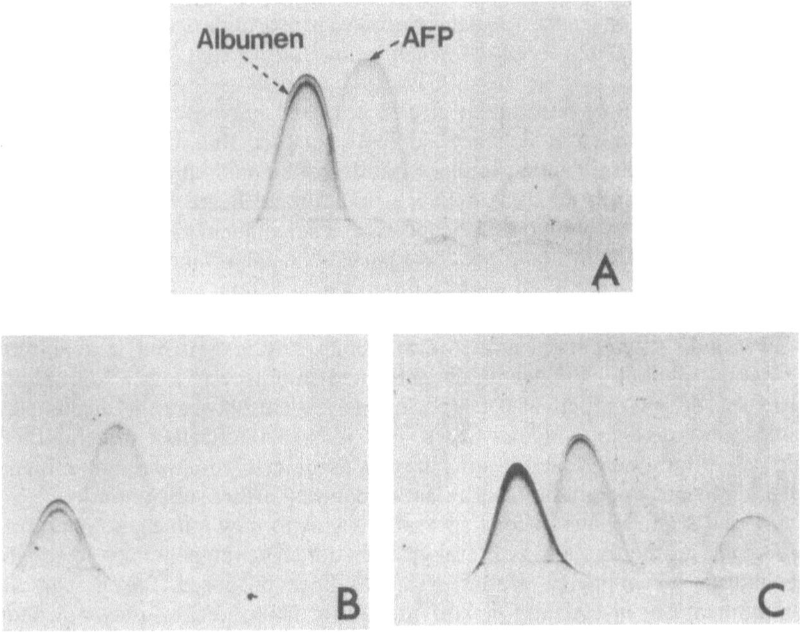

Figure 16.2 Immunoelectrophoretic patterns of fetal plasma proteins in mice showing α-fetoprotein and albumin peaks from: A, normal 18-day mouse fetus; B, 18-day mouse fetus from mother infected with Coxsackievirus B3 on the eighth day of pregnancy; C, 18-day mouse fetus from mother infected with Coxsackievirus B3 on the eighth day of pregnancy and given dietary casein hydrolysate to overcome pancreatic insufficiency

ORTHO- AND PARAMYXOVIRUSES

These are heterogeneous groups of medium sized RNA viruses which include influenza and parainfluenza, mumps and measles which have been implicated variously in the causation of anomalous fetal growth in man and other mammals. Much has been learned regarding the teratogenicity of influenza in the course of the several severe epidemics in this country, but less is available relating to mumps or measles infections. Measles is rare in pregnant women probably as a consequence of the increasing use of vaccination in childhood. Mumps although more common, occurs sporadically and is frequently inapparent in pregnancy. Also, the delayed hypersensitivity test which is used to diagnose mumps is not a reliable guide and may be misinterpreted[124].

Influenza

The clinical problem in human pregnancy

Unlike most other viruses, the antigenicity of influenza is highly variable such that the complications arising in the course of epidemics have been inconsistent. This inconsistency is possibly attributable to differences in the modes of collection and analysis of data. However, in respect of the ability of influenza viruses to cause fetal deformities, prospective studies so far have been small and the number of deformed infants insufficient to establish a clear relationship[125].

Influenza is a common infection and is a well known cause of abortion, stillbirth and premature delivery[2]. Infections occurring in the second trimester are more likely to cause placental damage and stillbirth, whereas later infections result in premature delivery or perinatal death[126]. Congenital deformities in human babies ascribed to maternal influenza infections involve the central nervous system, circulatory system, palate and lymphoreticular tissues[27,28], but the implications of influenza as a human teratogen are controversial.

Two main studies have shown teratogenic relationships but in each case, the actual number of congenitally deformed infants was low[27,28]. In subsequent surveys no statistical correlation was evident between the incidence of central nervous system deformities and maternal influenza infections[29,30]. The risk of lymphoreticular lesions may be associated with maternal influenza infection in early pregnancy, but this too requires further substantiation[125].

Clinical studies have yielded several clues as to how influenza infection in a pregnant mother may impair fetal growth, but as yet no generally acceptable mechanism for influenza-teratogenesis has been proposed. Saxén (personal communication) in Helsinki, is convinced that fetal infection does not occur usually and that any impaired fetal growth that results is primarily attributable to maternal illness, e.g. fever, disinclination to feed and metabolic upset. This aspect has received considerable attention in experimental studies and is discussed below. Surprisingly however, in a recent review article, it was clearly stated that '... influenza virus *can* penetrate the placenta and enter a developing human foetus – that much we know'[127]. To achieve this, virus would have to escape from the respiratory tract to reach the urinogenital tract and fetal tissues through the maternal circulation[129]. Although this may occur, viraemia has rarely been reported[129]. Viruria is also infrequent[130], but might conceivably set up an ascending infection resulting in an infection of the fetus by way of its membranes[131,132]. Alternatively, transplacental infection may arise through placental damage caused either by infection or inappropriate choice of drugs. Although the human uterine endometrium supports influenza replication[128], there is seemingly no evidence to show that infection in the preimplantation embryo is an important route of prenatal infection in man.

If fetal infection does occur in man and as a consequence, deformities do arise, then one might expect that susceptible tissues such as the central nervous system or lymphoreticular tissues which are supposed to be damaged, would support virus in culture. This has *not* been observed[128] and thus one is led to the conclusion that, if influenza is teratogenic, it must function by an indirect mechanism as suggested by Saxén.

Animal studies of influenza teratogenesis

Numerous experimental studies have been reported on the pathogenesis of influenza in the embryo or fetus and a very complex picture has emerged. One of the earliest studies was that of Hamberger and Habel[133] who injected virus of the PR8 strain into hens' eggs at 48 h incubation and demonstrated deformities of the central nervous system and spinal column. These effects were not present in embryos infected at 90 h incubation. Virus was isolated from these tissues and from the alimentary tract, myocardium and lens placode[134].

Abortion is a commonly observed effect in pregnant mammals infected with influenza. Two of eight rhesus monkey fetuses injected directly through the uterine wall at 105–111 days pregnancy aborted. The remainder were apparently normal at birth but were found to be hydrocephalic at autopsy[135]. The pathogenesis of this condition is not known but may result from an inflammation of the ependyma or choroid with an obstruction of the third and fourth ventricles or the foramen of Monro. In this work, virus persisted in neural tissues for up to 11 days after infection and thus was probably directly responsible for the damage seen.

Experimental studies in rodents on the teratogenesis of parainfluenza type 1 (Sendai) virus did not yield gross abnormalities, but fetal losses were increased and survivors were smaller than normal at birth, and showed prolonged gestational periods[136–138]. Virus was not regularly recovered from conceptuses when mothers were infected intranasally but maternal respiratory distress was a common finding. Thus a reduced state of maternal health leading to an 'unfavourable' intra-uterine environment was a probable cause for the impaired fetal growth reported. In contrast, when influenza virus was administered to rats intravenously, intra-uterine growth retardation and runting were prominent but virus *was* present in fetal tissues[139]. This infection persisted into post-natal life in at least 20–30% of those examined. Viral infection was seen in the central nervous system but was not associated with histological abnormalities.

Histological abnormalities were not seen in guinea pig or ferret fetuses following intracardial infection of pregnant mothers. Nevertheless, some interesting observations were made on the propensity of the virus to replicate in fetal brain, umbilical cord, placenta and fetal membranes[140,141]. Low titres of virus were recovered from fetal tissues and membranes after intracardiac infection of pregnant mothers but not when infection was administered intranasally. In the fetal ferret, virus was regularly found in the central nervous system, heart and lymphatic tissues, all of which failed to support viral replication in culture[141]. In the guinea pig, influenza occurred sporadically in fetal tissues, yet in culture the tissues showed a similar pattern of susceptibility to that seen with ferret and human tissues. That is, a viral replication occurred in respiratory, alimentary and urinogenital tract tissues[128,140]. Fetal membranes in the three species differed in their ability to support viral replication in culture. In the ferret, the amnion, placenta and umbilical cord yielded virus whereas in human tissues examined only the uterine endometrium yielded significant amounts of virus. In the guinea pig only the chorion exhibited evidence of permitting viral replication. The similar insusceptibility of guinea pig and human placentae to influenza replication may reflect morphological similarities.

Rubeola (measles)

Rubeola infection as discussed here refers to illnesses characterized by a morbilliform rash. German measles is discussed above under rubella.

Limited information is available implicating maternal measles infection in pregnancy as a cause of fetal risk. Infants may be smaller than normal at birth but are unlikely to be malformed[142]. Fetal infection does occur and may be as high as 88% when mothers contract the disease in the second month of pregnancy[143]. A fetal infection rate of about 34% was presumed with infections in the third month but after five months, the risk is negligible. However, an infection acquired late in pregnancy may be a cause of pneumonia in the child shortly after birth[144].

Congenital deafness can result from congenital rubeola infection but this complication is rare[145]. In these instances, it is probable that the mechanism for damage is by a direct pathogenesis since virus has been isolated from infants whose mothers were infected in pregnancy.

Mumps

Teratogenicity and pathology of mumps infection in human pregnancy

Mumps infection in pregnancy is frequently inapparent and for many years was not considered to be responsible for fetal risk[146–148]. The disease rarely occurs in sweeping epidemics as does rubella or influenza and is seldom seen in pregnancy[149]; the incidence is probably of the order of one case in 30 000 pregnancies[150, 151].

Although early epidemiologists believed mumps to be without teratogenic activity, it is now held to be a cause for abortion, stillbirth and possibly congenital deformity[94, 124, 152, 153]. Occurring in the first trimester of pregnancy mumps may lead to abortion without fetal infection[94]. Congenital deformities associated with mumps later in pregnancy include auditory and optic defects, urinogenital disease, spina bifida and the specific cardiac defect of endocardial fibroelastosis[154–156].

Endocardial fibroelastosis is a primary lesion which occurs in the absence of other cardiac defects and is often fatal[157]. A sclerosis or hardening of endocardial surfaces, particularly in the right ventricle is characteristic. Cardiac contractility is impaired by a thick layer of whitish tissue which extends between papillary muscles (Figure 16.3). Although mumps may be a cause of this condition, limited supporting evidence is available[124]. The delayed hypersensitivity reaction used in the diagnosis of mumps has been used in the identification of only eight infants with suspected endocardial fibroelastosis[154, 155]. If a correlation does exist, then more epidemiological evidence is obviously necessary supported where possible by virological studies and

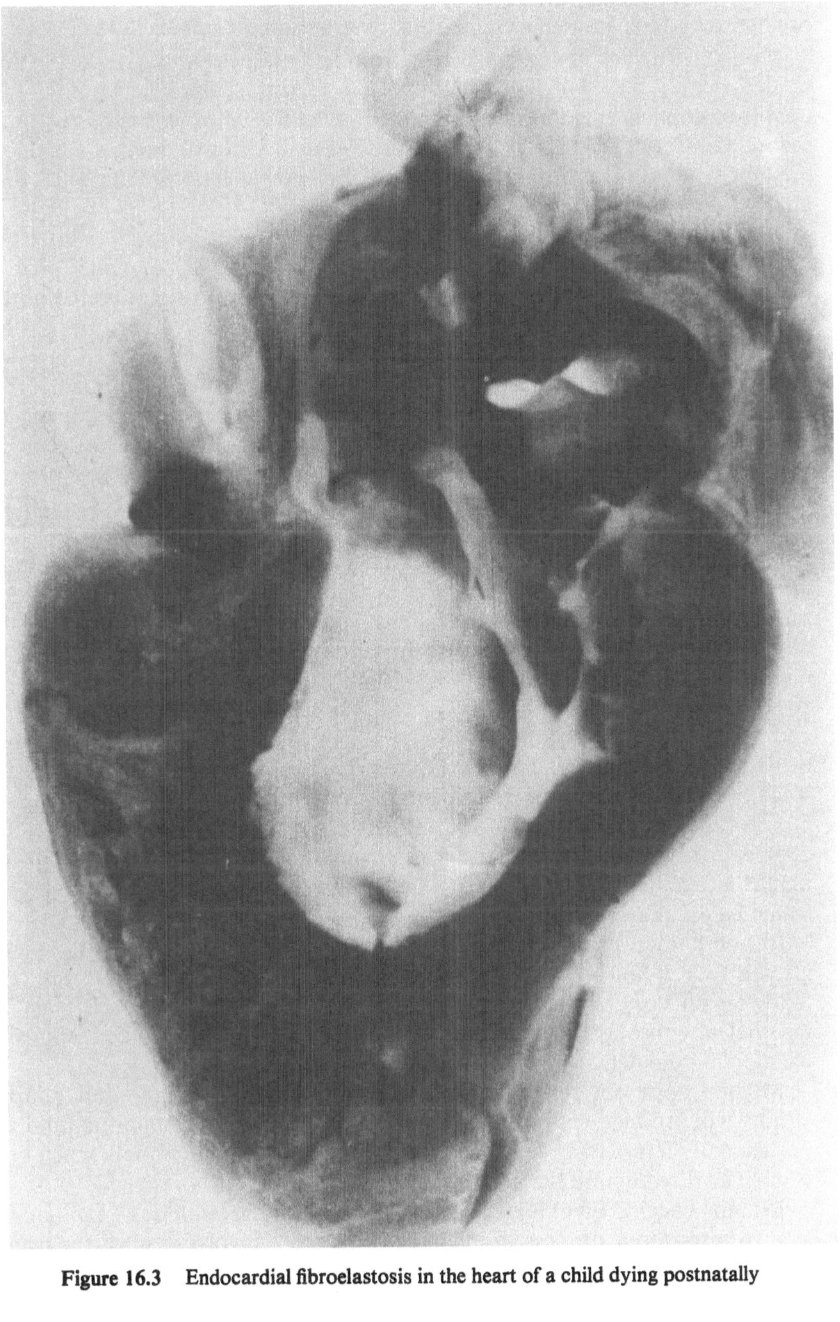

Figure 16.3 Endocardial fibroelastosis in the heart of a child dying postnatally

isolation of virus from affected tissues[15]. Current views are that endocardial fibroelastosis may be caused by an infectious agent such as mumps, but other predisposing factors such as lymphatic obstruction are not excluded[158].

Mumps virus has not been isolated from fetal tissues although specific antibodies were detected in cord blood of 12 Eskimo babies[152]. Monif[159] was unable to confirm this in three pregnancies where mumps was acquired in the second trimester. Further, in 45 patients injected with live mumps vaccine in pregnancy, virus was found in placentae but not in fetal tissues 7–10 days afterwards[156].

Animal studies on the pathogenicity of mumps in the embryo and fetus

Numerous experimental studies have been undertaken to examine the ability of mumps to induce endocardial fibroelastosis, but so far a relationship has been reported in only one study using chick embryos[160]. Myocarditis was present in chicks examined at 18 days, but endocardial fibroelastosis had developed in those killed one year after hatching. In other studies using the chick embryo, mumps infection resulted in lens abnormalities, myocarditis and growth retardation[161, 162]. Growth retardation in the infected embryo was related to a diminished cell size. Although the DNA content in these embryos was normal, the brain and carcase contained less RNA.

A clear relationship has been demonstrated between the time of mumps infection in chick embryos and the type of growth defect seen[163]. Organ sensitivity and time of infection has been studied in detail in the development of virus-induced lenticular damage in culture[164]. As in the case of rubella, mumps virus induced lens deformities only in explants made at the open vesicle stage. Virus could be demonstrated in the lens by immunofluorescence and 50–60% of the explants developed a central cataract. Lenticular damage was not seen in explants cultured in the presence of virus at the closed vesicle stage, when virus was present in neural ectoderm only. Thus, the lens capsule protects the lens from the injurious effects of the virus at this later stage. The observations extrapolated to the *in utero* mammalian fetus suggest that lenticular anomalies can only be expected when the amniotic fluid becomes infected before the lens vesicle closes.

In non-human primates, gestational mumps may cause fetal or post-natal growth retardation or result in abortion[165, 166]. Virus is present in the fetus following an early pregnancy infection but not later. In a single case, placental insufficiency through virally-induced damage was presumed to be responsible for a premature delivery.

Mumps does not readily infect mice or hamsters directly but adapted strains will produce viraemia with infection of the uterus, amniotic fluid and occasionally fetuses[167–170]. Fetal infections occur more commonly when virus is injected directly into the amniotic cavity.

Mumps vaccine (live) injected into hamsters on the tenth day of pregnancy lead to respiratory distress in neonates with some involvement of the central nervous system[171]. Viral inclusion bodies persisted at these sites for up to 21 days after infection. When these neonates were examined histologically, hydrocephalus with ependymal involvement and possibly aqueductal stenosis was recognised.

HERPESVIRUSES

This is a large group of medium sized DNA viruses which includes Herpesvirus simplex (HVS), varicella (herpes zoster) and cytomegalovirus (CMV). These infections are commonly inapparent in pregnant women but have been associated with abortions, stillbirth, intra-uterine growth retardation and congenital deformities. Infections occurring late in pregnancy may also be a cause of neonatal illness and disabilities evident later in life.

Herpesvirus simplex (HVS)

Two main types of HVS are recognized, type 1 which is responsible for the common cold sore or blisters in the oral region, and type 2 which produces genital lesions including cervicitis and vaginitis[172]. HVS type 1 may account for up to 10% of genital infections.

HVS infections occur in up to 50% of pregnancies and they may lead to fetal infection and abortion. However in studies carried out so far, only seven cases of congenital deformity have been recorded. These were children exhibiting microcephaly, cerebral calcifications, microphthalmos and retinal dysplasia[7,173,174]. Naib and his colleagues[175] examined the incidence of abortions in HVS infections amongst 228 patients and noted that in about one-third, abortion was associated with infections that were well established before conception. Thus, HVS can be a latent infection. As many as 95% of these patients were infected with HVS type 2.

HVS is known to infect the human conceptus but there is some dispute as to the actual route of infection. Transplacental infection is possible but current opinion is that the fetus becomes involved by an ascending infection from the lower genital tract. Alternatively, infection may be acquired perinatally as the child passes through an infected birth canal[176]. Fatal neonatal illness is more likely to develop from this latter mode of infection.

Non-human primates injected intravaginally with HVS type 2 develop a vaginitis as in man but this has been shown to lead to paralysis and a fatal generalized disease[177]. Cebus monkeys however, recover uneventfully from the infection and exhibit no permanent damage[178]. Vaginitis with a severe generalized disease has also been recorded in guinea pigs, mice and hamsters following intravaginal injection of adapted strains of virus[179,180].

Equine HVS type 2 infection appears to be non-fatal. In the pregnant mare, fetal infection occurs but not with congenital deformity[181]. However, when this virus was adapted and injected into pregnant hamsters, a severe hepatic degeneration developed and the fetuses aborted[182]. This observation was related to placental damage manifest by degenerative changes in the trophoblast cells.

Cytomegalovirus (CMV)

Cytomegaloviruses were first identified in 1956[183] even though the disease caused in man was recognized thirty years earlier[184]. Like rubella, CMV presents a serious risk of fetal malformation and consequently has received considerable attention from epidemiologists, clinicians and experimentalists[185–188].

In man, CMV shows a predilection for the fetal central nervous system and is a well known cause of mental retardation in infants following gestational infection. However, since CMV infection in pregnancy is commonly inapparent, the true extent of the teratogenic problem is not fully appreciated. Often infants exposed to infection in pregnancy appear normal at birth but are less viable postnatally and become mentally backward. Estimates in the United Kingdom indicate that 0.5–1.0 per 1000 pregnancies are complicated by CMV each year and that between 400 and 800 infants are affected[22].

Infants showing congenital CMV usually excrete virus in urine, saliva and other body secretions for months or even years after birth[189,190]. However, they are often asymptomatic at birth or show a mild illness manifest by jaundice, respiratory distress and hepatosplenomegaly. Microcephaly and mental disorders are well known complications of CMV but deafness and cerebral blindness have been recorded[187,191].

Fetal CMV infection is known to occur in man but important routes are not well defined[193]. As with several other viral infections, maternal infection in the first or second trimester presents a greater risk to the fetus but a late gestational infection may lead to postnatal illness. Clinical and pathological studies indicate that fetal infection may arise by an ascending infection from the lower genital tract as in the case of HVS or is acquired at birth as the infant passes through the birth canal. This route of infection is suggested by the observation of CMV lesions in the cervical region and isolation of virus from secretions at this site[194,195].

Isolation of CMV from semen in a healthy male is a probable indication as to the origin of these infections[196]. Fetal infection is clearly possible through virus-induced placental damage which is well known[197]. Characteristically placental lesions in congenital CMV consist of inflammatory changes with focal necrosis and villous degeneration and atrophy[198–200].

CMV infection in pregnancy may be accompanied by an acute maternal illness with fever, hepatosplenomegaly and lymphoses[197]. This can result in fetal infection at any time but the resulting fetal damage seems to relate to the severity of the maternal illness as well as the stage in pregnancy at which the infection is acquired and the overall period of exposure. Infection from the fourth week of pregnancy leads to a generalized dissemination of virus in the fetus with inflammatory changes and an increase in IgM levels but with no obvious abnormality in growth or survival[200,201]. On the other hand, infection at 13 weeks' pregnancy caused widespread fetal involvement with severe congenital damage[202]. However, retrospective examination of case reports indicate that the degree of fetal involvement varies enormously and that there is no simple correlation between time of infection and the extent of virus dissemination[185]. Also, there seems to be very limited information as to how CMV may cause fetal defects. Selective cell death, mitotic inhibition and chromosomal damage have been suggested but these views are largely speculative[190,203].

The pathogenesis of CMV has been examined in several animal models but owing to the extreme species specificity of the various strains of CMV used, the observations made in these studies may be of limited value in assisting our understanding of the human problem.

In man, fetal infection is appreciated and there is increasing evidence that fetal damage is probably the result of a direct viral effect. In comparison, in the murine strain which has frequently been used to study the consequence of CMV infection in pregnancy, fetal infection has not been demonstrated and increased fetal wastage and retarded growth have been ascribed to impaired maternal health[204–206]. Fetal infection may occur in the guinea pig and bovine strains of CMV but no studies appear to have been reported where fetal damage occurred in these species[207–209]. Murine CMV differs from the human strain in that it does not regularly produce placental damage even though inclusion bodies may be formed at that site[205]. Virus injected directly into the mouse fetus did lead to infection of the fetal brain but this observation was not related to neurological damage or subnormal mental development.

Maternal ill-health was held to be responsible for fetal loss through implantation failure when mice were infected with CMV at about ovulation[210]. However, when pre-implantation embryos from infected mothers were explanted into culture media, they developed as normal, suggesting that an 'unfavourable' intra-uterine environment created by the sick mother was primarily responsible for the effect seen. When mice were infected later in pregnancy, retarded growth was a prominent feature of fetal development[206]. This too was attributed primarily to maternal ill-health characterized by frank inflammatory changes in the liver, heart, salivary glands and adrenal cortex (Figure 16.4)[206]. Lesions comprising a vacuolar degeneration in the adrenal cortex indicated severe stress which was particularly marked in those animals infected intraperitoneally where widespread peritonitis and splenic

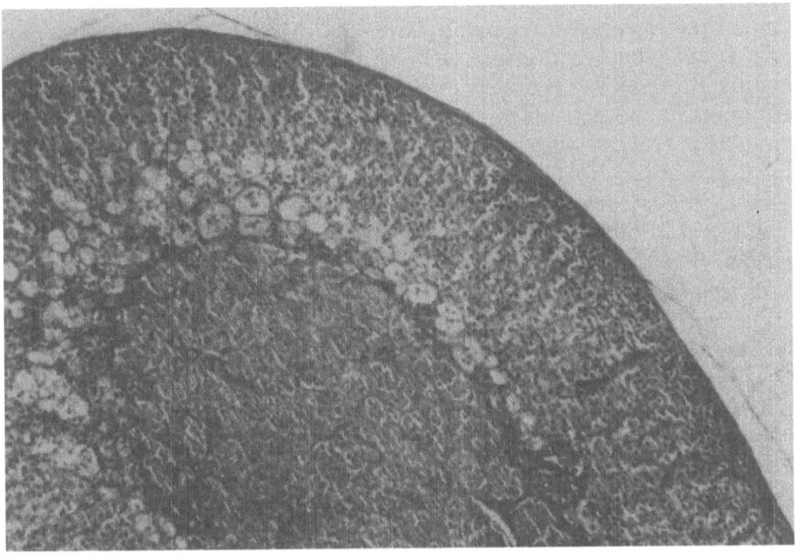

Figure 16.4 Degenerative changes in the adrenal zona reticularis of a pregnant mouse infected with cytomegalovirus, H&E (× 105)

necrosis developed. In contrast, when CMV was administered by intramuscular route, peritonitis did not arise and the spleens were profoundly enlarged with multiple centres of reactive hyperplasia. Hepatic inflammatory and necrotic changes were considered to be important causes of retarded fetal growth through disturbed protein metabolism, possibly of the type seen experimentally in mice infected with Coxsackieviruses B[115,120,121]

Varicella (Herpes zoster)

Congenital chicken pox has been known for many years[211] but evidence implicating it as a cause of fetal teratogenesis in man is limited to five cases of children showing a consistent pattern of limb hypoplasia accompanied by typical varicelliform skin lesions[212]. Virus-induced neural damage may underlie such defects as chorioretinitis, cataract, microphthalmos, nystagmus and optic atrophy which have been recorded[212,213]. Maternal infection with varicella from 11 weeks of pregnancy was associated with premature delivery, muscle lesions and respiratory disorders[214]. However, infection later in pregnancy may lead to a fatal illness in up to 20% of neonates within the first 10 days after birth[145].

Fetal infection results from a maternal illness early in pregnancy and thus a direct viral pathogenicity is a likely mechanism for abortion or congenital deformities seen[5,145,215]. Maternal pneumonitis seen as an occasional complication of varicella infections may be contributory in causing premature births.

OTHER VIRUSES TERATOGENIC IN MAN

Limited clinical information supplemented by experimental studies in animal species is available to show that other viral infections are teratogenic in man. Hepatitis B, smallpox (variola) and vaccinia virus used in its prevention, Western equine encephalomyocarditis (WEE) virus and lymphocytic choriomeningitis (LCM) viruses are known causes of abortion, stillbirth and neonatal death but their ability to evoke congenital deformity is still not clear.

Due to active vaccination campaigns in many countries, smallpox is now rare but fetal infection is recorded and may be a cause of disease in the newborn child[216]. Structural defects in congenital smallpox or vaccinia virus infection in pregnancy appear to be rare, but the risk of abortion does exist. In one study where 14 mothers were vaccinated before 24 weeks' gestation only one child survived, nine aborted and the remainder died in early infancy[217]. The risk indicated here is possibly high in view of subsequent studies using larger population samples where no fetal damage or fatality occurred and only two babies showed evidence of congenital vaccinia[218,219]. Experimentally, vaccination in pregnant baboons was seen to lead to maternal and fetal infection but not abnormal fetal growth or lowered survival[220].

Hepatitis viruses are of two main types, type A which rarely leads to fetal infection except in cases of extreme malnutrition, and type B which is a potential cause of abortion or neonatal disease[221]. Approximately 80% of children infected congenitally with hepatitis B survive for one year after birth and half of these exhibit signs of liver disease[94, 222]. Mongolism may be a complication of prenatal hepatitis B but this is not clearly resolved and requires further epidemiological and experimental examination[223, 224].

Rabies is known to infect the human fetus following maternal infection but the limited information available indicates that the mother is probably at greater risk than her unborn infant[225]. Kuru, a rare disease found mainly in New Guinea, also presents greater risk to a pregnant mother but the fetus is not infected and development is normal[226].

VIRAL TERATOGENS IN ANIMALS

Dimmock and Edwards' description of the so-called abortion agent' of mares which did not overtly influence maternal health but led to widespread fetal infection, is an early report of viral teratogenicity in animals[1].

Since the 1930s, a wide range of animal viruses has been identified which result in fetal death or congenital abnormality. These viruses fall into two main categories, those which affect farm animals causing abortion or subnormal infants in cattle, pigs and sheep and which are of economic importance and secondly, those infections which affect non-commercial species such as rats, ferrets, cats and hamsters. In certain cases, as with some reoviruses, vesicular stomatitis virus, rabies and Colorado tick fever virus, humans can be infected and infection in pregnancy may lead to impaired fetal development. Western equine encephalomyocarditis virus is an example of this.

Amongst the teratogenic animal viruses of economic importance in various parts of the world, Akbane virus of sheep, bovine ephemeral fever, bovine viral diarrhoea, infectious rhinotracheitis, parainfluenza-3 and syncytial virus are significant in cattle breeding. Hog cholera virus, bluetongue, swine fever and enterovirus infections, are known causes of fetal death and mummification in pigs. With many of these infections, abortion is a common occurrance[227], but some infections are known to cause gross fetal deformities particularly involving the central nervous system. Cerebellar hypoplasia in fetal pigs exposed in early gestation to hog cholera virus and cerebral hydrencephaly in fetal sheep infected with bluetongue virus are examples[228, 229]. Bovine diarrhoea virus injected into pregnant heifers lead to meningeal inflammation in the fetuses with neuronal necrosis and focal haemorrhage in the cerebellum[230]. Fetal infection was demonstrated by specific immunofluorescence studies.

In cases of abortion however, the pathogenesis is less clear. Maternal illness may be contributory, but with some bovine infections in particular, overt signs of maternal infection are not always present. Placental damage or generalized fetal infection is possibly responsible but changes have rarely been reported.

To understand more fully, the implications of viruses in causing fetal loss or deformity in farm animals, some laboratory studies have been undertaken using smaller species. For example, foot-and-mouth disease virus was injected into pregnant mice[231]. Fetal infection was inconsistent but the virus persisted in the placenta for up to six days. Although in cattle, foot-and-mouth disease is associated with abortion, in mice infection at 0–12 days of pregnancy did not lead to increased fetal loss or impaired growth in surviving offspring. In contrast, bluetongue virus, a known cause of central nervous system defects in sheep[228, 232], lead to cerebral malformations in mouse fetuses after injection into pregnant dams at susceptible stages of pregnancy[233].

Several viruses of small mammals lead to fetal defects following maternal infection in pregnancy. Well known examples that may be given here are: the H-1 strain of rat virus[234], reovirus type 3 in rats and hamsters[235, 236], and feline panleucopenia virus[237]. Fetal involvement varied greatly with the different viruses and the teratogenic responses ranged from intra-uterine wastage, growth retardation and reduced postnatal viability to gross deformity with defects affecting the central nervous system and musculoskeletal tissues. As in other infections discussed above, maternal illness was not always apparent and defects seen were possibly attributable to direct viral damage in susceptible fetal tissues.

DISCUSSION

Evidence reviewed indicates that viral infections are a significant source of teratogenic risk in man and animal species. Although the different teratogenic viruses differ greatly in their propensities to impair fetal growth and survival, several distinct patterns for teratogenesis have emerged from the present comparative survey. Direct fetal infection through transplacental passage, ascending infection from the genital tract or infection in the birth process is the principle cause of abnormality or fetal disease. Maternal ill-health and resulting placental damage are important factors in fetal loss through implantation failure, abortion and retarded intra-uterine development. For most of the infections discussed here, there is a clear correlation between the type and severity of the fetal damage and the time in pregnancy when maternal infection is acquired. This is undoubtedly related to the stage of development of the fetus and the susceptibility of its tissues to injury.

The influence of infection in very early pregnancy at pre-implantation states is unclear for many teratogenic viruses but available information on influenza and CMV indicate that failure in implantation is a likely event at this stage. Abortion, stillbirth and congenital deformity are common consequences of maternal viral infections occurring at slightly later stages of the first trimester when tissue differentiation and organogenesis are active. Tissues including the eye, heart and central nervous system which have been associated with rubella, mumps, CMV and HVS infections, are clearly more susceptible to damage at this stage in pregnancy than later. Infections occurring in the second and third trimesters are less frequently associated with gross congenital deformity but may be accompanied by risks of stillbirth, fetal growth retardation and possibly neonatal illness depending on the nature of the infecting virus and the severity of the illness. Neonatal illness through perinatal infection as has been seen with rubella and CMV, appears to be related to a persistence of virus in the tissues.

Cytological damage leading to defective organ growth are common causes of impaired growth and survival in viral teratogenesis. Good evidence for direct viral cytolysis has been produced in respect of rubella, CMV and several of the animal viruses. Less clear is the role of viruses in causing fetal deformities or defective mental development and mongolism through genetical or non-cytolytic changes. The production of a specific growth inhibitor substance in virally infected cells from small babies is an interesting speculation.

Where fetal infection is not seen or where fetal distress forms part of a generalized disease process as in the case of influenza, the mechanism for fetal damage is often unclear. Features such as maternal fever resulting in intrauterine hyperthermia, respiratory distress leading to hypoxia and general metabolic upset are possibly causes of abnormal fetal growth or survival irrespective of whether fetal infection occurs or not.

Much information as to the importance of these non-specific changes in viral infection as a cause of fetal deformity has been produced by experimental studies in animals. For example, Edwards, recognizing that influenza infections which are frequently accompanied by high fever and were associated with fetal nervous system defects, examined the type of effect produced by hyperthermia in pregnant guinea pigs and upon cells cultured *in vitro*[238–240]. Cell growth in culture was disrupted and in guinea pigs retarded and central nervous system defects of the type associated with maternal influenza were present. These guinea pig fetuses also exhibited cardiac defects of patent ductus arteriosus, confluent ventricles and auricular defects similar in type to those reported in human babies exposed to congenital rubella.

Hypoxia resulting from severe respiratory illness in pregnant mothers infected with influenza or rhinovirus infections may also be a cause of aberrant fetal growth. The critical level of blood oxygen necessary for normal fetal growth probably varies according to geographical region but epidemiological evidence from studies carried out in Peru shows that congenital heart disease of patent ductus arteriosus and septal defects are more common in children born from high altitude areas with lower oxygen levels[241]. Laboratory animals exposed to low oxygen concentrations in pregnancy have shown similar cardiac defects[242]. Jackson and others have suggested that as a result of their experimental studies, closure of the ductus arteriosus requires an increased systemic oxygen level[243]. Thus, high systemic oxygen levels lead to an early obliteration or reduction in the bore of the ductus shunt whereas lower than normal oxygen significantly increases the shunt.

Maternal anorexia leading to undernutrition is a further non-specific effect liable to result in subnormal fetal growth. Abundant experimental evidence has shown that dietary deprivation in pregnancy is a cause of fetal wastage and retarded growth, but congenital deformities have also been reported[245].

Despite the obvious advantages of using animal species in studying the consequence of human viral infections in pregnancy, limited information has been gained from this approach to further our understanding of how many viruses impair fetal growth. With rubella for example, which is probably the best known human viral teratogen, animal responses have been inconsistent and in some cases not reproducible. In influenza studies in the guinea pig, tissues in culture exhibited similar patterns of susceptibility to that seen for some human tissues but not central nervous system or lymphoreticular organs which are supposed to be susceptible to deformity in congenital infections. The relationship between Coxsackievirus B infection and congenital heart disease indicated by two epidemiological studies in man has not been confirmed by experimental studies in mice. These differences in species susceptibility no doubt reflect variations in pathogenicity, or alternatively they indicate a lack of teratogenic risk.

So far, only two human viral infections are clearly liable to cause gross structural or functional deformity in the human fetus following maternal infection at susceptible periods of pregnancy, i.e. rubella and CMV. Influenza, HVS and others are possibly detrimental for fetal survival but further epidemiological information is necessary to substantiate their role in the aetiology of congenital malformations. Fetal tissues are clearly more susceptible than adult tissues to injury but variations in reported abnormalities may be attributed to genetical, environmental and therapeutic factors and are thus not simply the result of the infection alone.

References

1. Goodpasture, E. (1942). Virus infection of the mammalian fetus. *Science*, **95**, 391
2. Bland, P. B. (1919). Influenza and its relation to pregnancy and labor. *Am. J. Obstet. Gynecol.*, **79**, 184
3. Gregg, N. M. (1941). Congenital cataracts following German measles in the mother. *Trans. Ophth. Soc. Aust.*, **3**, 35
4. Swan, C., Tostevin, A. L., Moore, B., Mayo, H. and Black, G. H. B. (1943). Congenital defects in infants following infectious diseases in pregnancy. *Med. J. Aust.*, **2**, 201
5. Dudgeon, J. A. (1968). Dysmorphogenesis: the role of infection. *Proc. Roy. Soc. Med.*, **61**, 1285
6. Catalano, L. W. and Sever, J. L. (1971). The role of viruses as causes of congenital defects. *Annu. Rev. Microbiol.*, **25**, 255
7. Sever, J. L. (1971). Virus infections and malformation. *Fed. Proc.*, **30**, 114
8. Coid, C. R. (1973). Comparative aspects of infection in pregnancy. In: *Intra-uterine Infections*, pp. 117–133, Ciba Foundation Symposium. (North Holland, Amsterdam: Elsevier, Excerpta Medica)
9. Harris, R. E. (1974). Viral teratogenesis: a review with experimental and clinical perspectives. *Am. J. Obstet. Gynecol.*, **119**, 996
10. Kilham, L. and Margolis, G. (1975). Problems of human concern arising from animal models of intra-uterine and neonatal infections due to viruses: a review. *Prog. Med. Virol.*, **20**, 113
11. Margolis, G. and Kilham, L. (1975). Problems of human concern arising from animal models of intra-uterine and neonatal infections due to viruses: a review. *Prog. Med. Virol.*, **20**, 144
12. Johnson, R. T. (1971). Viral infections and malformations of the nervous system. *Birth Defects*, **7**, 56
13. Johnson, K. P. (1974). Viral infections of the developing nervous system. *Adv. Neurol.*, **6**, 53
14. Johnson, R. T. (1975). Hydrocephalus and viral infection. *Dev. Med. Child Neurol.*, **17**, 807
15 Overall, J. C. (1972). Intra-uterine virus infections and congenital heart disease. *Am. Heart J.*, **84**, 823
16. Lerner, A. M. and Wilson, F. M. (1973). Virus cardiomyopathy. *Prog. Med. Virol.*, **15**, 63
17. Lansdown, A. B. G. (1978). Viral infections and diseases of the heart. *Prog. Med. Virol.*, **24**, 70
18. Dimmick, J., Mahmood, K. and Altshuler, G. (1976). Antenatal infection: adequate protection against hyaline membrane disease. *Obstet. Gynecol.*, **47**, 56

19. Coid, C. R., Lansdown, A. B. G. and MacFadyen, I. R. (1977). Fetal growth retardation and low birthweight following infection in pregnancy. In: C. R. Coid (ed.), *Infections and Pregnancy*, pp. 289–305. (London: Academic Press)
20. Mims, C. A. (1968). Pathogenesis of viral infections of the fetus. *Prog. Med. Virol.*, **10**, 194
21. Sever, J. L. (1970). Viruses and the fetus. *Int. J. Gynecol. Obstet.*, **8**, 763
22. Dudgeon, J. A. (1975). Intra-uterine infection. *Proc. Roy. Soc. Med.*, **68**, 365
23. Grist, N. R. and Bell, E. J. (1974). A six year study of Coxsackievirus B infections in heart disease. *J. Hyg. Cambridge*, 73, 165
24. Whitehead, J. E. M. (1973). Silent infections and the epidemiology of viral carditis. *Am. Heart J.*, **85**, 711
25. Saxén, L., Hjelt, L., Sjöstedt, J. E., Hakosalo, J. and Hakosalo, J. (1960). Asian influenza during pregnancy and congenital malformations. *Acta Path. Microbiol. Scand.*, **49**, 114
26. Droughet, V. and Rouquette, C. (1970). Serological screening for inapparent Coxsackie B virus infections during pregnancy in mothers of malformed children and in those of normal children. *Bull. INSERM*, **25**, 29
27. Coffey, V. P. and Jessop, W. J. E. (1959). Maternal influenza and congenital deformities. A prospective study. *Lancet*, ii, 935
28. Hakosalo, J. and Saxén, L. (1971). Influenza epidemic and congenital effects. *Lancet*, ii, 1346
29. Doll, R., Hill, A. B. and Sakula, J. (1960). Asian influenza in pregnancy and congenital defects. *Br. J. Prev. Soc. Med.*, **14**, 167
30. Laurence, K. M., Carter, C. O. and David, P. A. (1968). Major central nervous systems malformations in South Wales. II Pregnancy factors, seasonal variation, and social class effects. *Br. J. Prev. Soc. Med.*, **22**, 212
31. Karkinen-Jääskelainen, M. and Saxén, L. (1974). Maternal influenza, drug consumption and congenital defects of the central nervous system. *Am. J. Obstet. Gynecol.*, **118**, 815
32. Krugman, S. (1965). Rubella: clinical and epidemiological aspects. *Arch. ges. Virusforsch.*, **16**, 477
33. Dudgeon, J. A. (1975). Congenital rubella. *J. Pediat.*, **87**, 1078
34. Ojala, P., Vesikau, T. and Elo, O. (1973). Rubella during pregnancy as a cause of congenital hearing loss. *Am. J. Epidemiol.*, **98**, 395
35. Karmody, C. S. (1968). Subclinical maternal rubella and congenital deafness. *N. Engl. J. Med.*, **278**, 809
36. Weil, M. L., Itabashi, H. H., Cremer, N. E., Oshiro, L. S., Lennette, E. H. and Carnay, L. (1975). Chronic progressive panencephalitis due to rubella virus simulating subacute sclerosing panencephalitis. *N. Engl. J. Med.*, **292**, 994
37. Townsend, J. J., Baringer, J. R., Wolinsky, J. S., Malamud, J., Mednick, J. P., Panitch, H. S., Scott, R. A. T., Oshiro, L. S. and Cremer, N. E. (1975). Progressive rubella panencephalitis, late onset after congenital rubella. *N. Engl. J. Med.*, **292**, 990
38. Higgins, I. T. T. (1965). The epidemiology of congenital heart disease. *J. Chron. Dis.*, **18**, 699
39. Overall, J. C. and Glasgow, L. A. (1970). Virus infection of the fetus and newborn infant. *J. Pediat.*, **77**, 315
40. Gumpel, S. M., Hayes, K. and Dudgeon, J. A. (1971). Congenital perceptive deafness: role of intra-uterine rubella. *Br. Med. J.*, **2**, 300
41. Töndury, G. and Smith, D. W. (1966). Fetal rubella pathology. *J. Pediat.*, **68**, 867
42. Singer, D. B., Rudolph, A. J., Rosenberg, H. S., Rawls, W. E. and Boniuk, H. (1967). Pathology of the congenital rubella syndrome. *J. Pediat.*, **71**, 665
43. Way, R. C. (1967). Cardiovascular defects and the rubella syndrome. *Can. Med. Ass. J.*, **97**, 1329
44. Easterly, J. R. and Oppenheimer, E. H. (1973). Intra-uterine rubella infection. *Perspect. Pediatr. Res.*, **1**, 313
45. Blattner, R. J. (1974). The role of viruses in congenital defects. *Am. J. Dis. Child.*, **128**, 781
46. Kibrick, S. and Loria, R. M. (1974). Rubella and cytomegalovirus. Current concepts of congenital and acquired infections. *Pediat. Clin. N. Am.*, **21**, 513
47. Forrest, J. H. and Menser, M. A. (1975). Recent implications of intra-uterine and postnatal rubella. *Aust. Paediat. J.*, **11**, 65

48. Lejarraga, H. and Peckham, C. S. (1974). Birthweight and subsequent growth of children exposed to rubella infection in utero. *Arch. Dis. Childh.*, **49**, 50
49. Brook, C. G. D. (1972). Evidence for a sensitive period of adipose cell replication in man. *Lancet*, ii, 624
50. Bellanti, J. A., Artenstein, H. S., Olson, L. C., Buescher, E. L., Luhrs, C. E. and Milstead, K. L. (1965). Congenital rubella: clinico-pathologic, virologic and immunologic studies. *Am. J. Dis. Child.*, **110**, 464
51. Töndury, G. (1953). Embryopathies caused by viruses in pregnancy. *Etudes neo-natale*, 2, 105
52. Menser, M. A. and Reye, R. D. K. (1974). The pathology of congenital rubella: a review written by request. *Pathology*, **6**, 215
53. Blattner, R. J. and Heyes, F. M. (1961). Role of viruses in the etiology of congenital malformations. *Prog. Med. Virol.*, 3, 311
54. Naeye, R. L. and Blanc, W. (1965). Pathogenesis of congenital rubella. *J. Am. Med. Ass.*, **194**, 109
55. Nusbacher, J., Hirschhorn, K. and Cooper, L. Z. (1967). Chromosome abnormalities in congenital rubella. *N. Engl. J. Med.*, **276**, 1401
56. Alford, B. R. (1968). Rubella – La bête noire de la medicine. *Laryngoscope, St. Louis*, **78**, 1623
57. Sekeles, E. and Ornoy, A. (1975). Osseous manifestations of gestational rubella in young human fetuses. *Am. J. Obstet. Gynecol.*, 122, 307
58. Heggie, A. D. (1977). Growth inhibition of human embryonic and fetal rat bones in organ culture by rubella virus. *Teratology*, **15**, 47
59. Lansdown, A. B. G. and Grasso, P. (1969). Histological changes in the skeletal system of the developing quail embryo treated with sodium salicylate. *Experientia*, **25**, 885
60. Lansdown, A. B. G. (1970). Histological changes in the skeletal elements of developing rat foetuses following treatment with sodium salicylate. *Fd. Cosmet. Toxicol.*, **8**, 647
61 Vermeij-Keers, C. (1975). Primary congenital aphakia and the rubella syndrome. *Teratology*, **11**, 257
62. Lindsay, J. S., Caruthers, D. G., Hemenway, W. G. and Harrison, M. S. (1953). Inner ear pathology following maternal rubella. *Ann. Otol. Rhin. Laryng.*, **62**, 1201
63. Monif, G. R., Sever, J. L., Schiff, G. M. and Traub, R. G. (1965). Isolation of rubella virus from products of conception. *Am. J. Obstet. Gynecol.*, **91**, 1143
64. Menser, M. A., Harley, J. D., Hertzberg, R., Dorman, D. C. and Murphy, A. M. (1967). Persistence of virus in lens for three years after prenatal rubella. *Lancet*, ii, 387
65. Karkinen-Jääskelainen, M., Saxén, L., Vaheri, A. and Leinikki, P. (1975). Rubella cataract in vitro: sensitive period of the developing human lens. *J. Exp. Med.*, **141**, 1238
66. Rawls, W. E. and Melnick, J. L. (1966). Rubella virus carrier derived from congenitally infected infants. *J. Exp. Med.*, **123**, 795
67. Plotkin, S. A. and Vaheri, A. (1967). Human fibroblasts infected with rubella virus produced a growth inhibitor. *Science*, **156**, 659
68. Töndury, G. (1962). *Embryopathien; uber die Wirkungsweise (Infektionsweg und Pathogenese) von Viren auf den menschlichen* Keimling. (Berlin: Springer-Verlag)
69. Töndury, G. and Töndury, T. A. (1972). Course of infection and pathogenesis of the human embryo. *Rev. Suisse Zool.*, **79**, 179
70. Marshall, W. C. (1975). Effects of rubella virus on the human fetus. *Proc. Roy. Soc. Med.*, **68**, 369
71. Giles, P. F. H. (1972). Rubella vaccination and termination of pregnancy, *Br. Med. J.*, **4**, 666
72. Bolognese, R. J., Corson, S. L., Fucillo, D. A., Sever, J. L. and Traub, R. (1973). Evaluation of possible transplacental infection with rubella vaccination during pregnancy. *Am. J. Obstet. Gynecol.*, 117, 939
73. Modlin, J. F., Herrmann, K., Brandling-Bennett, A. D., Eddings, D. L. and Hayden, G. F. (1976). Risk of congenital abnormality after inadvertant rubella vaccination of pregnant women. *N. Engl. J. Med.*, **294**, 972
74. Fleet, W. F., Benz, E. W., Kazon, D. T., Lefkowitz, L. B. and Herrmann, K. L. (1974). Fetal consequences of maternal rubella immunization. *J. Am. Med. Ass.*, **227**, 621
75. Parkman, P. D., Phillips, P. E., Kirschstein, R. L. and Meyer, H. M. (1965). Experimental rubella virus infection in the rhesus monkey. *J. Immunol.*, **95**, 743

76. Sever, J. L., Meier, G. W., Windle, W. F., Schiff, G. M., Monif, G. R. and Fabiyi, A. (1966). Experimental rubella in pregnant rhesus monkeys. *J. Infect. Dis.*, **116**, 21
77. DeLaHunt, C. S. and Reiser, N. (1967). Rubella induced embryopathies in monkeys. *Am. J. Obstet. Gynecol.*, **99**, 580
78. Hendricks, A. G. (1966). Teratological findings in a baboon colony. FDA Conference on Non-human Primate Toxicology, Airlie House, Virginia, USA
79. Oxford, J. S. and Schild, G. C. (1966). Growth of rubella virus in the hamster (*Mesocricetus auratus*). *Virology*, **28**, 780
80. Sato, H., Hishiyama, M. and Shishido, A. (1976). Growth of rubella, measles and mumps viruses in rat fetus. *Jap. J. Med. Sci. Biol.*, **29**, 39
81. Cotlier, E., Fox, J., Bohigian, G., Beaty, C. and Du Pree, A. (1968). Pathogenic effects of rubella virus on embryos and newborn rats. *Nature (Lond.)*, **217**, 38
82. Kleinman, H., Ramras, D. G., Cooney, M. K. and Boyd, L. (1962). ECHO-9 virus infection and congenital abnormalities; a negative result. *Pediatrics*, **29**, 261
83. Landsman, J. B., Grist, N. R. and Ross, C. A. C. (1964). ECHO-9 virus infection and congenital malformations. *Br. J. Prev. Soc. Med.*, **18**, 152
84. Brown, G. C. and Karunas, R. S. (1972). Relationship of congenital anomalies and maternal infection with selected enteroviruses. *Am. J. Epidemiol.*, **95**, 207
85. Dudgeon, J. A. (1970). Prenatal virus infections: In: Debre and R. Celers (eds.) *The Evaluation and Management of Human Viral Infections*. (Philadelphia: Saunders)
86. Freeman, D. W. and Barno, A. (1959). Deaths from Asian influenza associated with pregnancy. *Am. J. Obstet. Gynecol.*, **78**, 1172
87. Siegal, M. and Greenberg, M. (1955). Incidence of poliomyelitis in pregnancy, its relation to maternal age, parity and gestational period. *N. Engl. J. Med.*, **253**, 841
88. Sever, J. L. and White, L. R. (1968). Intra-uterine viral infections. *Anna. Rev. Med.*, **19**, 471
89. Reynolds, D. W., Stagno, S., Stubbs, K. G. (1974). Inapparent congenital cytomegalovirus and mental deficiency. *N. Engl. J. Med.*, **290**, 291
90. Schaeffer, M., Fox, M. J., and Li, C. P. (1954). Intra-uterine poliomyelitis infection: report of a case. *J. Am. Med. Ass.*, **155**, 248
91. Bates, T. (1955). Poliomyelitis in pregnancy, fetus and newborn. *Am. J. Dis. Child.*, **90**, 189
92. Ringe, M. E. (1957). Poliomyelitis in pregnancy: a report of 79 cases in Connecticut. *N. Engl. J. Med.*, **256**, 281
93. Czeizel, A. (1967). Coxsackievirus and congenital malformation. *J. Am. Med. Ass.*, **210**, 156
94. Fucillo, D. A. and Sever, J. L. (1973). Viral teratology, *Bacteriol. Rev.*, **37**, 19
95. Sussman, M. L., Struass, L. and Hodes, H. L. (1959). Fatal Coxsackie group B virus infection in the newborn: report of a case with necropsy findings and a brief review of the literature. *Am. Med. Ass. J. Dis. Child.*, **97**, 488
96. Hall, C. B. and Miller, D. G. (1969). The detection of a silent Coxsackie B5 virus perinatal infection. *J. Pediat.*, **75**, 124
97. Brown, G. C. and Evans, T. N. (1967). Serologic evidence of Coxsackievirus etiology of congenital heart disease. *J. Am. Med. Ass.*, **199**, 151
98. Burch, G. E., Sun, S. C., Chu, K. C., Sohol, R. S. and Colcolcough, H. L. (1968). Instetstitial and Coxsackievirus B myocarditis in infants and children. *J. Am. Med. Ass.*, **203**, 1
99. Kibrick, S. and Benirschke, K. (1956). Acute aseptic myocarditis and meningo-encephalitis in the newborn child infected with Coxsackievirus group B, type 3. *N. Engl. J. Med.*, **255**, 883
100. Kibrick, S. and Benirschke, K. (1958). Severe generalized disease (encephalo-hepatomyocarditis) occurring in the newborn period and due to infection with Coxsackievirus, group B. *Pediatrics*, **22**, 857
101. Fruhling, L., Korn, R., Lavallaureix, J., Surjus, A. and Fousseraux, S. (1962). La myoendocardite chronique fibroelastique du nouveau-né et du nourrisson (fibroélastose). Donnees morphologiques, etiologiques et pathogéniques nouvelles. Rapports avec certaines malformations cardiaques. *Ann. Anat. Pathol.*, **7**, 227
102. Plager, H., Beebe, R. and Miller, J. K. (1962). Coxsackie B-5 pericarditis in pregnancy. *Arch. Int. Med.*, **110**, 735

103. O'Shaunessey, W. J. and Buechner, H. A. (1962). Hepatitis associated with a Coxsackie B5 virus infection during late pregnancy. *J. Am. Med. Ass.*, **179**, 71
104. Dalldorf, G. and Sickles, G. M. (1948). An unidentified filtable agent isolated from the faeces of children with paralysis. *Science*, **108**, 61
105. Lansdown, A. B. G. (1976). Influence of Coxsackievirus B infections on fetal growth in mice. *Teratology*, **13**, 291
106. Dallorf, G. and Gifford, R. (1954). Susceptibility of gravid mice to Coxsackievirus infection. *J. Exp. Med.*, **99**, 21
107. Came, P. E. and Crowell, R. L. (1964). Studies of resistance of fetal mouse tissues in culture to Coxsackie A viruses. *Virology*, **23**, 542
108. Selzer, G. (1969). Transplacental infection of the mouse fetus by Coxsackie viruses. *Israel J. Med. Sci.*, **5**, 125
109. Suptel, E. A. and Maximovich, N. A. (1964). The possibility of intra-uterine infection with Coxsackie virus in mice inoculated by different routes. *Acta Virol.*, **8**, 46
110. Solov'ev, V. D., Khesin, Y. A. and Gutman, N. R. (1970). On the disruption of intra-uterine development in mice in pregnant females inoculated with Coxsackie viruses. *Vopr. Virusol.*, **15**, 60
111. Meyer, J. and Löffler, H. (1968). Infection of gravia mice with Coxsackievirus. *Path. Microbiol.*, **32**, 139
112. Soike, K. (1967). Coxsackie B-3 virus infection in the pregnant mouse. *J. Infect. Dis.*, **117**, 203
113. Droughet, V. and Levantis, F. (1968). Le passage transplacentaire du virus Coxsackie B3 chez la souris gestante et la contamination foetale *in utero*. *Ann. Inst. Pasteur*, **114**, 249
114. Surjus, A. (1961). Effets du virus Coxsackie B3 sur la souris gestante et sa transmission transplacentaire. *Ann. Inst. Pasteur*, **100**, 825
115. Lansdown, A. B. G. (1975). Influence of time of infection during pregnancy with Coxsackievirus B3 on maternal pathology and foetal growth in mice. *Br. J. Exp. Pathol.*, **56**, 119
116. Lansdown, A. B. G. (1975). Histological observations on epidermal development in fetal mice subject to intra-uterine growth retardation due to maternal infection with Coxsackievirus B3. *Biol. Neonat.*, **27**, 368
117. Lansdown, A. B. G. (1977). Histological observations on thymic development in fetal and newborn mammals subject to intra-uterine growth retardation. *Biol. Neonat.*, **31**, 252
118. Coid, C. R. and Ramsden, D. B. (1973). Retardation of foetal growth and plasma protein development in foetuses from mice injected with Coxsackie B3 virus. *Nature* (*Lond.*), **241**, 460
119. Lansdown, A. B. G. (1977). Coxsackievirus B3 infection in pregnancy and its influence on foetal heart development. *Br. J. Exp. Pathol.*, **58**, 378
120. Lansdown, A. B. G. and Coid, C. R. (1974). Pathological changes in pregnant mice infected with Coxsackie B3 virus as a possible cause of retarded foetal development. *Br. J. Exp. Pathol.*, **55**, 101
121. Lansdown, A. B. G. and Ellaby, S. J. (1974). Histochemical demonstration of changes in liver cell enzymes in pregnant mice infected with Coxsackie B3 virus. *Histochemistry*, **40**, 175
122. Lansdown, A. B. G. (1975). Pathological changes in pregnant mice infected with Coxsackievirus B3 and given dietary casein hydrolysate supplement. *Br. J. Exp. Pathol.*, **56**, 373
123. Lansdown, A. B. G., Coid, C. R. and Ramsden, D. B. (1975). Mitigation of virus induced foetal growth retardation in mice by dietary casein hydrolysate. *Nature* (*Lond.*), **254**, 599
124. Gersony, W. M., Katz, S. L. and Nadas, A. S. (1966). Endocardial fibroelastosis and mumps virus. *Pediatrics*, **37**, 430
125. McKenzie, J. S. and Houghton, M. (1974). Influenza virus infections during pregnancy: association with congenital malformation and with subsequent neoplasms in children, and potential hazards of live virus vaccines. *Bacteriol. Rev.*, **38**, 356
126. Zhukovskij, A. N. (1972). The effect of influenza on the course of pregnancy and intra-uterine pathology of the foetus. *Vopr. Virusol.*, **5**, 515
127. Monitor, (1975). Pregnancy hazard of influenza virus. *New Scientist*, **66**, 596
128. Rosztoczy, I., Sweet, C., Toms, G. L. and Smith, H. (1975). Replication of influenza virus

in organ cultures of human and simian urogenital tissues and human foetal tissues. *Br. J. Exp. Pathol.*, **56**, 322

129. Lehmann, N. I. and Gust, I. D. (1971). Viraemia in influenza. A report of two cases. *Med. J. Aust.*, **2**, 1166
130. Davenport, F. M. (1961). Pathogenesis of influenza. *Bacteriol. Rev.*, **25**, 294
131. Greenberg, M., Jacobziner, H., Pakter, J. and Weisl, B. A. G. (1958). Maternal mortality in epidermic Asian influenza. *Am. J. Obstet. Gynecol.*, **76**, 897
132. Yawn, D. H., Pyeatte, J. C., Joseph, J. M., Eichler, S. L. and Garcia-Banuel, R. (1971). Transplacental transfer of influenza virus. *J. Am. Med. Ass.*, **216**, 1022
133. Hamberger, V. and Habel, K. (1947). Teratogenic and lethal effects of influenza A and mumps viruses on early chick embyros. *Proc. Soc. Exp. Biol. Med.*, **66**, 608
134. Heath, H. D., Shear, H. H., Imagawa, D. T., Jones, M. H. and Adams, J. M. (1956). Teratogenic effects of herpes simplex, vaccinia, influenza A (NWS), and distemper virus on early chick embryos. *Proc. Soc. Exp. Biol. Med.*, **92**, 675
135. London, W. T., Fucillo, D. A., Sever, J. L. and Kent, S. G. (1975). Influenza virus as a teratogen in rhesus monkeys. *Nature (Lond.)*, **255**, 483
136. Coid, C. R. and Wardman, G. (1971). The effect of para-influenza type 1 (Sendai) virus infection on early pregnancies in the rat. *J. Reprod. Fertil.*, **24**, 39
137. Coid, C. A. and Wardman, G. (1972). The effect of maternal respiratory disease induced by para-influenza type 1 (Sendai) virus on foetal development and neonatal mortality in the rat. *Med. Microbiol. Immunol.*, **157**, 181
138. Ohba, N. (1958). Formation of embryonic abnormalities of the mouse by a viral infection of mother animals. *Acta Path. Jpn.*, **8**, 127
139. Tucker, M. J. and Stewart, R. B. (1976). Intra-uterine transmission of Sendai virus in inbred mouse strains. *Infect. Immun.*, **14**, 1191
140. Sweet, C., Collie, M. H., Toms, G. L. and Smith, H. (1977). The pregnant guinea pig as a model for studying influenza virus infection *in utero*: infection of foetal tissues in organ culture and *in vivo*. *Br. J. Exp. Pathol.*, **58**, 133
141. Sweet, C., Toms, G. L. and Smith, H. (1977). The pregnant ferret as a model for studying the congenital effects of influenza virus infection *in utero*: infection of foetal tissues in organ culture and *in vivo*. *Br. J. Exp. Pathol.*, **58**, 113
142. Siegel, M. and Fuerst, H. T. (1966). Low birth weight and maternal virus diseases: a prospective study of rubella, measles, mumps, chickenpox and hepatitis. *J. Am. Med. Ass.*, **197**, 680
143. Wenner, R. and Flammer, P. (1950). Rubeolen und Schwangerschaft. *Gynecologia*, **130**, 436
144. Dyer, I. (1940). Measles complicating pregnancy, report of 24 cases with 3 instances of congenital measles. *South. Med. J.*, 33, 601
145. Menser, M. A. and Forrest, J. M. (1973). Maternal infections and the developing foetus. *Med. J. Austr.*, **1**, 448
146. Schwartz, H. A. (1950). Mumps in pregnancy. *Am. J. Obstet. Gynecol.*, **60**, 875
147. Ylinen, O. and Järvinen, P. A. (1953). Parotitis during pregnancy. *Acta Obstet. Gynecol.*, **32**, 121
148. Hyatt, H. W. (1961). Relationship of maternal mumps to congenital defects and fetal deaths, and to maternal morbidity and mortality. *Am. Pract.*, **12**, 359
149. Sever, J. L. (1968). The prevention of mental retardation through control of infectious diseases. *Public Health Service Publication*, no. 1692 p. 37
150. Greenhill, J. P. (1933). Acute (extragenital) infection in pregnancy, labor and puerperium. *Am. J. Obstet. Gynecol.*, **25**, 760
151. Greenberg, M. W. (1948). Parotitis complicating the early puerperium. *Am. J. Obstet. Gynecol.*, **55**, 340
152. Asse, J. M., Noren, G. R., Reddy, D. V. and St. Geme, J. W. (1972). Mumps-virus infection in pregnant women and the immunological response of their offspring. *N. Engl. J. Med.*, **286**, 1379
153. St. Geme, J. W., Noren, G. R. and Adams, P. (1966). Proposed embryopathic relation between mumps virus and primary endocardial fibroelastosis. *N. Engl. J. Med.*, **275**, 339
154. Noren, G. R., Adams, P. and Anderson, R. C. (1963). Positive skin reactivity to mumps virus antigen in endocardial fibroelastosis. *J. Pediat.*, **62**, 604

155. Vosburgh, J. B., Diehl, A. M., Liu, C., Lauer, R. M. and Fabiyi, A. (1965). Relationship of mumps to endocardial fibroelastosis. *Am. J. Dis. Child.*, **109**, 69
156. Yamouchi, T., Wilson, C. and St. Geme, J. W. (1974). Transmission of live attenuated mumps virus to the human placenta. *N. Engl. J. Med.*, **290**, 710
157. Goodwin, J. F. (1973). Cardiomyopathies in England. In: E. Bajusz and G. Rona (eds.). *Cardiomyopathies*, pp. 79–93. (Baltimore, Maryland: University Park Press)
158. Miller, A. T., Pick, R. and Katz, L. N. (1963). Ventricular endo-myocardial changes after impairment of cardiac lymph flow in dogs. *Br. Heart J.*, **25**, 182
159. Monif, G. R. G. (1974). Maternal mumps infection during gestation: observations on the progeny. *Am. J. Obstet. Gynecol.*, **119**, 549
160. St. Geme, J. W., Peralta, H., Farias, E., Davis, C. W. C. and Noren, G. R. (1971). Experimental gestational mumps virus infection and endocardial fibroelastosis. *Pediatrics*, **48**, 821
161. Davis, C. W. C., St. Geme, J. W., Dufour, F. D. and Martin, H. L. (1975). In vitro analysis of experimental avian viral myocarditis. *J. Infect. Dis.*, **132**, 125
162. Yamouchi, T., St. Geme, J. W., Oh, W. and Davies, C. W. C. (1975). The biological and biochemical pathogenesis of mumps virus-induced embryonic growth retardation. *Pediat. Res.*, **9**, 30
163. Williamson, A. P., Blattner, R. J. and Simonsen, L. (1957). Cataracts following mumps virus in early chick embryos. *Proc. Soc. Exp. Biol. Med.*, **96**, 224
164. Karkinen-Jääskelainen, M. (1973). Spatial and temporal restriction of mumps virus induced lesions in the developing chick lens. *Acta Path. Microbiol. Scand.*, **Suppl. no. 243**, pp. 5–52
165. St. Geme, J. W. and Van Pelt. (1974). Fetal and postnatal growth retardation associated with gestational mumps virus infection of the rhesus monkey. *Lab. Anim. Sci.*, **24**, 895
166. London, W. T., Curfman, B. and Sever, J. L. (1971). Mumps infection of the rhesus monkey fetus. *Teratology*, **4**, 234
167. Kilham, L. and Overman, J. R. (1953). Natural pathogenicity of mumps virus for suckling hamsters on intra-cerebral injection. *J. Immunol.*, **70**, 147
168. Kilham, L., Murphy, H. W. and Overman, J. R. (1953). Performance of mumps neutralization tests in suckling mice. *J. Immunol.*, **71**, 183
169. Ferm, V. H. and Kilham, L. (1963). Mumps virus infection in the pregnant hamster. *J. Embryol. Exp. Morphol.*, **11**, 659
170. Kilham, L. and Margolis, G. (1974). Intra-uterine infections induced by mumps virus in hamsters. *Lab. Invest.*, **31**, 34
171. Kilham, L. and Margolis, G. (1975). Induction of congenital hydrocephalus in hamsters with attenuated and natural strains of mumps virus. *J. Infect. Dis.*, **132**, 462
172. Josey, W. E., Nahmias, A. J. and Naib, Z. M. (1971). Genital herpes simplex in historical perspective. *Bull. N.Y. Acad. Med.*, **52**, 935
173. Florman, A. L., Gershon, A. A., Blackett, P. R. and Nahmias, A. J. (1973). Intra-uterine infection with herpes simplex virus, resultant congenital malformations. *J. Am. Med. Ass.*, **225**, 129
174. Hanshaw, J. B. (1973). Herpesvirus hominis infections in the fetus and the newborn. *Am. J. Dis. Child.*, **126**, 546
175. Naib, Z. H., Nahmais, A. J., Josey, W. E. and Wheeler, J. H. (1970). Association of maternal genital herpetic infection with spontaneous abortion. *Obstet. Gynecol.*, **35**, 260
176. Tobin, J. O'H. (1975). Herpesvirus hominis infection in pregnancy. *Proc. Roy. Soc. Med.*, **68**, 371
177. Felsburg, P. J., Heberling, R. L., Brack, M. and Kalter, S. S. (1973). Experimental genital herpes infection in the marmoset. *J. Med. Prim.*, **2**, 50
178. Felsburg, P. J., Heberling, R. L. and Kalter, S. S. (1972). Experimental genital infection of cebus monkeys with oral and genital isolates of herpesvirus hominis type 1 and type 2. *Arch. ges. Virusforsch.*, **39**, 223
179. Renis, H. E. (1975). Genital infection of female hamsters with herpesvirus hominis type 2. *Proc. Soc. Exp. Biol. Med.*, **150**, 723
180. Lukas, B., Wiesendanger, W. and Schmidt-Ruppin, K. H. (1975). Herpes genitatis in guinea pigs. I. Kinetic study in infection with herpesvirus hominis type 2. *Arch. Virol.*, **49**, 1

181. Gleeson, L. J. and Studdert, H. J. (1977). Equine herpesviruses: experimental infection of a foetus with type 2. *Aust. Vet. J.*, **53**, 360
182. Burek, J. D., Roos, R. P. and Narayan, O. (1975). Virus-induced abortion: studies of equine herpesvirus 1 (Abortion virus) in hamsters. *Lab. Invest.*, **33**, 400
183. Rowe, W. P., Hartley, J. W., Waterman, S., Turner, H. C. and Huebner, R. J. (1956). Cytopathogenic agent resembling human salivary gland virus recovered from tissue cultures of human adenoids. *Proc. Soc. Exp. Biol. Med.*, **92**, 418
184. Goodpasture, E. W. and Talbot, F. B. (1921). Concerning the nature of 'protozoan-like' cells in certain lesions in infancy. *Am. J. Dis. Child.*, **21**, 415
185. Plummer, G. (1973). Cytomegoloviruses of man and animals. *Prog. Med. Virol.*, **15**, 92
186. Hanshaw, J. B. (1970). Developmental abnormalities associated with congenital cytomegalovirus infection. *Adv. Teratol.*, **4**, 64
187. Hayes, K. (1974). Prenatal virus infection with particular reference to cytomegaloviruses. *Aust. Paediat. J.*, **10**, 56
188. Stern, H. and Tucker, S. M. (1973). Prospective study of cytomegalovirus infection in pregnancy. *Br. Med. J.*, **2**, 268
189. Krech, U. and Jung, M. (1973). Epidemiologie, Virologie and virologische Diagnose der Cytomegalie. *Klin. Wschr.*, **51**, 529
190. Reynolds, D. W., Stagno, S., Stubbs, K. G., Dahle, A. J., Livingston, M. H., Saxon, S. S. and Alford, C. A. (1974). Inapparent congenital cytomegalovirus infection with elevated cord IgM. levels. *N. Engl. J. Med.*, **290**, 291
191. Hanshaw, J. B. (1966). Cytomegalovirus complement-fixing antibody in microcephaly. *N. Engl. J. Med.*, **275**, 476
192. Stern, H., Elek, S. D., Booth, J. C. and Fleck, D. G. (1969). Microbiological causes of mental retardation: the role of prenatal infections with cytomegalovirus, rubella virus and toxoplasma. *Lancet*, **ii**, 443
193. Weller, T. H. (1971). The cytomegaloviruses, ubiquitous agents with protean clinical manifestations. *N. Engl. J. Med.*, **285**, 203
194. Numazaki, Y., Yano, N., Morizuka, T., Takai, S. and Ishida, N. (1970). Primary infection with human cytomegalovirus: virus isolation from healthy infants and pregnant women. *Am. J. Epidemiol.*, **91**, 410
195. Varga, A. and Browell, B. (1960). Viral inclusion bodies in vaginal smears. *Obstet. Gynecol.*, **16**, 441
196. Lang, D. J. and Kummer, J. F. (1972). Demonstration of cytomegalovirus in semen. *N. Engl. J. Med.*, **287**, 756
197. Hayes, K. and Gibas, H. (1971). Placental cytomegalovirus infection without fetal involvement following primary infection in pregnancy. *J. Pediat.*, **79**, 401
198. Lelong, M., Lepage, F., Le Tan Vinh, Tournier, P. and Chany, C. (1960). Le virus de la maladie des inclusions cytomégaliques. *Arch. Franc. Pediat.*, **17**, 1
199. Cochard, A. M., Le Tan Vinh and Lelong, M. (1963). Le placenta dans la cytomegalie congenital: études anatomoclinique de 3 observations personelles, *Arch. Franç. Pédiat.*, **20**, 35
200. Benirschke, K., Mendoza, G. R. and Bazeley, P. L. (1974). Placental and fetal manifestations of cytomegalovirus infection. *Virchows Arch. B. Cell. Path.*, **16**, 121
201. Davis, L. E., Tweed, G. V., Stewart, J. A., Bernstein, T. A., Miller, G. L., Gravelle, C. R. and Chin, T. D. Y. (1971). Cytomegalovirus mononucleosis in the first trimester pregnant female with transmission to the fetus. *Pediatrics*, **48**, 200
202. Altshuler, G. (1973). Implications of two cases of human placental plasma cells. Reports of cytomegalic inclusion disease in a thirteen week's fetus and of ascending herpes infection in the newborn. *Am. J. Pathol.*, **70**, 18a
203. Peckham, C. S. (1972). A clinical and laboratory study of children exposed *in utero* to maternal rubella. *Arch. Dis. Child.*, **47**, 571
204. Medearis, D. N. (1964). Mouse cytomegalovirus infection. III. Attempts to produce intrauterine infections. *Am. J. Hyg.*, **80**, 113
205. Johnson, K. P. (1969). Mouse cytomegalovirus: placental infection. *J. Infect. Dis.*, **120**, 445
206. Lansdown, A. B. G. and Brown, J. D. (1978). Pathological observations on experimental cytomegalovirus infections in pregnancy. *J. Pathol.*, **125**, 1

207. Plowright, W., Ferris, M. D. and Scott, G. R. (1960). Blue wildebeest and the agent of bovine catarrhal fever. *Nature (Lond.)*, **188**, 1167
208. Plowright, W., Kalunda, M., Jessett, D. M. and Herniman, K. A. J. (1972). Congenital infection in cattle with the herpesvirus causing malignant catarrhal fever. *Res. Vet. Sci.*, **13**, 37
209. Lam, K. M. and Hsiung, G. D. (1971). Herpesvirus infection of guinea pigs. II. Transplacental transmission. *Am. J. Epidemiol.*, **93**, 308
210. Neighbour, A. P. (1976). The effect of maternal cytomegalovirus infection on preimplantation development in the mouse. *J. Reprod. Fertil.*, **48**, 83
211. Hubbard, T. W. (1878). Varicella occurring in an infant twenty-four hours after birth. *Br. Med. J.*, **1**, 822
212. Williamson, A. P. (1975). The varicella-zoster virus in the etiology of severe congenital defects. *Clin. Pediat.*, **14**, 553
213. Duehr, P. A. (1955). Herpes zoster as a cause of congenital cataract. *Am. J. Ophthalmol.*, **39**, 157
214. McKendry, J. B. J. and Bailey, J. D. (1973). Congenital varicella associated with multiple defects. *Can. Med. Ass. J.*, **108**, 66
215. Brunell, P. A. (1967). Varicella-zoster infections in pregnancy. *J. Am. Med. Ass.*, **199**, 315
216. Harley, J. D. and Gillespie, A. M. (1972). A complicated case of congenital vaccinia. *Pediatrics*, **50**, 150
217. Green, D. K., Reid, S. M. and Rhaney, K. (1966). Generalised vaccinia in the human foetus. *Lancet*, i, 1296
218. Friart, G. M. (1963). Vaccination antivariolique *in utero*. *Arch. Belges Derm.*, **19**, 191
219. Liebeschuetz, H. J. (1964). The effects of vaccination in pregnancy on the foetus. *J. Obstet. Gynecol. Br. Cwth.*, **71**, 132
220. Kalter, S. S., Heberling, R. L. Panigel, M., Brack, M. and Felsburg, P. J. (1973). Fetal infection of the baboon, (*Papio cynocephalus*) with vaccinie virus. *Proc. Soc. Exp. Biol. Med.*, **143**, 1022
221. Haemmerli, U. P. (1966). Jaundice during pregnancy. *Acta Med. Scand.*, **179**, Suppl. 444, 17
222. Zuckerman, A. J. (1974). Maternal transmission of hepatitis. *Nature (Lond.)*, **249**, 105
223. Stoller, A. and Collmann, R. D. (1965). Incidence of infective hepatitis followed by Down's syndrome nine months later. *Lancet*, ii, 1221
224. Stoller, A. (1968). Virus-chromosome interaction as a possible cause of cases of Down's syndrome (mongolism) and other congenital anomalies. In: D. M. H. Woolham (ed.) *Advances in Teratology*, 3, 97. (London: Logos)
225. Spence, M. R., Davidson, D. E., Dill, G. S., Boonthai, P. and Sagartz, J. W. (1975). Rabies exposure during pregnancy. *Am. J. Obstet. Gynecol.*, **123**, 655
226. Zigas, V. (1973). Effect of Kuru on pregnancy. *Trop. Geogr. Med.*, **25**, 262
227. Hubbert, W. T., Bryner, J. H., Fernelius, A. L., Frank, G. H. and Estes, P. C. (1973). Viral infection of the bovine fetus and its environment. *Arch. ges. Virusforsch.*, **41**, 86
228. Osburn, B. I., Silverstein, A. M., Prendergast, R. A., Johnson, R. T. and Parshall, C. J. (1971). Experimental viral induced congenital encephalopathies. I. Pathology of hydranencephaly and porencephaly caused by blue tongue virus. *Lab. Invest.*, **25**, 197
229. Johnson, K. P., Ferguson, L. C., Byington, D. P. and Redman, D. R. (1974). Multiple fetal malformations due to persistent viral infection. *Lab. Invest.*, **30**, 608
230. Brown, T. T., DeLahunta, A., Bistner, S. I., Scott, F. W. and McEntee, K. (1974). Pathogenetic studies of infection of the bovine fetus with bovine viral diarrhoea virus. *Vet. Pathol.*, 11, 486
231. Anderson, A. A. and Campbell, C. H. (1976). Experimental placental transfer of foot-and-mouth disease virus in mice. *Am. J. Vet. Res.*, **37**, 585
232. Richards, W. P. C. and Cordy, D. R. (1967). Blue tongue virus infection: pathologic responses of nervous systems in sheep and mice. *Science*, **156**, 530
233. Narayan, O. and Johnson, R. T. (1972). Effects of viral infection on nervous system development. I. Pathogenesis of blue tongue virus infection in mice. *Am. J. Pathol.*, **68**, 1
234. Ferm, V. H. and Kilham, L. (1965). Histopathologic basis of teratogenic effects of H-1 virus on hamster embryos. *J. Embryol. Exp. Morphol.*, 13, 151

235. Kilham, L. and Margolis, G. (1973). Pathogenesis of intra-uterine infections in rats due to reovirus type 3. I. Virologic studies. *Lab. Invest.*, **28**, 597
236. Kilham, L. and Margolis, G. (1973). Pathogenesis of intra-uterine infections in rats due to reovirus type 3. II. Pathologic and fluorescent antibody studies. *Lab. Invest.*, **28**, 605
237. Kilham, L., Margolis, G. and Colby, D. E. (1967). Congenital infections of cats and ferrets by feline paneukopenia virus manifested by cerebellar hypoplasia. *Lab. Invest.*, **17**, 465
238. Edwards, M. J. (1972). Influenza, hyperthermia and congenital malformation. *Lancet*, **i**, 320
239. Edwards, M. J. (1969). Congenital defects in guinea pigs: fetal resorption, abortions and malformations following induced hyperthermia during early gestation. *Teratology*, **2**, 313
240. Wanner, R. A., Edwards, M. J. and Wright, R. G. (1975). The effect of hyperthermia on the neuro-epithelium of the 21-day guinea pig foetus: histologic and ultrastructural study. *J. Pathol.*, **118**, 235
241. Alzamora, V., Rotta, A., Battilana, G., Abugattas, R., Rubio, C., Bouroncle, J., Zapata, C., Santa-Maria, E., Binder, T., Subiria, R., Paredes, D., Pandio, B. and Graham, G. (1953). On the possible influence of great altitudes on determination of certain cardiovascular anomalies: preliminary report. *Pediatrics*, **12**, 259
242. Clemmer, T. P. and Telford, I. R. (1966). Abnormal development of rat heart during prenatal hypoxic stress. *Proc. Soc. Exp. Biol. Med.*, **121**, 800
243. Jackson, B. T. (1968). The pathogenesis of congenital cardiovascular anomalies. *N. Engl. J. Med.*, **279**, 80
244. Moss, A. J., Emmanouilides, G. C., Adams and Chuang, K. (1964). Response of ductus arteriosus and pulmonary and systemic arterial pressure to changes in oxygen environment in newborn infants. *Pediatrics*, **33**, 937
245. Runner, H. M. and Miller, J. R. (1956). The congenital deformity in the mouse as a consequence of fasting. *Anat. Rec.*, **124**, 437

17
Environmental effects on normogenesis and teratogenesis, with special regard to noise and vibration

J. FANGHÄNEL AND G.-H. SCHUMACHER

THE INTERACTION OF GENETIC AND PERISTATIC FACTORS AS MALFORMATION CAUSES

Formerly, teratological literature dealt chiefly with the development of different malformations, but research into their causes has attracted increasing interest during recent decades. Teratological research received a major impetus when the Australian physician Gregg[1] discovered that rubella infection of the mother during early pregnancy can cause severe malformations in the fetus. Research into the causes of malformations was further stimulated by the thalidomide catastrophe in the Federal Republic of Germany. The latter in particular forcefully showed the previously ignored necessity of thoroughly checking new drugs for teratological effects and simultaneously indicated the complexity of the whole problem. However, important new knowledge was brought to light by the genetic techniques of chromosome analysis introduced in 1956.

More recent publications[2–8] especially have underlined the fact that malformations are caused by complicated interactions between genetic and environmental factors. Some malformations are determined by genetic, and others by environmental factors. However, both groups of factors are involved in the large majority of congenital malformations (Figure 17.1).

According to Rosenbauer[7], exogenic factors are responsible for some 90% of all malformations, whereas only 10% can be attributed to genetic causes. Rosenbauer[7] assumes, furthermore, that 10% of the malformations are caused by the embryonal environment, and that the remaining 80% are caused by some interplay of exogenic and endogenic factors of which we still do not have any detailed knowledge.

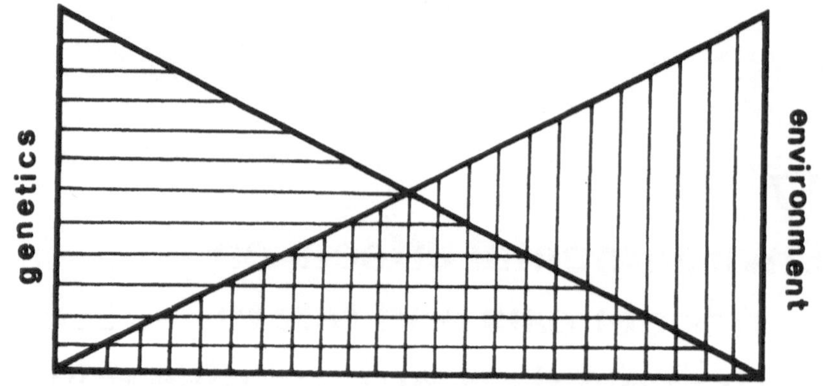

Figure 17.1 The interaction between genetic and environmental factors after K. H. Degenhardt[3] (simplified). The horizontal shaded portion represents genetic and the vertically shaded portion environmental factors. Where these areas overlap, both groups of factors act jointly

Von Kreybig[9] describes three possible groups of malformation sources which frequently interact but may also take effect separately:

— endogenic malformation sources which must be sought in the developing embryo itself,
— peripheral or physiological malformation sources such as anomalous fetomaternal relations or metabolic anomalies of the mother, and
— exogenic malformation sources located in the environment and acting on the embryo through the maternal organism or metabolism.

Figure 17.2 shows how closely genotype and phenotype are related in teratological experiments, largely as a result of the complex interaction between endogenic and exogenic factors mentioned above.

The development of all teratogenic damage, irrespective of the causal factor, is based on a common principle: similar disturbances in the biomechanism of the cell. These disturbances include interruption of reaction chains producing energy at cellular level, inhibition of cell division, and inhibition of ferment activity and synthesis during specific metabolic processes. Freundt[10] states in this connection that, where exogenic noxae are involved, the borders between inactive, active but tolerable, harmful, and lethal are not clearly defined.

Mortality due to malformations is relatively high during the first half of pregnancy, and a large proportion of spontaneous abortions are probably the result of malformations. During the second half of pregnancy, the mortality of malformed fetuses drops substantially. It then rises again slightly during the final weeks prior to birth, and thereafter declines steadily in the course of the first few months (Figure 17.3). The shape of the mortality curve can be explained by considering the different intra-uterine and extra-uterine conditions to which the fetus or infant, respectively, is exposed. A fetus nourished through the placenta by its mother is fully able to develop despite, for example, malformations of the respiratory, digestive, or excretory organs, and even in the presence of brain or heart malformations. However, such offspring

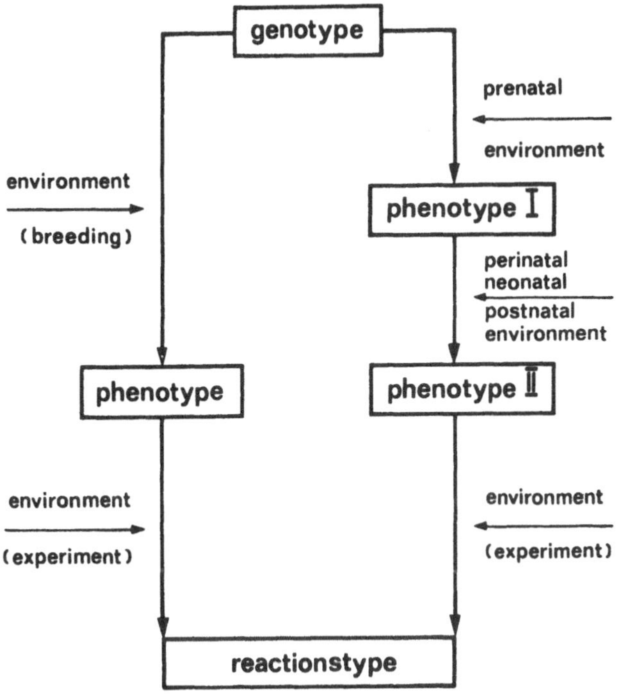

Figure 17.2 The relationship between genotype and phenotype

die upon exposure to extra-uterine conditions unless the congenital defects can be repaired by surgery. Newborns with malformations are viable only if the malformation is slight or affects areas or organs which are not essential for extra-uterine life. In view of the fact that several malformations can be rectified by current surgical practice, teratology is also of major practical importance[11].

PERISTATIC FACTORS

Environmental effects are assuming increasing importance in the industrialized society. Many of the peristatic factors involved in the development of malformations are already a major environmental problem. This explains the increasing importance of experimental teratology and applied teratological research because these activities show what should be done to prevent malformations.

Goerttler[5] defines the environment as everything existing outside of the chromosomes (Figure 17.4). The chromosomes are subject to nucleus–cytoplasmic interactions. The oocyte and sperm are exposed to a variety of influences in the extracellular and intraorganismic areas. The uterus is for a long time the environment of the developing embryo and fetus, the placenta performing an important mediator function during this period. Although the intramaternal environment provides a high level of protection for the embryo against teratogenic noxae, it simultaneously exposes it to disturbances arising

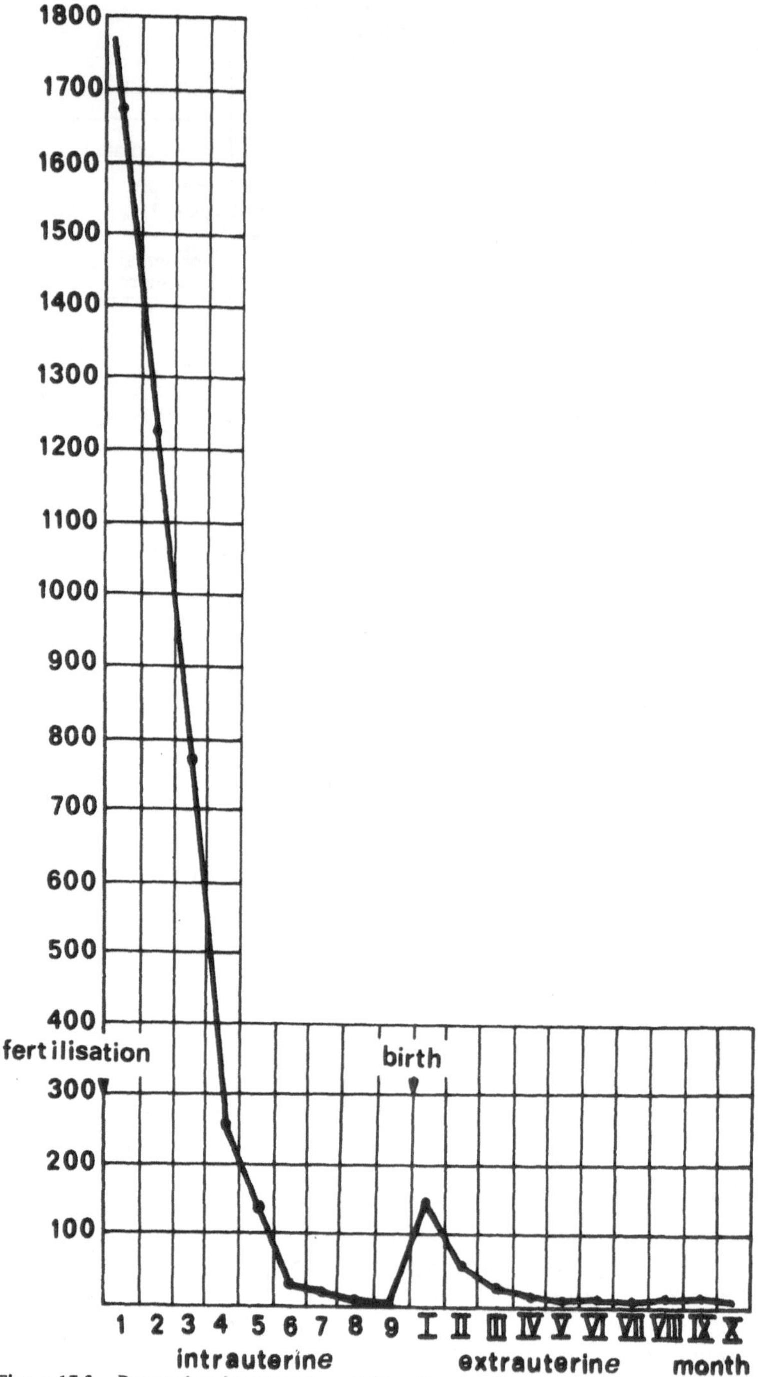

Figure 17.3 Prenatal and postnatal mortality after von Pfaundler and Schlossmann, in M. Wrete[84]

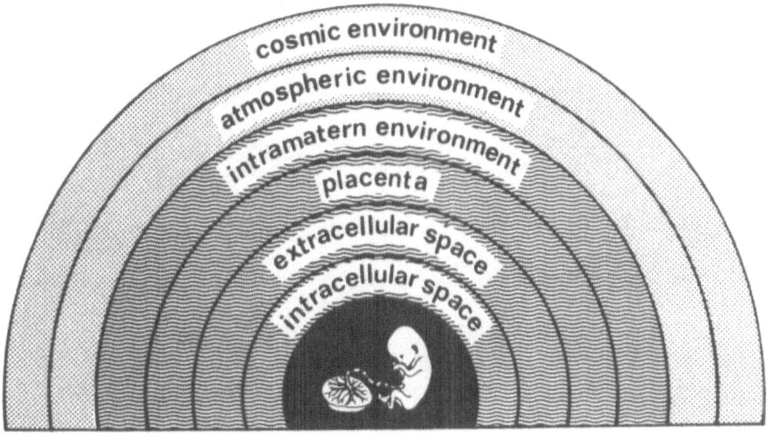

Figure 17.4 Environmental areas, modified after Goerttler[82]

in the maternal organism itself. Outside of the maternal organism lies the atmospheric environment with its physical, mechanical, chemical, and infectious factors. Radiation penetrating into the atmosphere, finally, forms part of the cosmic environment.

Although environmental teratogenic factors have long been sought, few specific noxae have so far been correlated with malformations. In contrast, the number of environmental factors is extraordinarily large. For example, Figure 17.5 summarizes the factors mentioned in the international literature as being responsible for clefts in the lip, palate and jaw. The majority of these factors were determined in animal experiments and therefore are not all applicable to man.

The period of teratogenic determination is decisive in connection with the development of malformations. It is initiated by several biochemical and fermentative processes and coincides with a phase of elevated mitotic activity in the organism concerned. For some known teratogenic substances, it is possible to derive a typical pattern, or malformation calendar, corresponding to the process of differentiation. Such a 'calendar' is shown in Figure 17.6 for the development of the human embryo.

It is well known that iatrogenic factors can also have teratogenic effects. Typical examples are ionizing radiation, ultrasonic vibration, short wave radiation, curettage, medicaments, narcosis, vaccination, undercooling, and overheating. A comprehensive list was published by Rosenbauer[6,7]. X-, neutron-, and gamma-rays are strongly teratogenic and cause brain damage, malformations of the skull, blindness, cleft palates and limb malformations. The natural dose of cosmic and radioactive radiation absorbed during a pregnancy is about 100 mR. The devastating effects of high radioactivity doses have been demonstrated by the atomic explosions above Hiroshima and Nagasaki, a topic dealt with especially by Japanese authors[12–14]. Of the Japanese women affected by these explosions, 28% had miscarriages, 25% of the children died during the first year of life, and 25% of the surviving children

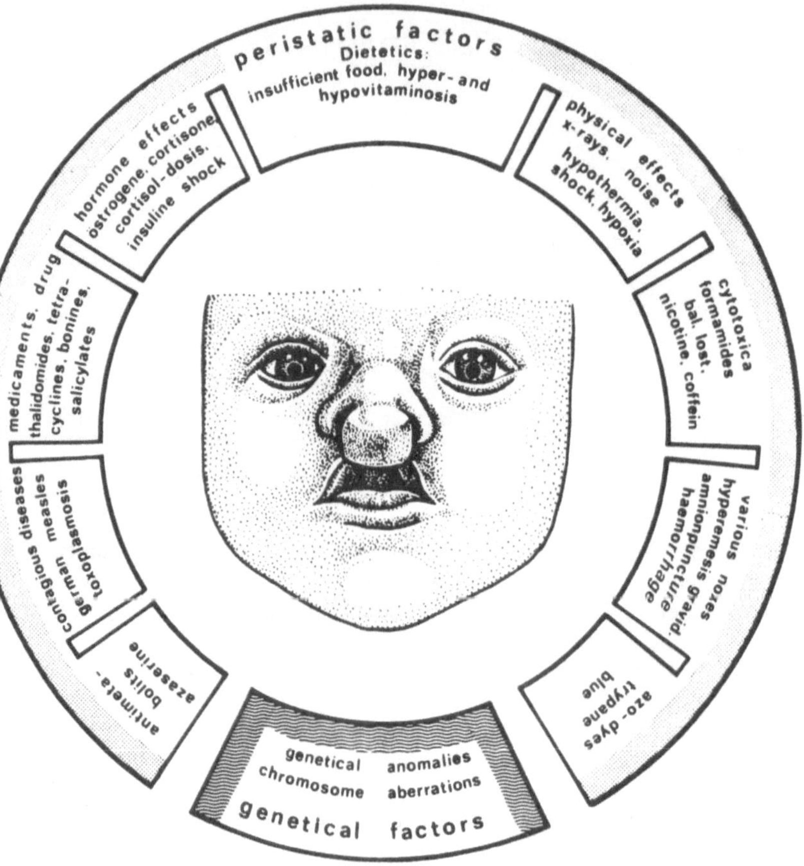

Figure 17.5 Teratogenic factors causing craniofacial dysplasia in man and animals, from Schumacher[83]

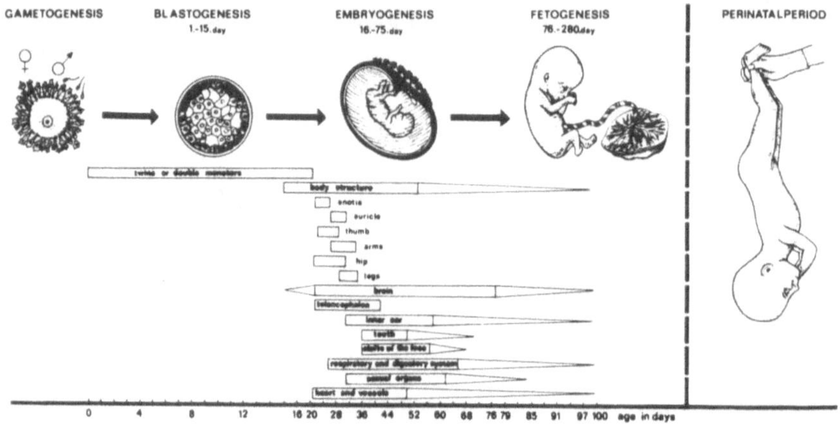

Figure 17.6 Phases in the development of the human embryo and a 'malformation calendar' compiled from data in Goerttler[82], from Schumacher[83]

suffered from defects of the central nervous system. Indirect effects on the reproductive cells must also be expected.

Among the medicaments, cytostatic preparations acting as mitosis inhibitors, alkylating substances, or antimetabolites are important malformation sources. Lathyrogenic substances have also attracted the interest of teratologists[15] on account of the skeletal and vascular damage they have been shown to cause. Although in placental animals a barrier exists between the maternal and embryonal organism, numerous drugs or their metabolic products pass from mother to embryo. Some authors[16] suspect that the majority of medicaments can cross the uteroplacental barrier. In Wendler's opinion[17], every individual represents a genetic entity and possesses an inherited response to medicaments, their conversion, and their decomposition products. Malformations from these sources affect particularly the head, brain, eyes, ears, skeleton, bowels, and reproductive organs.

The teratogenic effects of oxygen deficiency have long been known. These may be due to modifications of the placenta or vessels in the umbilical cord, vascular disorders of the mother, narcosis during pregnancy, and several other causes. The literature dealing with primary and secondary oxygen deficiency as teratogenic noxae has reached almost insurmountable proportions. Of the numerous authors, we shall mention only Becher[18], and Degenhardt[19], and Büchner[20] whose fundamental research on chicken and rabbit embryos showed that reduced atmospheric pressures and oxygen deficiency during embryonal development lead to CNS and heart defects, and to retarded growth.

Numerous malformations can be induced by hormone application in animal experiments, and analogous observations have also been made in man. It is already known that hormones from the adrenal gland (Table 17.1),

Table 17.1 Teratogenic effects of the suprarenal gland

	Species	*Malformations*	*References*
Adrenalin	Rat	Nervous system	Jost, A. (1953). *Arch. Franc. Pediatr.*, **10**, 865
	Rat	Extremities, skeleton, lower jaw	Jost, A. (1964). *Brux. Med.*, **44**, 245
Corticosteroids	Mouse	Cleft palate	Kalter, H. (1954). *Genetics*, **39**, 185
	Mouse	Cleft palate, exencephalia, eyes	Miller, J. R. (1962). *Nature* (*Lond.*), **104**, 891
	Rat	Suprarenal gland	Lohmeyer, H. (1961). *Geburtsh. u. Frauenheilk.*, **21**, 560
	Rabbit	Cleft palate	Fainstat, T. (1954). *Endocrinology*, **55**, 502
	Man	Cleft palate	Harris, J. W. S., u. Mitarb. (1956). *Lancet*, 1045

After Rosenbauer[7]

reproductive organs (Table 17.2), pancreas (Table 17.3), thyroid gland, and hypophysis can have teratogenic effects. Shock and stress probably act directly through the endocrine apparatus to produce malformations.

Among the nutritional and metabolic disturbances, malnutrition, vitamin

Table 17.2 Teratogenic effects of hormones from the gonads

	Species	*Malformations*	*References*
♂	Rat	Intersexuality	Greene, R. R., u. Mitarb. (1939). *Am. J. Anat.*, **65**, 414
	Man	Intersexuality	Wilkins, L. (1960). *J. Amer. Med. Ass.*, **172**, 1028
	Mouse	Extremities	Kameyama, Y., u. Mitarb. (1963). *Proc. Ann. Res. Ass. Japan*, 3rd Ann. Meet
	Rabbit	Skeleton, umbilicus	Courrier, R., u. Mitarb. (1939). *Compt. Rend. Sc. Soc. Biol.*, **130**, 726
♀	Toad	Extremities, gonads	Takahashi, H. (1957). *Ann. Zool. Jap.*, **30**, 199
	Chicken	Kidneys	Landauer, W., u. Mitarb. (1939). *Proc. Soc. Exp. Biol. Med.*, **41**, 80
	Rat	Intersexuality	Greene, R. R., u. Mitarb. (1940). *Am. J. Anat.*, **67**, 305
	Mouse	Skeleton	Gardner, W. U., u. Mitarb. (1937). *Proc. Soc. Exp. Biol. Med.*, **37**, 678
	Mouse	Eyelids	Raynaud, A. (1943). *C. Rend. Sc. Soc. Biol.*, **136**, 337
	Rabbit	Skeleton	Courrier, R., u. Mitarb. (1939). *Compt. Rend. Sc. Soc. Biol.*, **130**, 726
	Man	Extremities, brain	Uhlig, H. (1959). *Geb. Hilf. u. Frauenheilkunde*, **19**, 346

After Rosenbauer[7]

Table 17.3 Teratogenic effects of pancreatic hormones

	Species	*Malformations*	*References*
Insulin	Chicken	Micromelia, skeleton, skull	Landauer, W. (1953). *Proc. Nat. Acad. Sci., USA*, **39**, 54
	Chicken	Brain, heart	Baroon, P., u. Mitarb. (1962). *J. Embryol. Exp. Morphol.*, **10**, 88
	Rat	Eyes, skeleton	Lichtenstein, H., u. Mitarb. (1951). *Proc. Soc. Exp. Biol. Med.*, **78**, 398
	Rabbit	Extremities, skull, brain	Brinsmade, A., u. Mitarb. (1951). *Naturw.*, **23**, 398
	Man	Microcephalia, eyes	Wickes, I. G. (1954). *Br. Med. J.*, **2**, 1029
Glucagon	Rat	Eyes, skeleton	Tuchmann-Duplessis, H., u. Mitarb. (1962). *Presse Méd.*, 1411

After Rosenbauer[7]

deficiency, and, for some vitamins, hypervitaminosis are known to have teratogenic consequences (Table 17.4). Vitamin A deficiency, for example, may cause the death or malformation of the embryo, affecting especially the CNS

Table 17.4 Teratogenic effects of nutritional disturbances

	Species	*Malformations*	*References*
Vitamin A hypovitaminosis	Rat	Aortic arch, eyes	Wilson, J. G., u. Mitarb. (1949). *Am. J. Anat.*, **85**, 113
	Pig	Extremities, eyes, heart, kidneys, hydrocephalia	Pallidan, B. (1961). *Acta Vet. Scand.*, 2, 32
	Chicken	Micromelia	Warkany, J., u. Mitarb. (1943). *J. Bone Joint Surg.*, **25**, 261
Vitamin B riboflavin deficiency	Rat	Extremities, jaw, tongue, ribs	Warkany, J., u. Mitarb. (1943). *Proc. Soc. Exp. Biol. Med.*, **54**, 92
Folic acid deficiency	Rat	Hydrocephalia	O'Dell, B. L., u. Mitarb. (1951). *Proc. Soc. Exp. Biol. Med.*, **76**, 349
Pantothenic acid deficiency	Rat	Anophthalmia, exencephalia	Biosselot, J. (1948). *Arch. Franc. Pédiat.*, 225
Vitamin E hypovitaminosis	Rat	Syndactylia, exencephalia, hydrocephalia	Runnder, M. N., u. Mitarb. (1956). *Anat. Rec.*, **124**, 437
Malnourishment	Mouse	Skeleton	Cheng, D. W., u. Mitarb. (1953). *Proc. Iowa Acad. Sci.*, **60**, 290

After Rosenbauer[7]

and the eyes. Administration of excessive quantities of vitamin A has produced similar effects in animals, as have the vitamin B complex and vitamin D. The continuous animal experimental studies undertaken by Persaud[21,22], who has shown in several publications that the use of herbs as medicaments or food during pregnancy in Third World countries, is an important factor in the aetiology of congenital malformations (Table 17.5) are noteworthy in this context.

Table 17.5 The effects of leucine, riboflavin phosphate, and carnitine on the teratogenic activity of hypoglycine A

	Number of maternal animals	*Implantations*	*Resorptions*	*Abnormal fetuses*
Hypoglycine A + leucine	6	52	17 (32.7%)	35 (100%)
Hypoglycine A + riboflavin phosphate	6	51	3 (5.9%)	21 (43.7%)
Hypoglycine A + carnitine	6	48	3 (6.3%)	36 (80%)
Hypoglycine A	6	53	4 (7.5%)	40 (81.6%)

After Persaud[22]

Having roughly outlined the problem, we now wish to report on the studies we have performed on the effects of extra-aural noise and vibration as teratogenic noxae at the Anatomical Institute of the Wilhelm Pieck University, Rostock, during the past few years.

The steadily increasing noise and vibration load to which all persons living in both town and country are exposed has resulted in acoustic (Figure 17.7) and vibration effects becoming one of the most topical environmental protection problems[23–26].

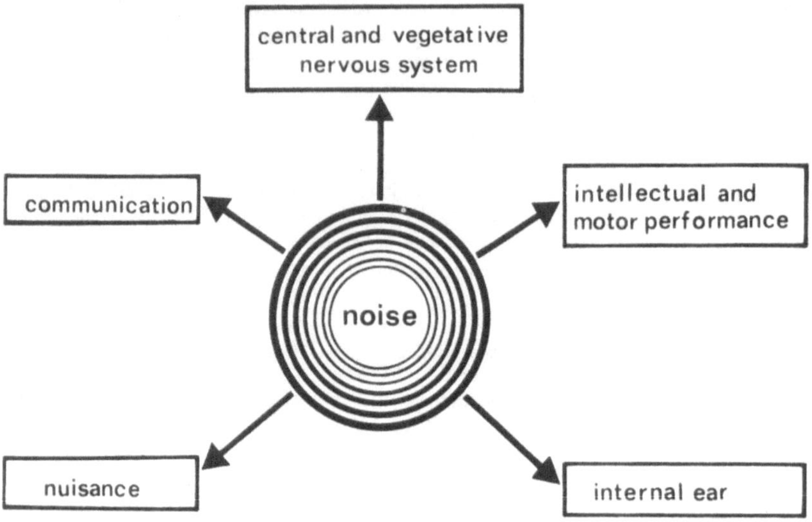

Figure 17.7 The effects of noise as an environmental factor, from Schumacher and Fanghänel[25]

Noise belongs to the environmental factors with the most widespread effects[27]. As a concept, noise is difficult to define because its perception is processed in the brain and is therefore subjectively influenced. Moldenhauer[28] differentiates between direct and indirect harmful effects, the former acting through the auditive apparatus and the latter through the CNS. The fact that the auditive organ is very sensitive and communicates with all organ systems by means of the central, vegetative, and peripheral nervous systems must be taken into special account when estimating noise damage. Physically, noise can be quantified by measuring the acoustic pressure or the volume. Lehmann[29] and Rosenau[30] distinguished four noise level ranges which produce different responses in the organism. At noise level range I from 30 to 65 phones psychic modifications can be observed. Vegetative responses occur at sound level range II from 65 to 90 phones, and at sound level stage III, from 90 to 130 phones, the two previously mentioned responses are accompanied by damage to the inner ear manifested as partial or total deafness[31–34]. Direct exposure of ganglia cells to sound level stage IV, i.e. above 130 phones, leads to the paralysis or destruction of the cells[29, 30, 35–37].

In the case of vibration, we are concerned with vibrations affecting the whole body, i.e. low-frequency mechanical oscillations at frequencies from one to 90 Hz transmitted through the limbs, the seat, and the back (so called z axis = along the axis of the body). The perception of vibration is rather complex and occurs when the nerve endings at the ends of the follicles and in Meissner's corpuscles are excited by at least 18 stimuli/s[38]. Vibration is physically defined by acceleration, velocity, and amplitude. Panzke[39] dealt with the physical fundamentals of mechanical oscillations. Oscillations arise due to the fluctuation of a mechanical quantity in alternate directions. They can be of a stochastic or a determinate nature. Stochastic vibration exposure is measured analogously to noise exposure by using an equivalent continuous acceleration[40]. In this way it is possible to determine the duration of exposure to oscillations with a strictly periodic acceleration which is equivalent to a defined duration of exposure to oscillations with aperiodic acceleration. The frequency-dependent effective acceleration value is used as a basis for measuring vibration. The specification of a maximum permissible duration of daily exposure to vibration of a certain frequency is a practical example for the use of the effective acceleration value. The pairs of limiting values calculated in this manner are used to divide places of work into four categories depending on the type of work involved. Category I contains limits for mechanical oscillations at work places. Categories II and III take into account activities requiring greater degrees of concentration, and category IV concerns intellectual work. A considerable amount of literature has been published on the use of these categories for industrial medical purposes[41].

MATERIALS AND METHODS

Our studies are based on about 4000 rat embryos and newborn from a strain ROZT: Wistar (*Rattus norvegicus* Berkenhout), the mothers of which were exposed to noise during pregnancy. About 2000 embryos and newborn from mothers which were not exposed to noise served as controls. The animals were obtained from the Central Institute for Nutrition of the Academy of Sciences of the GDR, Potsdam-Rehbrücke. They were bred further at the Central Experimental Animal Unit of the Institute of Theoretical Medicine of the Medical Faculty at the Wilhelm Pieck University, Rostock, and then kept at the Anatomical Insititute in Rostock. The animals were bred and kept as described in Hagemann[42] and Schumacher, Fanghänel, and Schultz[43]. The animals were kept separately according to sex (10 animals per cage; wire cages measuring 35 × 35 × 34 cm; with wooden shavings, changed once a week). They were fed on pellets (Rezeptur K 10 mm) *ad libitum*. The animals were paired at an age of 90 days. Climatic conditions: temperature 22–24 °C, relative humidity 45–50%.

The animals were mated at an age of 100 to 120 days in the evening and were separated in the morning on the second day after mating. This day was considered as the first day after conception (1st p.c.). Since the number and bodyweight of the newborns are causally related[44], the animals intended for our experiment were taken from litters of various sizes. Primiparae were exposed to sound emitted by a loudspeaker (VEB Funkwerk Leipzig, Type L 3 354 PBK, 6.30 ohms, 12.5 VA) connected to an appropriate amplifier (VEB Funkwerk Kölleda, Type 4008, Work's No. 699). The AF signal representing the exposure tone was produced by a beat oscillator (VEB Funkwerk Erfurt, Type 2620 K, Work's No. 1310), whereas industrial noise was fed to the amplifier by a tape recorder (Tesla, Type B 4) by means of an endless tape loop. Figure 17.8 shows the experimental design. The room in which sound exposure took place measured 5 × 3.3 × 2.9 m. In order to avoid vibration, the

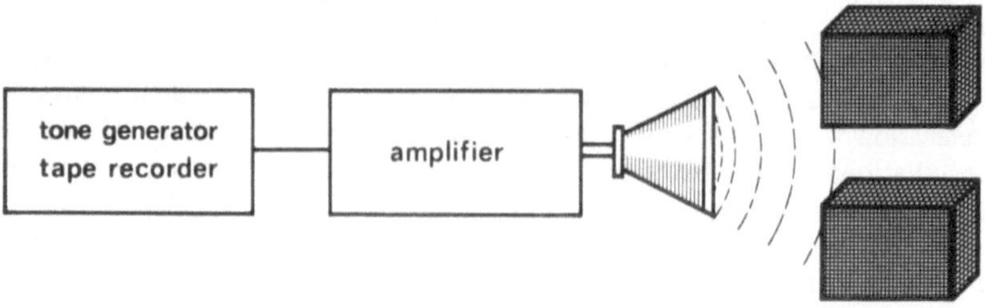

Figure 17.8 Experimental design for exposure to sinus tones and pulsed noise

loudspeaker was suspended at a height of 2.50 m and situated 2.6 m from the cage. The animals were exposed to sound in the late evening and during the night.

Exposure to sinus tones: single and double exposure; frequencies 4 kHz, 6 kHz, and 10 kHz; sound level: 100 dB (SPL); exposure on 8th, 10th, 12th, 13th, 14th, 15th, and 16th p.c.; duration of exposure: 10, 12, 14, 16, 20, 24, 28 h.

Pulsed noise exposure (noise from forge at VEB Schiffswerft Neptun Rostock): single and double exposure; pulse duration: 360 ms; pulse frequency: 52/mp/min; sound level: 115 dB (SPL); frequency range: 390–9500 Hz; principal frequency: ca. 840 Hz; exposure on 8th, 9th, 10th, 11th, 12th, 13th, 14th, 15th, 16th p.c.; duration of exposure: 2, 3, 4, 5, 6, 7, 8, 10, 12 h. The unit of volume is the phone and equates to one decibel (dB). 10 phones correspond to one Bel (B). Figure 17.9 compares the volumes of different noise sources.

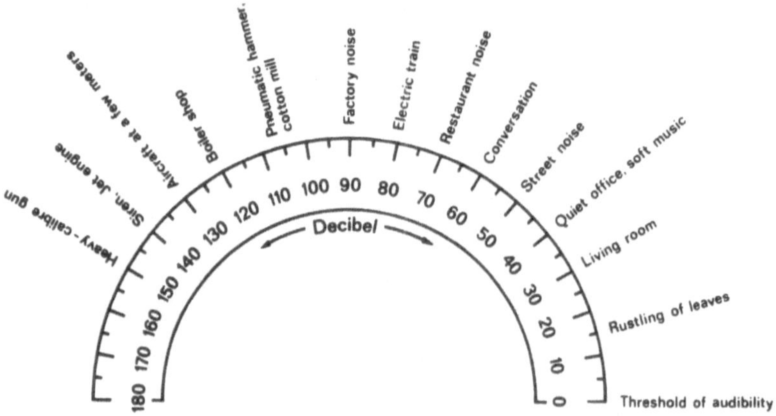

Figure 17.9 Noise levels

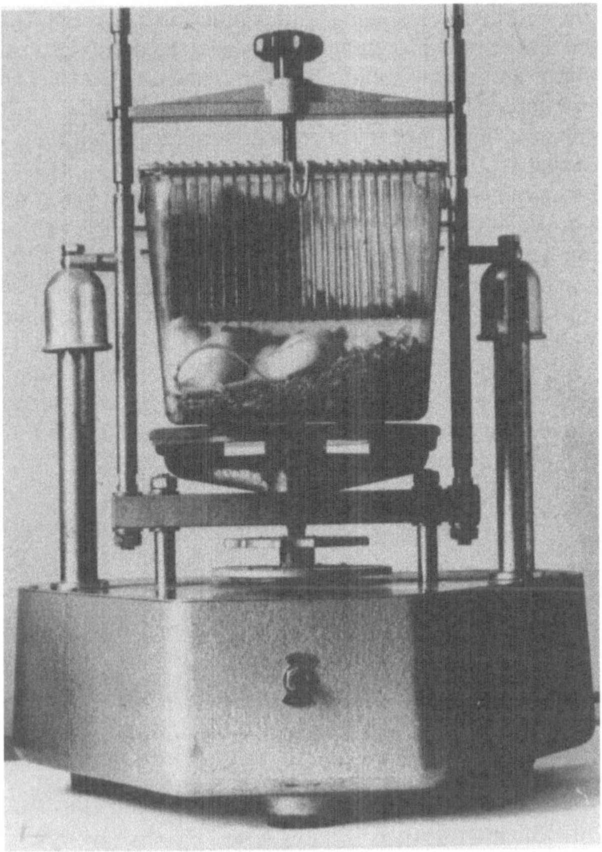

Figure 17.10 Agitation table for vibration exposure

Vibration: single exposure on a vibrating table (Figure 17.10) (VEB Rapido Radebeul); acceleration: ca. 30 ms^{-2}; exposure on 8th, 9th, 10th, 11th, 12th, 13th, 14th, 15th, 16th day p.c.; duration of exposure: 4–8 h. Table 17.6 summarizes the permissible values for oscillatory acceleration in the z axis.

Combined sinus tone and vibration: sinus tone: 4 kHz at 100 dB, 10 h; vibration: 4 h; exposure on 13th p.c.

Table 17.6 Effects of vibration

Time of exposure	*Acceleration* (ms^{-2})
1 min	5.6
10 min	4.7
30 min	3.1
1 h	2.4
5 h	0.9
10 h	0.5

Acceleration: ca. 30 ms^{-2}

Some of the mothers were killed immediately before completion of pregnancy and the fetuses and placentae were investigated biometrically (parietococcygeal length, gross body weight, median placenta diameter) and morphologically for visible malformations. Student's *t*-test was used to determine the level of significance. The numbers of resorptions (resorption rate) and malformations were also ascertained.

In other series of experiments, the young animals were killed at intervals of five days from the 40th to the 140th day after birth. They were investigated gravimetrically and the data were used to calculate growth curves. We found that homogeneous linear differential equations (Figure 17.11) form a particularly good approach for such work[45–48]. The coefficients were obtained by the difference method. Integration of the differential equation led to a solution in the form of the general integral. The graphs of the special solutions of the general integral produce the growth curves, and the growth rate and increase in growth rate are the first and second derivatives of the special solutions. All

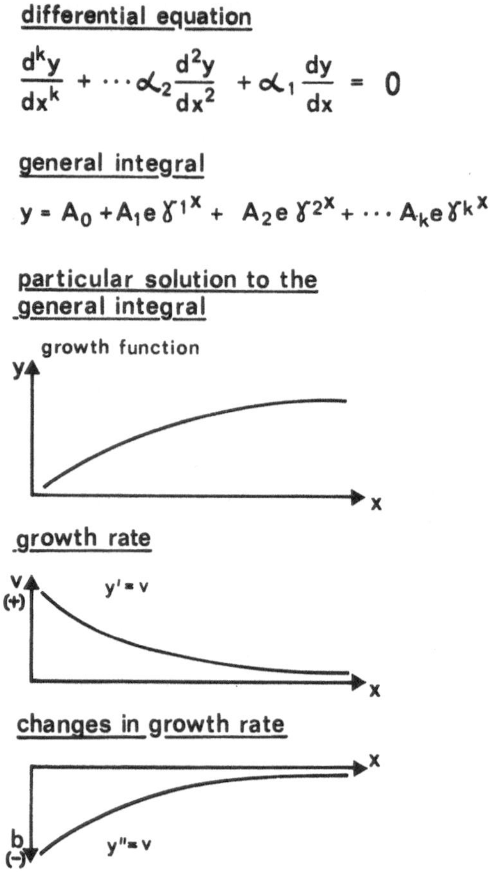

Figure 17.11 The mathematical approach for the biometric analysis of postnatal growth of offspring from sound-exposed maternal animals, after Fanghänel and Schumacher[45]

calculations were performed on ZRA-1 and R-300 computers at the computer centre of the Wilhelm Pieck University, Rostock, In order to identify significant differences, we determined confidence intervals for the curve segments from morphological measurements. Differences are significant at the point where the intervals diverge. In test series, it is thus possible to ascertain the degree of a noxa from curves plotted on the basis of length measurements[45].

We shall not reproduce here the different weights and distances measured. The authors are willing to provide these upon request.

FINDINGS

Exposure to sinus tones

In earlier studies[25,26,49], no changes in growth or malformation frequency were observed after exposure to sinus tones (4 kHz and 6 kHz) for less than 10 h (Figure 17.12). The resorption rate was about 5%. Sinus tones thus have no teratogenic effect under these conditions. This was checked biometrically on series of newborn and young rats killed at intervals. All parameters such as gross body weight and parieto-coccygeal length exhibited the same growth trends as those of the controls. The duration of pregnancy of the animals exposed to noise was normal.

A second period of exposure lasting 10 h and more between the 10th and 14th p.c. resulted in more frequent morphological peculiarities such as retardation (80%), increased resorption rates (25%), and a greater variety of placental anomalies (Figure 17.12). The mean significant reduction in parieto-coccygeal length of newborns was 6 mm. The corresponding gross body weight was about 1 g below normal. Some of the placentae were significantly

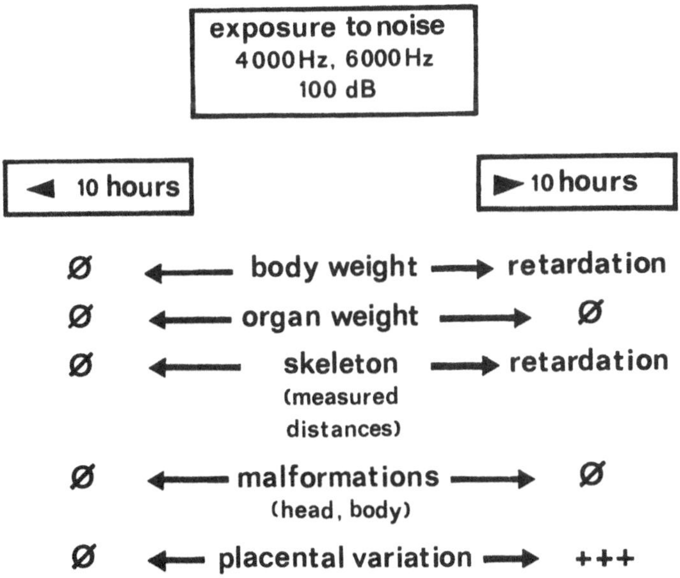

Figure 17.12 The effects on fetuses and newborn of exposing pregnant rats to sinus tones

smaller than normal (70% had a diameter of 2–3 mm compared with the normal diameter of 6 mm). Some 50% of the resorptions had a diameter of 2 mm. Classification of resorptions into early, median, and late stages as recommended by Heinecke[50] can give some indication of the causal noxa. The malformation rate was not elevated at 4 kHz and 6 kHz. A significant rise in the occurrence of malformations such as craniofacial dysplasia (Figure 17.13) and spina bifida was observed at 10 kHz. The malformation rate was 9.8%. In 40% of the mothers, pregnancy leasted 20 days under the experimental conditions described.

Double exposure to 4 kHz and 6 kHz for more than 10 h did not affect the malformation rate but produced a larger variety of placental anomalies.

The morphological changes we found were most pronounced when the animals were exposed on the 13th and 14th p.c.

As gravimetric studies[49] show, postnatal growth in rats from mothers which have been exposed to noise is slower than in the controls (Figure 17.14). The gross body weights (W_B) of male (Figure 17.14C) and female controls and of male (Figure 17.14E) and female offspring from noise-exposed mother animals increase rapidly at first, and then more slowly. This increase in weight can be interpreted by functions with e terms. We based our considerations on the following system of differential equations (derivation by time):

$W_B'' + 0.01012\, W_B' = 0$ (control, ♂)
$W_B'' + 0.01012\, W_B' = 0$ (rats from noise-exposed mothers, ♂)
$W_B'' + 0.01010\, W_B' = 0$ (controls, ♀)
$W_B'' + 0.01010\, W_B' = 0$ (rats from noise-exposed mothers, ♀)

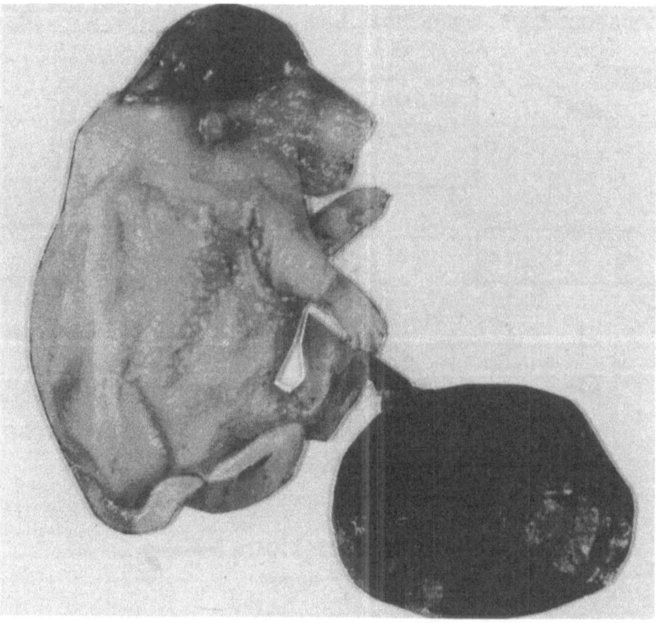

Figure 17.13 Anencephalus after exposing pregnant maternal animal to 10 kHz at 100 dB for 10 h, 14th p.c.

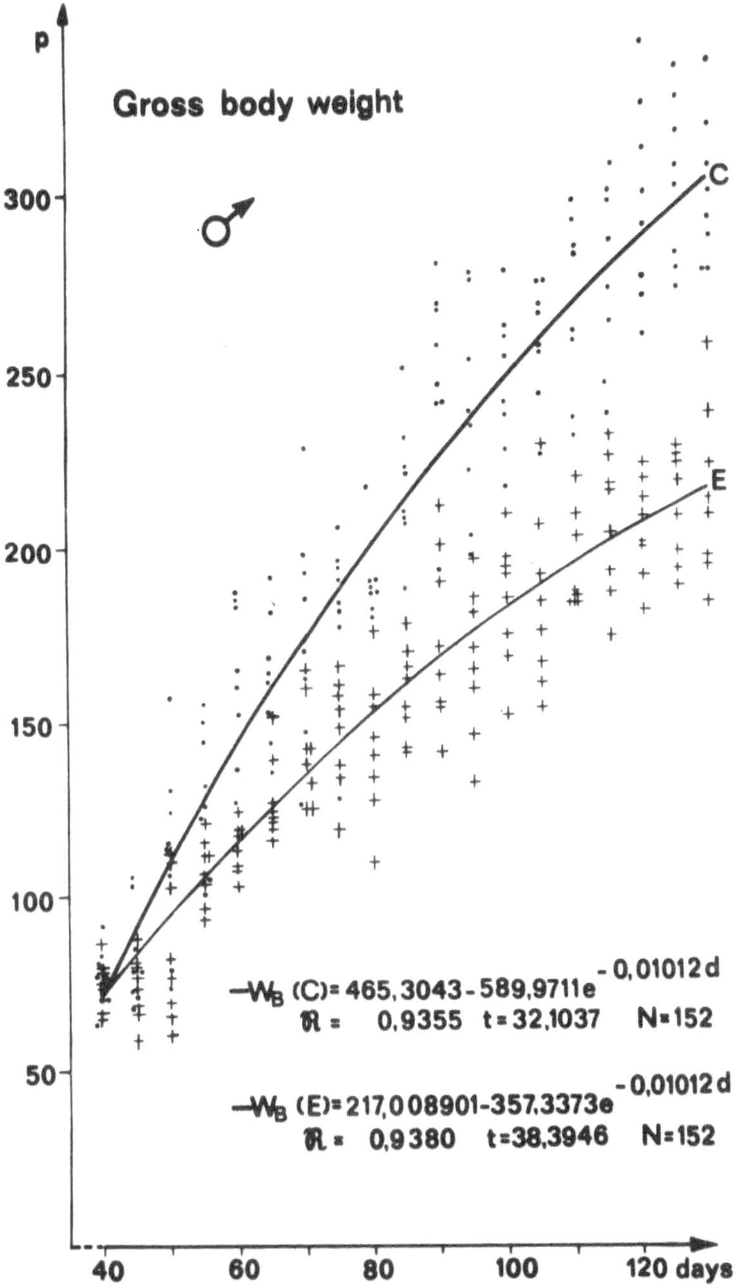

Figure 17.14 Postnatal body growth (W_B) of male rats from sound-exposed maternal animals (E) compared with controls (C), shown as special solutions of the general integral of a homogeneous linear differential equation. Sinus tone, 4 kHz, 100 dB, 10 h exposure, 14th p.c. $\mathscr{R}$ = non-linear correlation coefficient after Pearson; t = test coefficient; N = number of measurements; d = days. From Schumacher and Fanghänel[25]

The special solutions of the general integrals of the second order differential equations plotted as graphs adequately fit the empirical measured data (ℛ!). The growth rate and increase in growth rate were not constant throughout our observations. The curves for these quantities, which are not shown here, reveal that development becomes slower as the time increases.

Exposure to pulsed noise

Exposure to this noise form for less than 4 h between 8th and 16th p.c. had no effect on growth or morphology. Exposure to pulsed noise for 4–7 h induced a significant reduction in parieto-coccygeal length by up to 8 mm in newborns compared with the controls. The gross body weight was about 1 g lower than that of the controls. Retardation was most pronounced in animals exposed on the 13th and 14th p.c. (Figure 17.15). Some placentae were of reduced size

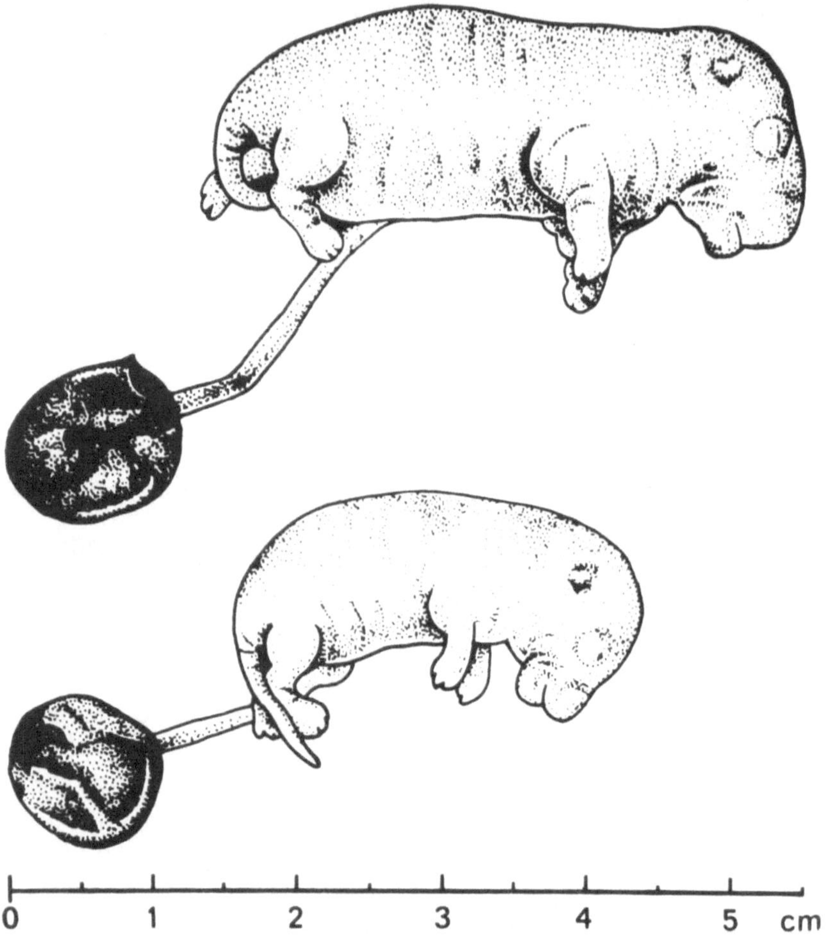

Figure 17.15 Newborn from sound-exposed mother (bottom) – exposure for 7 h, 14th p.c. – compared with newborn from control mother (top). From Schumacher and Fanghänel[26]

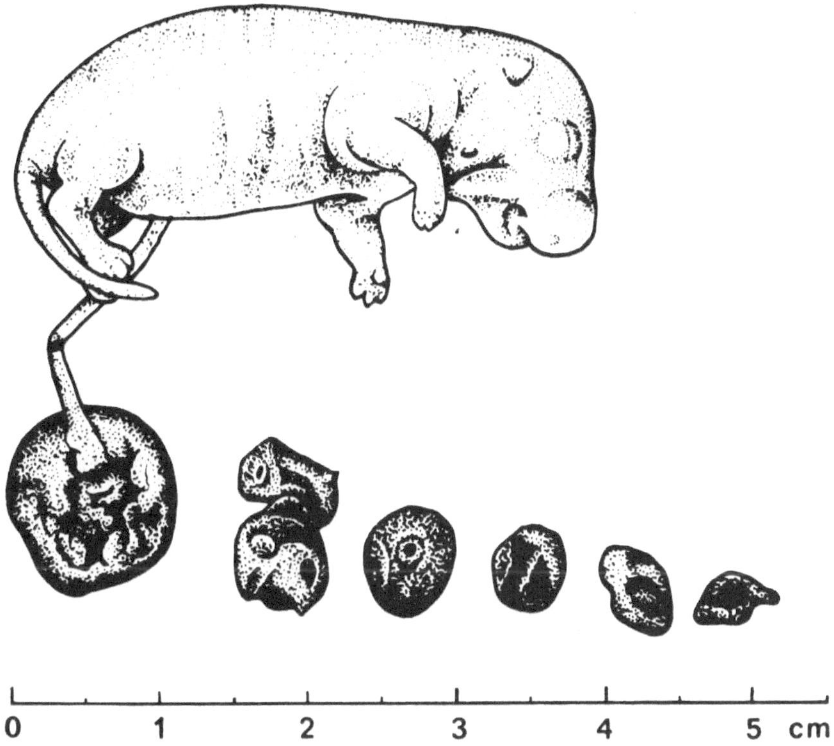

Figure 17.16 Morphological changes in placentae after exposure of maternal animals to pulsed noise. Exposure for 7 h, 14th p.c. From Schumacher and Fanghänel[26]

(65% of the placentae had a diameter of 2 mm), exhibited cystic modifications, or carried atrophied embryos (Figure 17.16). A large variety of resorptions were found, and these were concentrated particularly around the 13th p.c. (Table 17.7). The resorption rates were, at 29%, slightly higher than for exposure to sinus tones. Cyanosis of the whole body and haematomas (neck, back, tail) were frequently observed in the newborn. The amniotic fluid was frequently green to black in colour. The pregnancy lasted 21 days in 4% of the mother animals.

Table 17.7 Number of fetuses and resorptions in pregnant animals after exposure to pulsed noise for 7 h

Day of pregnancy	*Number of pregnant animals*	*Number of fetuses*	*Number of resorptions*	*% fetuses/ resorptions*
9th p.c.	19	159	5	3.1
10th p.c.	13	115	7	6.1
11th p.c.	28	238	11	4.6
12th p.c.	22	174	16	9.2
13th p.c.	17	113	18	15.9
14th p.c.	15	115	11	9.6

Table 17.8 Correlations between morphological aberrations and exposure to different noise sources

	Sinus tone	*Industrial noise*
Retardation	+	++++
Fetus resorptions	+	++
Placenta degenerations	+	++
Shortened pregnancy	+	+

Double exposure produced the same results, but the morphological modifications were more pronounced than those resulting from exposure to sinus tones (Table 17.8).

The mother exhibited abnormal physiological behaviour such as increased aggressiveness, flight reactions, and poor motor discoordination during exposure to both sinus and pulsed noise, particularly as the noise sources were switched on and off. Nitschkoff and Kriwitzkaja[51] observed the same response patterns in the F_1 generation (Figure 17.17). Such behaviour was more pronounced during exposure to pulsed noise.

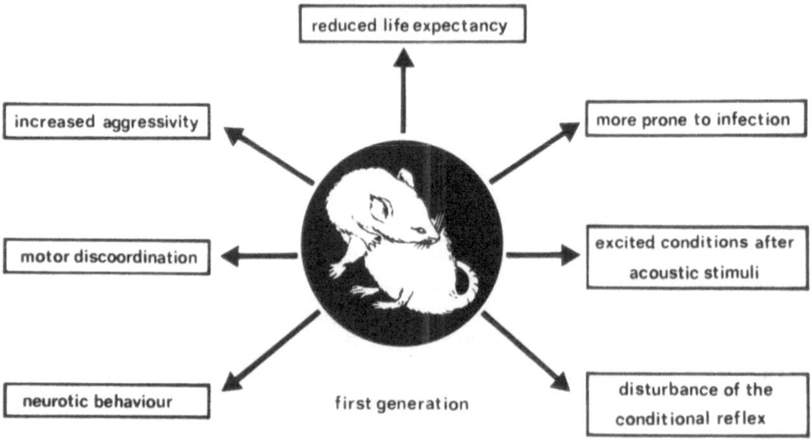

Figure 17.17 Effect of sinus tone and pulsed noise exposure on maternal animals and the F_1 generation. From Schumacher and Fanghänel[25]

Exposure to vibration

A higher resorption rate (70%) was observed following vibration exposure for 4–8 h. No retardation was observed, and the variety of placental anomalies remained within normal levels. Haematomas were found to be more frequent at exposure times of 8 h. The total mortality of the maternal animals (40%) was very high.

Additional exposure to sinus tones aggravated the effects of vibration exposure (Table 17.9), even at sinus tone loads which would not normally produce morphological changes. Retardations exceeded 10 mm and reached

Table 17.9 Effects of sinus tone and vibration exposure on the morphology of fetuses from exposed maternal animals

	Fetus resorptions	*Fetus retardations*	*Fetus haematomas*
Sinus tone, 4 h	Ø	Ø	Ø
Vibration, 4 h	+++	Ø	Ø
Vibration, 8 h	+	Ø	+++
Vibration, 4 h, and sinus tone, 4 h	+++	+++	+++

a frequency peak under these experimental conditions (Figure 17.18). The resorption rate was 70%. The number of haematomas observed in the newborn of mothers exposed to both sound and vibration was greater than in the controls. Vibration and combined exposure produced their greatest effects on the 13th and 14th p.c.

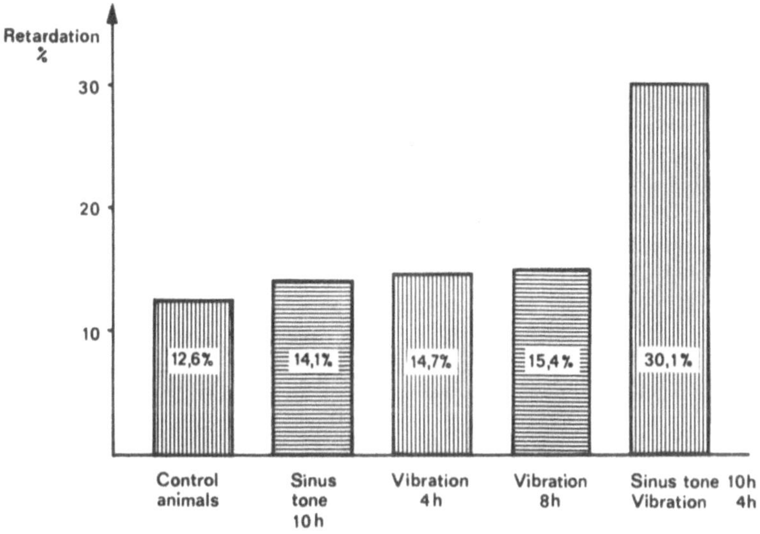

Figure 17.18 Occurrence and prevalence of retardation after exposing pregnant animals to sinus tones and vibration

DISCUSSION

Our studies using an experimental animal model have shown the effects of noise and vibration on embryonal development with respect to various physical parameters and the phenocritical phase.

Few similar teratological studies have been published so that no related results are available for comparison. Peters and Strassburg[52] observed that exposure to noise can cause malformations (cleft palates) in animals. After noise exposure (rattling crockery, beat music; 103 dB; 2–10 kHz) on the 10th day of pregnancy, the percentage of cleft palates in mice was not significantly higher than normal. Exposure on the 11th day of pregnancy, however, produced cleft palates in 31.7% of the animals compared with 2.4% for the controls. Heinecke and Klaus[33] found that exposure to pulsed noise (pneumatic drill) caused significant retardations in the F_1 generation of various laboratory mouse strains. The results of histological studies on nerve tissue following exposure to sinus tones have been published by Kuhlenkampff[53], Wüstenfeld and Schilling[54], Wüstenfeld and Gleiss[55] and Grosse, Lindner and Schneider[56]. These authors report morphological modifications to the medullary acusticus apparatus and the anterior horn cells of the myelin, and alterations in ganglia cells in *in vivo* tissue cultures.

Bredberg, Ades, and Engström[57] studied by means of scanning electron microscopy Corti's spiral organ in the guinea pig cochlea before and after exposure to noise (up to 100 dB, 4 kHz). After exposure, the sensory cells were irregularly positioned. Ades and coworkers[58] found a relationship between a shift in the auditory threshold and morphological changes in Corti's organ in rabbits. It was found that losses of hearing shown in audiograms coincided with the necrosis of hair cells in the spiral organ. Dieroff and Beck[59] report on metabolic disturbances and damage to the hair cells in the rabbit cochlea after exposure to industrial noise (100 dB, 4 kHz, 80 h). Biedermann and Meyer[60] found that the pulsed noise trauma in particular is a mechanical cell lesion which causes numerous leakages in the side of the ductus cochlearis.

Since noise affects, through the vegetative nervous system, the functions of all organs (Figure 17.19), as illustrated by the behaviour of the mothers during sound exposure, it is initially impossible to state specifically how it acts as a teratogenic factor. The vegetative responses of the organism to noise can be both sympathicotonic and vagotonic. However, the symptoms are said to be characteristic in 90–95% of the cases, such as heart circulation reactions involving precapillary vasoconstriction and peripheral blood supply disturb-

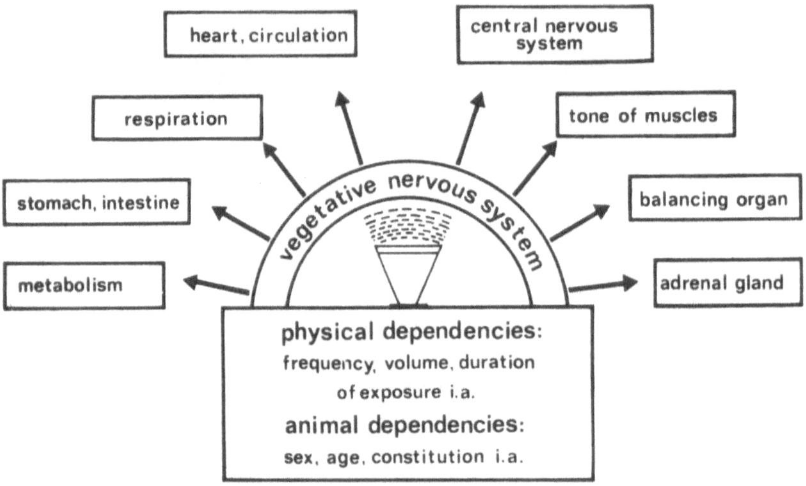

Figure 17.19 The effect of sound exposure on the organism via the vegetative nervous system. From Schumacher and Fanghänel[25]

ances, and a reduction in the systolic volume[29,61]. According to Lehmann[29], the systolic volume of the heart is reduced by half upon exposure to noise at 90 phones. Increased occurrence of extrasystoles, myocardial damage, chest disorders, and hypertonia were reported in workers exposed to noise[30]. Noise exposure is also known to induce peristaltic disturbances and disturbances in secretion into the digestive tract, changes in respiration frequency, rises in muscle tone, and dilation of the pupils[30,37,61,62]. Among others, Karrasch[33], Tamm[63], Bode[61], Rosenau[30], Nitschkoff and Kriwitzkaja[51], Jansen[64] and Baumann[65] have also reported on the increased prevalence of central nervous disorders, such as reduced performance, impaired concentration, premature tiredness, headaches, and irritability among those working in noisy environments. Spiegel[66] observed increases in the weight of hormone-producing organs such as the adrenal and thyroid glands as a result of noise stress.

As shown by the fact that exposure to pure tones must be of substantial duration to cause morphological changes, the noise quality is of major significance[49]. The occurrence of retardations and the resorption frequency show that, for *Rattus norvegicus*, the phenocritical phase occurs on the 13th and 14th days of pregnancy.

If we attempt to interpret the results of our experimental studies, it can be assumed that vasoconstriction in the maternal animals affects the diaplacentary function, and that this must affect the supply of nutrients, etc. to the fetuses. As stated earlier, oxygen deficiency and hypoglycaemia during the teratogenetic period of determination are factors which may induce malformations[18–20]. Our assumption is corroborated by the fact that short periods of exposure had, in contrast to lengthy exposure times, no detectable effect on fetal development. On the other hand, the different frequency-dependent sensitivities of the auditory organs of different species (Figure 17.20) show how difficult it is to generalize our assumptions[67]. As Klosterkötter[68] pointed out, furthermore, we are scarcely aware of the information conveyed to the experimental animals by a particular noise.

Recently, increasing interest has been attracted by the problem of vibration exposure. Since the organism is not a rigid structure but consists of numerous components which are more or less capable of oscillating, it is particularly susceptible to vibration. Vibration loads also act through the vegetative nervous system, as shown by the numerous angiopathic cases[69–71], changes in the corticosteroid balance[72] hyporeflexia and areflexia prevalence[73], and cases of disturbed sense of balance[74] among those exposed to vibration in the industrial and agricultural spheres. Klimkova-Deutschova[75] coined the term 'vibration neurodolesis' to describe the complex subjective and objective symptoms observed within the framework of a neurotic syndrome.

In our experimental animal model, disturbance of the diaplacentary function appears to be the primary cause[69]. From the aetiological standpoint, modification of the sympathetic ganglia, the peripheral nerves, and vessels play a certain role. As in the case of noise exposure, the phenocritical phase is on the 13th and 14th p.c. The mutually aggravating effects of noise and vibration exposure are, in view of day to day environmental phenomena, particularly noteworthy: Brust[76] states that the use of pneumatic drills, chain

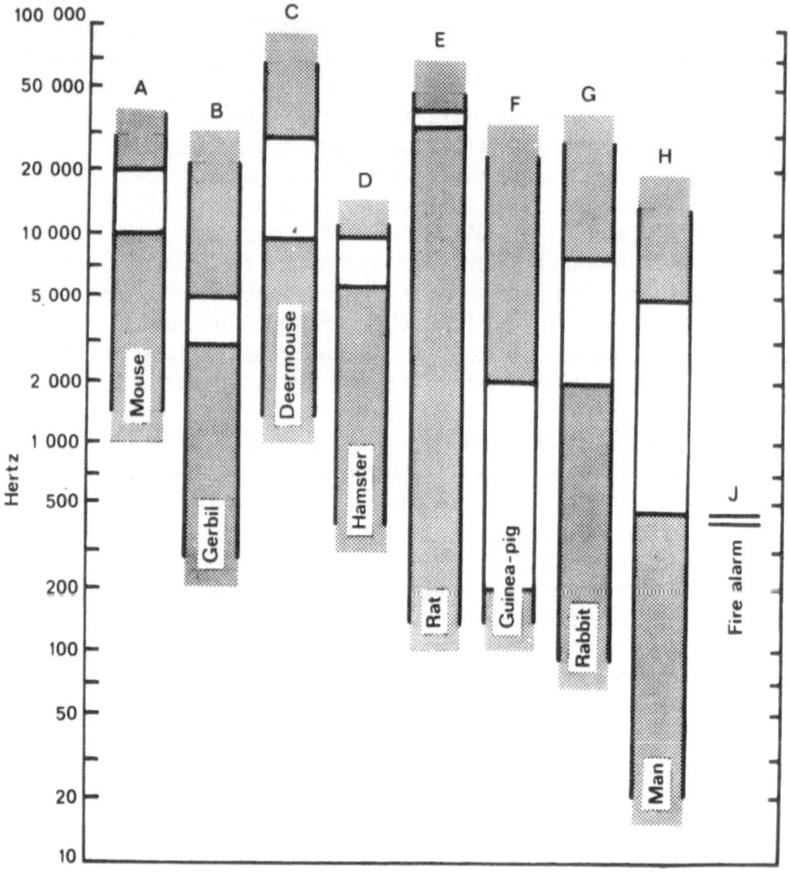

Figure 17.20 Frequency-dependent sensitivity of the auditive organ in animals and man (A–H) compared with the frequency spectrum of a fire alarm (J). Shaded areas: absolute sensitivity range, dotted areas: relative sensitivity ranges. From Clough and Fasham[67]

saws, etc., represents a dangerous combination of dust, noise, and vibration which is of extraordinary significance for industrial hygiene since it belongs to the most important multiple load group (cf. Berensci[77], a.o.).

Finally we wish to discuss a few generally valid remarks of a fundamental nature concerning the use of experimental animal models for teratological research which must be taken into account when analysing and assessing our results.

Differences in diaplacentary function and the specificity of phases during blastogenesis and embryogenesis in different animals and man lead to broadly divergent results. It is therefore difficult to compare results obtained with different animals, and to apply them to man. Freye[78] states that only genetically tested animals with placentae of the same type as found in man, i.e. a placenta haemochorialis, and with similar implantation phenomena should be used in experimental animal models. Furthermore, only virginal maternal animals should be used after a control pregnancy and after analysing, the offspring. Heinecke and Klaus[79] point out that the use of several species can balance out major physiological differences, and that outstanding genetic differences can be compensated for by the use of several strains. For this reason, the results obtained so far are being supplemented by experiments performed on mice. Heinecke and Klaus[79] also state that, when selecting suitable experimental animals, the most important parameters (resorption rate, fetus weight, and malformation rate) and some quantitative relationship between dose and effect must be known. According to Jörgensen[80], there is no ideal experimental animal. The greater the number of animal species which respond to teratogenic noxae with elevated malformation rates, the greater will be the probability that these noxae will also induce malformations in man. Rearing conditions, seasonal variations, and differences between primiparae and multiparae also affect the experimental animals[66]. In this connection, it must also be stressed that circadian rhythms have a substantial effect on responses during biological experiments and on the sensitivity of organisms to exogenic noxae[81]. It is therefore important that, in our experiments, the animals were exposed to noise and vibration at the same time of day.

We may conclude from all of these remarks that animal tests of new substances and noxae may still fail to reveal teratogenic effects in future.

The impressive evidence of the effects of exogenic factors on fetal development explain the great interest which teratology has attracted in the past few decades. We believe that further experimental activities in this field are justified not only because of the need for more information but that it is also an ethical obligation in order that all experience gained may be applied in practical medicine.

References

1. Gregg, N. M. (1941). Congenital cataract following German measles in the mother. *Trans. Ophthal. Soc. Austr.*, 3, 35
2. Blechschmidt, E. and Gassner, R. F. (1978). *Biokinetics and Biodynamics of Human Differentiation. Principles and Applications* (Springfield, Ill.: Charles C. Thomas)
3. Degenhardt, K. H. (1964). Mißbildungen des Kopfes und der Wirbelsäule. In: P. E. Becker (Hrsg.). *Humangenetik*. Bd. 2. S. pp. 489–604 (Stuttgart: Thieme Verlag)
4. Gebhardt, E. (1977). *Chemische Mutagene* (Stuttgart, New York: G. Fischer)
5. Goerttler, Kl. (1964). Kyematopathien. In: P. E. Becker (Hrsg.). *Humangenetik*. Bd. 2., S. 1–62. (Stuttgart: Thieme Verlag)
6. Rosenbauer, K. H. (1969). Exogene Mißbildungen. In: K. H. Rosenbauer (Hrsg.). *Entwicklung, Wachstum, Mißbildungen und Altern bei Mensch und Tier*, S. 99–141. (Stuttgart: Wissenschaftliche Verlagsgesellschaft)
7. Rosenbauer, K. H. (1973). Exogene Mißbildungsfaktoren bei Mensch und Tier. *Naturwiss. Rundsch., Stuttgart*, 26, 229
8. Shepard, Th. H. (1976). *Catalog of Teratogenic Agents* (Baltimore: Johns Hopkins Press)
9. Von Kreybig, Th. (1975). *Entstehung von Mißbildungen aus inneren und äußeren Ursachen.* (München, Berlin: Urban & Schwarzenberg)
10 Freundt, K. J. (1976). Schadstoffe in Lebensmitteln. *Med. Welt, Stuttgart*, 27, 1864
11. Gray, S. W. and Skandalakis, J. E. (1972). *Embryology for Surgeons.* (Philadelphia, London, Toronto: W. B. Saunders Company)
12. Hayashi, J. (1975). *Pathological Research on Influences of Atomic Bomb Exposure upon Fetal Development.* (Reprinted from the Research in the Effects and Influences of the Nuclear Bomb Test Explosions)
13. Ikeda, T., Okamoto, N. and Satow, Y. (1973). Biomechanical and cytogenetics study on

the teratogenicity of fast-neutron in the rat. *Acta Univ. Carolinae, Medica – Monographia, Praha*, 57/58, Part I, pp. 295–297
14. Okamoto, N., Ikeda, T. and Satow, Y. (1969). Effects of 14.1 – MeV Fast – Neutron Irradiation on the Cardiovascular System of the Rat Fetus. *Radiation Biol. of the Fetal and Juvenile Mammal.*, pp. 325–340
15. Merker, H. J., Zimmermann, B. and Günther, Th. (1972). Elektronenmikroskopische Untersuchungen über die D-Penicillaminwirkung am Knorpel embryonaler Ratten (Tag 16) in vitro. *Virchows Arch., Abt. B Zellpathol., Berlin (West)*, **12**, 51
16. Siegmund, G. and Degenhardt, K. H. (1965). Veränderte Eugenik. *Bild der Wissenschaft*, **11**, 917
17. Wendler, D. (1975). Stand und Perspektiven des teratologischen Screening. *Zbl. Pharmazie, Berlin*, **114**, 675
18. Becher, H. (1939). Über die Embryonalentwicklung bei verschiedenen atmosphärischen Druckverhältnissen. *Anat. Anz., Erg. H., Jena*, **88**, 144
19. Degenhardt, K. H. (1956). Mißbildungskorrelationen durch Sauerstoffmangel im Tierexperiment. *Naturwissenschaften, Berlin*, **43**, 525
20. Büchner, F. (1959). Die Bedeutung peristatischer Faktoren für die Entstehung der Mißbildungen und Mißbildungskrankheiten. *Verh. dt. Gesellsch. inn. Med., München*, **64**, 13
21. Persaud, T. V. N. (1971). Mechanism of teratogenic action of hypoglycin-A. *Experientia, Basel*, **27**, 414
22. Persaud, T. V. N. (1976). Zu Problemen der angeborenen Mißbildungen im Raum der englisch sprechenden Karibien. *Anat. Anz., Jena*, **140**, 345
23. Schumacher, G. H. and Fanghänel, J. (1973). Normo- und Teratogenese unter Schalleinwirkung. In: *Mensch und Umwelt aus der Sicht der Anthropologie.* Veröff. von H. Bach. (Wiss. Beiträge Friedrich-Schiller-Universität Jena.) Univ. Jena, pp. 192–200
24. Schumacher, G. H. and Fanghänel, J. (1974). Experimentelle Untersuchungen über den Einfluß von Lärm auf die Normound Teratogenese. *Anat. Anz., Erg. H., Jena*, **136**, 595
25. Schumacher, G. H. and Fanghänel, J. (1977a). Normo- und Teratogenese unter dem Einfluß von Umweltfaktoren. In: J.-H. Scharf und H. v. Mayersbach (Hrsg.) *Die Zeit und das Leben. Nova Acta Leopoldina.* NF 225, Bd. 46, S. 723–740. (Leipzig: J. A. Barth)
26. Schumacher, G. H. and Fanghänel, J. (1977b). Postnatale Entwicklung von Ratten nach Lärmstreß der schwangeren Muttertiere. *Labirint, str., Sarajevo*, pp. 57–66
27. Schindler, M. and Neuhofer, R. (1976). Industrie- und Verkehrslärmarten der Umgebung des VEB Kombinat Leuna-Werke 'Walter Ulbricht' – zwei Beispiele der Beeinflussung des Territoriums durch einen großen Industriebetrieb. *Z. ges. Hyg., Berlin*, **22**, 711
28. Moldenhauer, G. (1968). Die Auswirkung von Betriebslärm und seine Bekämpfung. *Holzindustrie, Leipzig*, **21**, 301
29. Lehmann, G. (1958). Nachweis und Bedeutung der vegetativ-nervösen Lärmbelastung. *Int. J. Prophyl. Med.*, **2**, 6
30. Rosenau, H. (1964). Lärm als medizinisches Problem. *Med. Welt, Stuttgart*, **15**, 816
31. Altvater, W. (1957). Lärm und Lärmschaden, ein gesundheitspolitisches Problem. *Öffentl. Ges. Dienst*, **19**, 47
32. Eckert-Möbius, A. (1968). *Lehrbuch der Hals-Nasen-Ohren-Heilkunde für Studenten und Praktische Ärzte.* (Leipzig: VEB G. Thieme)
33. Karrasch, K. (1952). Die Wirkung des Lärms auf den menschlichen Organismus. *Zbl. Arbeitswiss.*, **6**, 177
34. Zöllner, F. (1971). *Hals-Nasen-Ohrenheilkunde.* (Stuttgart: Thieme Verlag)
35. Jansen, G. (1960). Über die nichtspezifischen Wirkungen des Lärms auf den Menschen. *Med. Welt, Stuttgart*, **1**, 35
36. Lehmann, G. (1957). Der Kampf gegen den Lärm – eine wichtige Aufgabe der prophylaktischen Medizin. *Dt. med. Wochenschr., Stuttgart*, **82**, 465
37. Schuschke, G. (1970). Lärmkarte von Magdeburg. *Wissenschaft u. Fortschritt, Berlin*, **20**, 216
38. Quass, M. and Renker, U. (1976). *Arbeitshygiene, Berlin*, **176**, 346
39. Panzke, K. J. (1970). Mechanische Schwingungen bei handgeführten Vibrationsgeräten im Bauwesen und in Einwirkung auf den Menschen. *Ergometrische Berichte, Berlin*, pp. 3–38
40. Henkel, W. (1976). Schwingungsbewertung mit Hilfe der äquivalenten Dauerschwingbeschleunigung. *Z. ges. Hyg., Berlin*, **22**, 330

41. Weisshaupt, S. (1975). Dosimeter zur Ermittlung der Vibrationsbelastung. *Z. ges. Hyg., Berlin*, **21**, 617
42. Hagemann, E. (1960). *Ratte und Maus. Versuchstiere in der Forschung.* (Berlin: W. de Gruyter u. Co.)
43. Schumacher, G. H., Fanghänel, J. and Schultz, E. (1970). Untersuchungen an Rattus norvegicus BERKENHOUT unter normalen und experimentellen statischen Bedingungen. *Wiss. Z. Univ. Rostock. Math.-naturwiss. R., Rostock*, **19**, 177
44. Schumacher, G. H. and Rehmer, H. (1961). Der syrische Goldhamster als Laboratoriumstier. *Z. med. Labortechnik, Berlin*, **2**, 205
45. Fanghänel, J. and Schumacher, G. H. (1973). Biometrische Untersuchungen über die Normogenese und Teratogenese des Schädels. *Acta Univ. Carolinae, Medica – Monographia, Praha*, 56–57, Part I, S. 87–92
46. Fanghänel, J. and Timm, D. (1970). Mathematische Modelle in der modernen Morphologie. *Wiss. Z. Univ. Rostock, Math. Naturwiss. Reihe*, **19**, 185
47. Fanghänel, J., Timm, D., Schumacher, G. H. and Lau, H. (1977). Zum Problem biologisch-mathematischer Modelle. *Verh. Anat. Ges., Jena*, **71**, 1485
48. Scharf, J.-H. (1970). Moderne Morphologie und höhere Analysis. In: L.-H. Kettler (ed.). *Die heutige Stellung der Morphologie in Biologie und Medizin.* S. 45–99. (Berlin: Akademie-Verlag)
49. Schumacher, G. H., Fanghänel, J. and Dahl, D. (1976). Tierexperimentelle Untersuchungen zum Problem der Entstehung von Mißbildungen nach Lärmbeeinflussung. *HNO-Praxis, Berlin*, **311**, 145
50. Heinecke, H. (1969). *Die Maus als Modell in der Teratologie.* (Habilitations-Schrift, Halle/Saale)
51. Nitschkoff, S. and Kriwitzkaja, G. (1968). *Lärmbelastung, akustischer Reiz, neurovegetative Störungen. – Eine morphophysiologische Studie.* (Leipzig: VEB G. Thieme)
52. Peters, D. and Strassburg, M. (1968). Erzeugung von Gaumenspalten durch Lärm und Hunger. *Dt. zahnärztl. Z., München*, **23**, 843
53. Kuhlenkampff, H. (1951). Das Verhalten der Vorderwurzelzellen der weißen Maus unter dem Reiz physiologischer Tätigkeit. *Z. Anat. u. Entwickl.-Gesch., Berlin (West)*, **116**, 143
54. Wüstenfeld, E. and Schilling, R. (1966). Veränderungen der Nisslsubstanz im Ganglion spirale cochleae unter Reintonbeschallung. *Z. Zellforsch. u. mikrosk. Anat., Berlin (West)*, Abt. A **71**, pp. 517–524
55. Wüstenfeld, E. and Gleiss, H. (1969). Über die Beeinflussung der Kerngröße von Ganglienzellen durch Reintonbeschallung im medullären Akustikuskomplex des Meerschweinchens. *Anat. Anz., Erg.-H., Jena*, **125**, 447
56. Grosse, G., Lindner, G. and Schneider, P. (1972). Über den Einfluß der Reintonbeschallung auf Kulturen des Ganglion trigeminale von Hühnerembryo. *Z. mikrosk.-anat. Forsch., Leipzig*, **85**, 273
57. Bredberg, G., Ades, W. and Engström, H. (1972). Scanning electron microscopy of the normal and pathologically altered organ of Corti. *Acta Oto-laryng., Uppsala*, **Suppl. 301**, 3
58. Ades, H. W., Trahiotis, C., Kokko-Cunningham, A. and Averbruch, A. (1974). Comparison of hearing thresholds and morphological changes in the chinchilla after exposure to 4 kHz tones. *Acta Otolaryng., Stockholm*, **78**, 192
59. Dieroff, H. G. and Beck, E. (1966). Experimentell-mikroskopische Studie zur Frage der Lokalisation der industrielärmbedingten Hörermüdung und des später resultierenden bleibenden Hörschadens. *Arch. Ohren-, Nasen- u. Kehlkopfheilkd., Berlin (West)*, **186**, 1
60. Biedermann, M. and Meyer, Chr. (1977). Reaktionen der Meerschweinchencochlea auf Einzelimpulsbeschallung. *Med. u. Sport, Berlin*, **17**, 200
61. Bode, H. (1959). Studien über den Lärm, seine Wirkungen auf den Organismus und Vorschläge für eine Normierung. *Med. Dissertation*, Rostock
62. Schneider, B., Schröder, J. and Wöhle, W. (1957). Bericht über den Lärmkongreß in Leningrad und über die gesetzlichen Lärmnormative der Sowjetunion. *Arbeit u. Soz.-Fürsorge, Berlin*, **12**, 77
63. Tamm, J. (1956). Über Lärmwirkungen auf den Menschen. *Zbl. Arb.-wiss., Frechen b. Köln*, **10**, 97
64. Jansen, G. (1970). Psychosomatische Lärmwirkungen und ihre Grenzwerte für die

vegetative Belastung durch Schall. *Arbeitsmedizin, Sozialmedizin, Arbeitshygiene, Berlin (West)*, 5, 256

65. Baumann, R. (1975). Streß-Sensibilität und Adaptation. *Z. ges. inn. Med., Leipzig*, **30**, 15
66. Spiegel, A. (1975). Über das Versuchstier 'pro analysi'. *Dt. med. Wochenschr., Stuttgart*, **88**, 1203
67. Clough, G. and Fasham, J. A. L. (1975). Silent fire alarm. *Lab. Anim. London*, **9**, 193
68. Klosterkötter, W. (1968). Lärmwirkungen: Ergebnisse und Aufgaben der medizinischen Lärmforschung. *Kampf dem Lärm, München*, **15**, 3
69. Coermann, R. (1969). Einfluß von Stößen und Vibrationen auf den Menschen. In: Proceeding of a Symposium, Prague, October 2–7, 1967, Vol. II, pp. 640–645. (Geneva: International Labour Office)
70. Horvath, F., Sztanskay, Cs. and Kakossy, I. (1970). Untersuchungen vibrationsbewirkter Gefäßverengungen. *Fortschr. Röntgenstr. u. Nukl.-Med., Stuttgart*, **115**, 164
71. Matzen, P. F. (1975). *Orthopädie für Studierende.* (Leipzig: J. A. Barth)
72. Williams, N. (1975). Biological effects of segmental vibration. *J. Occupat. Med., Chicago*, **17**, 37
73. Agrarwal, G. C. and Gottlieb, L. (1976). Effects of vibration on human spinal reflexes. In: Schahani, M. (ed.). *The Motor System: Neurophysiology and Muscle Mechanics.* pp. 181–187 (Amsterdam: Elsevier)
74. Lackner, J. R. and Graybiel, A. (1974). Elicitation of vestibular side effects by regional vibration of the head. *Aviation, Space & Environm. Med., Washington*, **45**, 1267
75. Klimkova-Deutschova, E. (1966). Neurologische Aspekte der Vibrationskrankheit. *Int. Arch. Gewerbepath. Gewerbehyg.*, **22**, 297
76. Brust, H. (1976). Mechanische Schwingungen. *Betriebssicherheit*, **2**, 6
77. Berencsi, G. (1976). Zu einigen Problemen des Gesundheitsschutzes der Werktätigen in der Landwirtschaft in der Ungarischen Volksrepublik. *Z. ges. Hyg., Berlin*, **22**, 557
78. Freye, H.-A. (1966). Die Verwendung von Säugetieren bei embryopathischen Untersuchungen. *Luynx, Prag*, **6**, 31
79. Heinecke, H. and Klaus, S. (1973). Zur Problematik der Prüfung von Teratogenen. *Acta Univ. Carolinae, Medica – Monographia, Praha*, 56–57, Part I, S. 93–96
80. Jörgensen, G. (1967). Vergleichende Pharmakogenetik des Menschen und der Säugetiere (II). *Med. Welt, Stuttgart*, **84**, 1
81. Mayersbach, H. von (1977). Die Bedeutung der Circadianrhythmik im medizinischen und biologischen Experiment. In: J.-H. Scharf and H. v. Mayersbach (eds.). *Die Zeit und das Leben. Nova Acta Leopoldina.* NF 225, Bd. 46, S. 115–129. (Leipzig: J. A. Barth)
82. Goerttler, Kl. (1971). Pathomechanik und Ablauf pränatalter Wachstumsstörungen. *Ergebn. exp. Med., Berlin*, **4**, 192
83. Schumacher, G. H. (1977). *Embryonale Entwicklung des Menschen.* 3. Aufl. (Berlin: VEB Verlag Volk und Gesundheit)
84. Wrete, M. (1955). *Die kongenitalen Mißbildungen, ihre Ursachen und Prophylaxe.* (Stockholm: Almqvist & Wiksell)

Index